Handbook of Biodiversity

Handbook of Biodiversity

Editor: Neil Griffin

STATES
ACADEMIC PRESS
www.statesacademicpress.com

States Academic Press,
109 South 5th Street,
Brooklyn, NY 11249, USA

Visit us on the World Wide Web at:
www.statesacademicpress.com

ISBN: 978-1-63989-252-5 (Hardback)

Cataloging-in-Publication Data

Handbook of biodiversity / edited by Neil Griffin.
p. cm.
Includes bibliographical references and index.
ISBN 978-1-63989-252-5
1. Biodiversity. 2. Biology. 3. Biocomplexity. 4. Ecological heterogeneity.
5. Numbers of species. I. Griffin, Neil.
QH541.15.B56 H36 2022
333.95--dc23

Table of Contents

Preface

The biological variety and variability of life on Earth is called biodiversity. It is generally a measure of variation at the genetic, species and ecosystem level. Some of the factors which influence biodiversity are altitude, soils, precipitation and temperature. Due to the warm climate and high primary productivity, terrestrial biodiversity is generally greater near the equator. The tropical forest ecosystem which has a rich biodiversity covers less than ten percent of the surface of the Earth, but contains about ninety percent of all the species on the planet. Marine biodiversity is higher along the coasts of the Western Pacific and the mid-latitudinal band. Biodiversity has been increasing through time and tend to cluster in hotspots, but is likely to slow down due to rapid deforestation and environmental changes. This book provides comprehensive insights into this field. It consists of contributions made by international experts. Researchers and students in the field of biodiversity will be assisted by this book.

This book unites the global concepts and researches in an organized manner for a comprehensive understanding of the subject. It is a ripe text for all researchers, students, scientists or anyone else who is interested in acquiring a better knowledge of this dynamic field.

I extend my sincere thanks to the contributors for such eloquent research chapters. Finally, I thank my family for being a source of support and help.

Editor

1

Biodiversity Restoration and Renewable Energy from Hydropower: Conflict or Synergy?

Wondmagegn Tafesse Tirkaso, Ing-Marie Gren, Leonard Sandin, Joel Segersten, David Spjut and Erik Degerman

Abstract

Hydropower plants have a negative impact on biodiversity by transforming stream habitat and hydrology and thereby affecting aquatic organisms negatively. The negative effects can be mitigated by releasing water into the old river bed. This study investigates if the measure of releasing water creates costs and if ecological conditions at the old river bed contribute to such an impact. To this end, we used the cost-minimization framework in economics for deriving hypotheses. Tests were made with data from a survey to 76 hydropower plants in Sweden with questions on existence of a cost, size of the plant, type of water release from reservoirs, characteristics of the dried downstream old river bed, and official statistics on ecological status of the downstream dried segments. The results showed that 42% of the plants reported no cost, measured as impact on electricity production, from release of water into downstream old river bed. We applied logit and probit models to explain the probability of a cost. Significant results were obtained were the electricity produced and program for minimum water discharges increase the probability of loss in electricity production, but favorable ecological conditions in the old river bed decrease the probability of a cost.

Keywords: hydropower, biodiversity, streams and rivers, restoration, old river bed, cost, survey data, econometrics, Sweden

1. Introduction

Similar to many other countries, Sweden needs to comply with national and international targets on renewable energy and biodiversity provision. Hydropower is important for the provision of renewable electricity production and accounts for approximately 47% of the total electricity production in the country [1]. Nuclear power is the second largest source of electricity and accounts for 34% of total energy production. Not only is hydropower a large source of electricity, but also acts a powerful regulatory device for the large fluctuations in demand and supply of electricity. Further, it is among the least expensive sources of energy as measured in SEK/kWh [2].

Establishments of hydropower plants change the hydrological conditions in the riverine landscape which affects habitats for animals and plants. Streams can be totally or partially dried and thereby destroying the habitats for several species and migration pathways for fish species. Although there is no national evidence on the extinction of species because of the hydropower production, the effects imply a degradation of habitats for red listed species [3], which goes against the national target of preserving biodiversity.

In order to mitigate these effects water power plants may be run with a release of water from the reservoir(s) into the downstream dry channel (the old natural river channel). However, this may only be achieved at a cost in terms of less electricity production and hence fulfillment of the target on renewable energy. This study investigates whether such a cost exists, and which factors contribute to the probability of its occurrence. To this end, we use the cost-minimization framework to derive testable hypotheses. Test are made with data from a survey on 76 hydropower plants in Sweden with questions on the existence of a cost in terms of negative impact on electricity production, the type of water release from reservoirs, and characteristics of the dried downstream channel and the plant. This data set was completed with official statistics on ecological status in the downstream segments. We use econometric methods to examine the impact of water discharges from the reservoir and other explanatory variables on electricity production. The dependent variable is a binary variable which equals 1 when electricity production is affected and 0 otherwise. We, therefore, use a probit model for the regression analysis, where we estimate how the explanatory variables affect the probability of losing electricity production.

There is a large body of literature on ecological effects of biodiversity restoration in freshwaters, such as wetland restoration (see reviews in Refs. [4, 5]). Despite this, the literature on the determination of costs of measures mitigating biodiversity degradation from hydropower plants is scant (e.g., [6–8]). The cost of restoration objects depends on the investment and management of the restoration measure as such, and on the ecological conditions at the site affecting the need and quality of restoration [5]. In our view, the main contribution of this study is the estimation of the explanatory power of ecological conditions and water release from reservoirs on the probability of a restoration cost in terms of reduction in electricity production.

2. Theoretical framework

The theoretical framework rests on the assumption that each power plant minimizes costs for restoring biodiversity. This is a common assumption in economics where firms are assumed to use inputs, such as labor and capital, at given prices to minimize costs for producing certain outputs. By applying the so-called duality theory a cost function can be derived which shows the relation between the output and production cost (e.g., [9]). The cost is then expressed as a function of given input prices, and output level. In our case, biodiversity improvement constitutes the output Q^i where $i = 1,.., n$ sites of the hydropower plants. The level of the output, or success of restoration, which can be measured as number of fish species or as a quality index, depends on ecological conditions at the site, E^{il} where $l = 1,.., m$ conditions such as length of the channel and natural water flow, and on restoration measures at the plant, M^{ig} where $g = 1,.., h$ different restoration measures such as water discharges from the dam. The biodiversity at the site is then written as $Q^i = Q^i(M^{i1},..,M^{ih}; E^{i1},.., E^{im})$.

A crucial assumption in our analysis is that the plant manager minimizes total cost for achieving a minimum level of biodiversity, Q^{*i}. Each restoration measure is then associated with a cost, $C^{ig}(M^{ig})$, and a maximum capacity of implementation, $\overline{M}^{ig}$. For example, there is a maximum limit of water discharges into the channel. Plant size, K^i, may also affect costs; a large plant can have more expertise for implementing restoration measures than a small plant. On the other hand, a larger plant may give rise to more damages in the downstream waters, the mitigation of which requires costly restoration measures. The decision problem for the plant manager is then written as:

$$\begin{aligned} &\text{Min}C^i = \sum\nolimits_g C^{ig}(M^{ig}; K^i) \\ &M^{ig} \\ &\text{Subject } to\, Q^i(M^{i1}, .., M^{ih}; E^{i1}, \ldots, E^{im}) \geq Q*^i \text{ and } M^{ig} \leq \overline{M}^{ig} \end{aligned} \tag{1}$$

By applying the so-called duality theory to Eq. (1) we can express the cost of restoration at the plant as a function of the chosen restoration target Q^{*i}, ecological conditions at the site, E^{ig}, and the restoration measures, $\overline{M}^{ig}$, which is written as follows:

$$C^i = C^i(Q*^i,\ E^{i1}, \ldots, E^{im}, \overline{M}^{i1}, .., \overline{M}^{ih}, K^i) \tag{2}$$

Our main interest is to investigate the impact on costs of a marginal increase in the restoration ambition, Q^{*i}. The hypothesis is that the cost, C^i, increases since more of the restoration measures need to be implemented. In this case, there is a conflict between biodiversity restoration and electricity provision since resources that could be used for electricity production are used for biodiversity restoration. On the other hand, a non-positive effect would imply the opposite interpretation. As shown in Eqs. (1) and (2), a test of this hypothesis requires data, not only on C^i and Q^{*i}, but also on E^{il}, $\overline{M}^{ig}$, and K^i.

3. Description of data

Unfortunately, the necessary data presented in Section 2 is not available for a sufficient number of plants. Therefore, a survey was distributed to hydropower plants with dried channels. It turned out that the plant managers were not able to answer questions on our output variable, diversity at downstream sections of the reservoirs. The survey was therefore completed with official data on fish species.

The questionnaire was sent to the four largest hydropower companies in Sweden with any kind of restoration measures, where downstream dry or nearly dry river and stream channels had been identified and where electrofishing data were available for the dry or nearly dry channel downstream of the reservoir. The questionnaire was filled out and returned for 76 hydropower plants where fish data (see below) from downstream sections of the reservoirs were available. These plants are located in the entire Sweden, see **Figure A1**.

The survey included questions on the variables presented in Section 2: costs, measures for the release of water into the channel, ecological conditions of the dry river beds, and annual electricity production. The latter was used as a description of the characteristic of the plant. However, the respondents were not able to assess the costs of the measures, but on whether there had been a cost in terms of loss in electricity production. Therefore, our cost variable is binary, where $Cost = 1$ when there is a loss in electricity production and $Cost = 0$ otherwise.

With respect to the choice of measures, a common strategy is to implement a program with minimum water discharge from the reservoir to the old river bed, which ensures that there is a minimum flow of water in order to potentially sustain downstream stream and river organisms. A question was included on the existence of such a program (*Mindisch*) (by court order or voluntarily) where $Mindisch = 1$ when the measure is in place and $Mindisch = 0$ otherwise. The plants can also implement other strategies for improving biodiversity, such as an even flow of water to downstream water. A question was therefore included on the existence of other measures (*Othmeas*) where $Othmeas = 1$ if such measures exist and $Othmeas = 0$ otherwise.

Sufficient length of dry channels and natural water flow in the dry channels provide favorable ecological conditions for restoring biodiversity. Questions were included on the length of the dry channels (*Length*), and natural water flows in the dry channels in m/s (*Msec*) as continuous variables. The size of the plant was measured as the annual electricity production (*Elprod*).

As shown in Section 2, the variable measuring biological conditions of the downstream dry or partly dry stream or river section should reflect the effects of restoration measures and ecological conditions in the channel. This would require data and analysis of biological conditions before and after the implementation of the measures. Such data is not available. Instead, we use data on measurements of biological conditions in the downstream river section, which is available as electrofishing data at the Swedish Electrofishing Register [10]. The Swedish stream and river fish index *VIX* was used to assess the ecological status of the downstream sections [11]. The *VIX* index ranges from 0 to 1, where high values denote high ecological status and low values denote bad ecological status according to the EU Water Framework Directive [12].

Descriptive statistics of the dependent variable *Cost* and the independent variables are displayed in **Table 1**.

The results from the survey showed that 58% of all plants report a cost in terms of a reduction in electricity production and 47% have implemented minimum flow discharges into the old

Variable	Observations	Mean	Standard deviation	Minimum	Maximum
Cost	66	0.576	0.498	0	1
Elprod, electricity production in kWh	66	90,400,000	202,000,000	900	1,120,000,000
Length in m	75	2113	2612	62	16,308
Msec, water flow in m^3/s	67	48	83	0	377
Mindisch	75	0.47	0.50	0	1
Othmeas	75	0.69	0.46	0	1
VIX	20	0.43	0.23	0	1

Table 1. Descriptive statistics.

natural downstream channel as a biodiversity restoration measure. It might be argued that mainly those dams with discharges of water from the reservoirs to the dry channel face a cost. However, the plants reporting a loss in electricity production is evenly distributed between plants with and without minimum discharges to the dried channels.

The range in dam size as measured by electricity production is large within the dataset, ranging from 900 to 1,120,000,000 kWh/year (**Table 1**). The continuous variables *Elprod, Length,* and *Msec* are highly skewed, and we, therefore, transformed them into the logarithms *Logelprod, Loglength,* and *Logmsec.*

4. Econometric model

We have employed the standard logit and probit models to estimate the explanatory power of the independent variables listed in **Table 1** on the probability of a cost. The difference between the logit and probit model is the distributions of the error terms. The former follows a cumulative standard logistic distribution, whereas the later follows a cumulative standard normal distribution (see e.g., [13, 14]). The probability functions in both the models are symmetric around zero and tend to give similar parameter estimate. Therefore, we have estimated parameters of interest applying both estimators. We know that our dependent variable *Cost,* can take only two values, i.e., 1 if there is a cost and 0 if there is no cost. The probability of $Cost = 1$ is p and the probability of $Cost = 0$ is $(1-p)$. Hence, the expectation of *Cost, E[Cost],* is given as follows:

$$E[Cost] = 1 * p + 0 * (1 - p) \Rightarrow p \tag{3}$$

Considering the probability that $Cost = 1$ is a function of different covariates presented in **Table 1**, denoted by vector X, and parameters of interest β, we can write the standard binary choice model as follows:

$$P(Cost = 1|X) = f(\beta X) \tag{4}$$

Consequently, the logit and probit models corresponding to Eq. (4) are given by Eqs. (5) and (6), respectively, as:

$$f\left(\beta' X\right) = \frac{e^{\beta' X}}{1 + e^{\beta' X}} \tag{5}$$

$$f\left(\beta' X\right) = \Pi\left(\beta' X\right) = \int_{-\infty}^{\beta' X} \frac{1}{\sqrt{2\pi}} e^{-\frac{u^2}{2}} du \tag{6}$$

Since probit and logit models are nonlinear in both parameter and variables, the usual ordinary least square (OLS) and weighted least square (WLS) estimators could not be plausible. For that reason, identification of parameters given by a vector β preferably should be obtained by applying the maximum likelihood (MLE) estimator. Generally, binary choice models can be derived from the latent variable model as it provides a link with standard linear regression models which makes interpretation of the parameters straightforward. Besides, the model illustrates the difference between logit and probit models. Suppose the binary outcome variable *Cost* and the corresponding latent variable *Cost** satisfies the single infix model as:

$$Cost* = \beta X + \varepsilon \tag{7}$$

Given that *Cost* is observable, it can be expressed as:

$$Cost = \begin{cases} 1 \ \ if \ \ Cost* > 0 \\ 0 \ \ if \ \ Cost* \leq 0 \end{cases} \tag{8}$$

Combining Eqs. (7) and (8), we can have the following response probabilities:

$$P(Cost = 1) = P(\beta X + \varepsilon > 0) = P(-\varepsilon < \beta X) = f(\beta X) \tag{9}$$

where, $f\left(\beta' X\right)$ is the cumulative density functions (CDF). In the case of probit model, the error term follows the standard normal distribution whereas it follows the logistic distribution in the case of a logit model.

The signs of parameters β are directly interpretable in both logit and probit models, but not the magnitudes. For that reason, deriving the marginal effects and discrete changes in the estimates is crucial in order to obtain the magnitudes of parameters. The marginal effect of the continuous covariate is given by partial derivative with respect to that variable, whereas the discrete changes of dummy covariate are given by the difference in predicted probabilities of the variable at 0 and 1, setting other covariates constant at their reference points. Mathematical notion of marginal effects and discrete changes in binary outcome models can be found in Ref. [13].

Recall from Section 3 that the availability of data is limited for the *VIX* variable with only 20 observations. We, therefore, estimated regression equations with and without this variable. The regression equation without *VIX* is specified as:

Model 1:

$$Cost = \beta_o + \beta_1 \text{Logelprod} + \beta_2 \text{Loglength} + \beta_3 \text{Logmsec} + \beta_4 \text{Mindisch} + \beta_5 \text{Othmeas} + \varepsilon \tag{10}$$

The estimates give information on the probability of a cost from introducing restoration measures *Mindisch* and *Othmeas*, and the impacts of *Logelprod*, *Loglength*, and *Logmsec*. The regression equation with *VIX* includes all variables and is specified as:

Model 2:

$$\begin{aligned} Cost &= \beta_6 + \beta_7\text{Logelprod} + \beta_8\text{Loglength} + \beta_9\text{Logmsec} \\ &\quad + \beta_{10}\text{Mindisch} + \beta_{11}\text{Othmeas} + \beta_{12}\text{VIX} + \varepsilon \end{aligned} \tag{11}$$

5. Results

The results showed that the independent variable *Othmeas* was never significant, and we therefore excluded this variable. Another result was that the inclusion of all explanatory variables in Model 2 gave poor statistical fit because of the low number of observations. We, therefore, excluded *Loglength* and *Logmse* in Model 2.

There might also be statistical problems associated with endogeneity in the included explanatory variables. Since the purpose of minimum discharge (*Mindisch)* is to sustain ecological conditions in the dry channels, this variable might be dependent on the ecological status in the stream channels, *Loglength*, and *Logmsec* in Model 1 and *VIX* in Model 2. If so, the ordinary least square (OLS) estimates will not give consistent estimates (e.g., [15]). Therefore, we tested for endogeneity in *Mindisch* by using *Loglength* and *Logmsec* as instruments in Model 1 and *VIX* as an instrument in Model 2. Wald tests of both models showed that exogeneity in *Mindisch* could not be rejected at the 10% level (see, e.g., [15] for a description of the test). This means that we can treat *Mindisch* as an independent variable.

We also tested for the existence of heteroscedasticity, which was not present in any model according to the results from Breach-Pagan tests (e.g., [16]). However, Pearson test of correlation among all explanatory variables showed significant associations at the 1% level between *Logelprod* and several other explanatory variables (**Table A1**). Despite these association, variance inflation factor (*VIF*) tests did not reveal problems of multicollinearity (mean $VIF = 1.29$ for Model 1 and mean $VIF = 1.36$ for Model 2).

The binary dependent variable denotes the likelihood of a cost of changes in any of the explanatory variables. We would expect *Mindisch* to increase the probability of a cost since this measure discharges water into the dry channel which could be used for electricity production. On the other hand, natural conditions in the dry channels, measured as channel length and natural water discharge, are likely to reduce the likelihood of a cost because there is less need for mitigation measures. As a measure of the size of the dam, *Logelprod* can increase the probability of a loss in electricity production. The regression results of Model 1 are presented in **Table 2**.

According to **Table 2**, the results from the logit and probit models are quite similar. All explanatory variables are significant and have the expected sign. The models are significant at the 0.01 level according to the model Chi-square statistic, and the predicted "$Cost = 1$" corresponds to 87% of the observed "$Cost = 1$." The statistical performance of the probit model is slightly better

Variable name	Logit		Probit	
	Coeff.	Prob.	Coeff.	Prob.
Constant	−3.891	0.331	−2.396	0.215
Logelprod	0.841***	0.000	0.496***	0.000
Loglength	−0.777*	0.061	−0.443**	0.041
Logmsec	−1.334***	0.001	−0.785***	0.000
Mindisch	1.826*	0.086	1.062*	0.053
Model significance[a] Pseudo R^2	$p = 0.000$		$p = 0.000$	
	0.440		0.444	
Predicted *Cost* = 1/observed *Cost* = 1	33/38		33/38	
AIC, BIC	54.177, 64.565		53.851, 64.238	

Notes: ***$p < 0.01$, **$p < 0.05$, *$p < 0.1$; [a]Chi-square(4).

Table 2. Regression results of Model 1 with different estimators, $N = 59$.

than the logit model as measured by pseudo R^2, Aikaike Information Criterion (*AIC*), and Bayesian Information Criterion (*BIC*) tests.

The results for the two estimators are similar when replacing *Loglength* and *Logmsec* with *VIX*, see **Table 3**.

The statistical performance of Model 2 as measured by the significance of explanatory variables, overall model significance pseudo R^2, *AIC*, and *BIC* was lower than for Model 1, which may be explained by the lower number of observations. A common result for Model 1 and Model 2 was the positive and significant effect of *Logelprod*. Although *Mindisch* has the expected negative sign in Model 2, it was not significant. The estimate of *VIX* has an unexpected negative sign. Since *VIX*

Variable name	Logit		Probit	
	Coeff.	Prob.	Coeff.	Prob.
Constant	−9.811	0.013	−5.825	0.004
Logelprod	0.731	0.013	0.430	0.002
VIX	−2.644	0.655	−1.597	0.496
Mindisch	0.750	0.754	0.464	0.659
Model significance[a] Pseudo R^2	$p = 0.004$		$p = 0.004$	
	0.408		0.407	
Predicted yes/observed yes	10/13		10/13	
AIC, BIC	23.339, 27.322		23.338, 27.321	

Notes: ***$p < 0.01$, **$p < 0.05$, *$p < 0.1$; [a]Chi-square(3).

Table 3. Regression results of Model 2 with different estimators, $N = 20$.

Variable	Logit				Probit			
	Model 1		Model 2		Model 1		Model 2	
	dy/dx	p-value	dy/dx	p-value	dy/dx	p-value	dy/dx	p-value
Logelprod	0.176***	0.000	0.152***	0.001	0.175***	0.000	0.154***	0.000
Loglength	−0.164*	0.053			−0.156**	0.035		
Logmsek	−0.280***	0.000			−0.277**	0.000		
VIX			−0.553	0.646			−0.572	0.486
Mindisch[a]	0.384***	0.065	0.157	0.750	0.375**	0.042	0.166	0.655

Notes: $^{***}p < 0.01$, $^{**}p < 0.05$, $^{*}p < 0.1$; [a]dy/dx is for discrete change of dummy variable from 0 to 1.

Table 4. Estimates of marginal effects of each of the explanatory variable at the mean value of all variables.

shows the fish habitat conditions at a downstream segment, a higher level of *VIX* should be associated with a higher probability of a cost according to the simple economic theory presented in Section 2. On the other hand, a negative sign indicates that there is no conflict in the achievement of biodiversity targets and energy production. However, the estimate is not significant and we cannot make conclusions about the effects of *VIX* on the probability of a cost.

Generally, coefficients of binary outcome models are in log-units and cannot directly be interpreted as marginal effects. This is due to the fact that the logit or probit transformation of the outcome variable has a linear relationship with the predictor variables. However, it is possible to derive the individual marginal effects or elasticities of covariates at their mean values (**Table 4**).

The probit and logit models give similar marginal effects of *Logelprod* for both Model 1 and Model 2 (**Table 4**). The probability of a loss in electricity production increases by 0.18 (Model 1) or 0.15 (Model 2). An increase in *Mindisch* has the largest impact on the probability, an increase by one unit raises the probability by 0.38 (Model 1). On the other hand, an increase in the natural conditions in the dry channel reduces the risk by 0.16 and 0.26 for *Loglength* and *Logmsek*, respectively.

6. Discussion and conclusions

The purpose of this study was to determine if restoration of biodiversity in dry channels at hydropower plants in Sweden can be costly for the plants and how the probability of a cost is affected by the size of the plant, site-specific factors in the dry channels, and ecological status in downstream regions of the river. The measure considered for restoration is the existence of a program for minimum releases of water from the reservoirs to the dry channel, and the cost is defined as a decrease in electricity production. The study rests on data from a survey of the largest hydropower plants in Sweden, which resulted in data for 76 plants with dry channels.

According to the responses in the survey, 58% of the plants with a program for minimum water discharges report a cost. The reasons for not reporting such a loss can be that it is considered as negligible or that the respondent has insufficient information. We cannot distinguish between these reasons, but it can be argued that impacts of releases of water from the reservoirs to the dry channels on electricity production would be shown in the continuous monitoring of electricity production. Nevertheless, we should be careful in interpreting the lack of reporting of a loss as the nonexistence of decreases in electricity production from programs on minimum discharges to dry channels.

The main results from our analysis of the different variables in explaining the probability of a reported cost are that the existence of a program for minimum water releases and a larger size of the plant as measured by kWh electricity production increase the probability. On the other hand, site characteristics as measured by the flow of natural water into the dry channel and length of the dry channels reduce the probability. These results point out potential cost savings for improving biodiversity in dry channels at hydropower plants by targeting water releases from reservoirs.

A cost-effective restoration policy requires that restoration measures are directed toward locations with high biodiversity impacts (e.g., [17]). Admittedly, due to lack of data on the impact of restoration measures on biodiversity, our results can give only partial guidance on the cost-effective restoration of biodiversity loss by means of water releases from reservoirs. Despite this limitation, the results can be useful when considering that current Swedish policy is to a large extent based on uniform regulations for all hydropower plants, such a maximum loss of 2.3% in the annual production of electricity [18]. Our results show that the probability of costs in terms of losses in electricity production is low for relatively small-sized plants, and where the natural flow of waters to the dry channels is high and the length of the channels is large. Thus, a comparison of costs and effects of current uniform policy with a policy targeting restoration measures toward plant sites with these characteristics can be of interest for economic analysis.

Acknowledgements

The research presented in this paper was carried out as a part of the R&D program Kraft och liv i vatten (KLIV). The partners behind the program are several hydropower companies in Sweden, Swedish Energy Agency, Swedish Agency for Marine and Water Management and Sweden's water authorities.

Appendix: Table A1 and Figure A1.

	Logelprod	Loglength	Logmsec	Mindisch	VIX
Logelprod	1				
Loglength	0.056	1			
Logmsec	0.279	−0.021	1		
Mindisch	−0.605	−0.013	−0.399	1	
VIX	−0.069	−0.031	−0.181	0.294	1

Table A1. Correlation matrix.

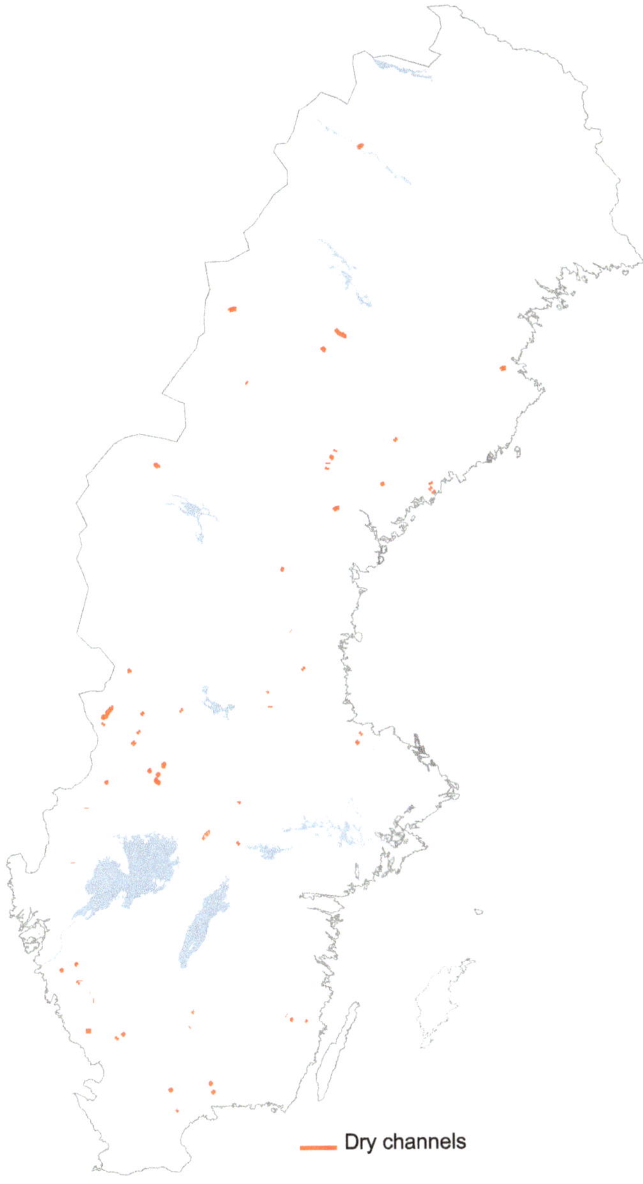

Figure A1. Locations of hydropower plants and dry channels included in the survey.

Author details

Wondmagegn Tafesse Tirkaso[1], Ing-Marie Gren[1]*, Leonard Sandin[2], Joel Segersten[2], David Spjut[2] and Erik Degerman[2]

*Address all correspondence to: ing-marie.gren@slu.se

1 Department of Economics, Swedish University of Agricultural Sciences, Uppsala, Sweden

2 Department of Soil and Water, Swedish University of Agricultural Sciences, Uppsala, Sweden

References

[1] Swedish Energy Agency. Electricity Production in Sweden 2015. [Internet]. 2016. Available from: http://www.energimyndigheten.se/nyhetsarkiv/2016/2015-var-ett-ar-med-stor-elproduktion-och-rekordstor-export-av-el/ [Accessed: November 2016]

[2] Sparrevik E, Viklands H, Bergsten P, Harju L. Ekologiska effekter och verksamhetspåverkan av förändrade produktionsvillkor i Vattenfalls storskaliga vattenkraftverk. Rapport, Vattenfall. [Internet]. 2011. Available from: https://corporate.vattenfall.se/globalassets/sverige/om-vattenfall/om-oss/var-verksamhet/vattenkraft/bilaga_10_ekologiska_effekter_och_verksamhetspaverkan_av_forandrade_produktionsvillkor.pdf No [Accessed: November 2016]

[3] Art Data banken. Rödlistade arter i Sverige. [Internet]. 2015. Available from: http://www.artdatabanken.se/media/2013/hela-boken.pdf. [Accessed: Februrary 2016]

[4] Brudvig LA. The restoration of biodiversity: Where has research been and where does it need to go? American Journal of Botany. 2011;**98**:549-558

[5] Meli P, Benayas JMR, Balvanera P, Ramos MM. Restoration enhances wetland biodiversity and ecosystem service supply, but results are context-dependent: A meta-analysis. PLoS One. 2014;**9**:e93507

[6] Rudberg P, Escobar M, Gantenbein J, Niiro N. Mitigating the adverse effects of hydropower projects: A comparative review of river restoration and hydropower regulation in Sweden and the United States. The Georgetown International Environmental Law Review. 2015;**27**:251-273

[7] Bullock JM, Aronson J, Newton AC, Pywell RF, Rey-Benayas JM. Restoration of ecosystem services and biodiversity: Conflicts and opportunities. Trends in Ecology and Evolution. 2011;**26**:542-549

[8] Nebhöver C, et al. Investing in ecological infrastructure. In: ten Brink P, editor. The Economics of Ecosystems and Biodiversity: National and International Policy Making. An output of TEEB. Earthscan; 2011. pp. 401-448

[9] Diewert WE. Duality approaches to microeconomic theory. In: Arrow KJ, Intriligator MD, editors. Handbook of Mathematical Economics. Vol. 2. Amsterdam: North-Holland; 1982

[10] SERS (Swedish Electrofishing Registry). [Internet]. 2017. Available from: http://www.slu.se/elfiskeregistret. [Accessed: November, 2016]

[11] Beier U, Degerman E, Sers B, Bergquist B, Dahlberg M. Environmental quality criteria to determine the status of fish in running waters – development and application of VIX. Finfo. 2007;**5**. Fiskeriverket

[12] European Commission. Directive 2000/60/EC of the European Parliament and of the Council of 23rd October 2000 Establishing a Framework for Community Action in the Field of Water Policy. Official Journal of the European Communities, L327/1. European Commission, Brussels. 2000

[13] Cameron AC, Trivedi PK. Microeconometrics Using Stata. Texas: Stata press College Station; 2009

[14] Hosmer DW, LemeshowS. Introduction to the logistic regression model. Applied Logistic Regression. 2nd ed. Hoboken, NJ, USA: John Wiley & Sons, Inc.; 2000

[15] Davidson R, MacKinnon JG. Estimation and Inference in Econometrics. New York: Oxford University Press; 1993

[16] Wooldridge JM. Introductory Econometrics: A Modern Approach. 5th ed. USA: South-Western, Mason; 2013

[17] Gren IM, Baxter P, Mikusinski G, Possingham H. Cost-effective biodiversity restoration with uncertain growth in forest habitat quality. Journal of Forest Economics. 2014;**20**:77-92

[18] Havs och vattenmyndighet. Nationell strategi för vattenkraft och vattenmiljö. [Internet]. 2017. Available from: https://www.havochvatten.se/hav/uppdrag--kontakt/publikationer/publikationer/2014-07-04-strategi-for-atgarder-inom-vattenkraften.html [Accessed: February 2017]

2

Usefulness of Plant Biodiversity in the Cities of Togo

Radji Raoufou and Kokou Kouami

Abstract

In an increasingly urbanized environment, the need for greenery and flowers is being more and more felt for esthetic reasons and ecological benefits. In Togo, more than six hundred species of horticultural plants are identified and composed of approximately 59% of dicotyledons (49 families, 145 genera, and 315 species) and 37.37% of monocotyledons. Pteridophytes and gymnosperms account for less than 6%. The spectrum of morphological types indicates that herbs are account for 55%, while trees and shrubs represent 15%. More than 50% of the species of this flora is exotic horticulture. A species distribution is made according to their decorative parts and their place of use. Across the country, 55 plantings are recorded and unevenly distributed in cities. Apart from their ornamental purpose, ornamental plants are used for feeding, traditional and industrial cosmetics in psychotherapy, horticultural therapy, and in traditional and conventional medicine. In this study, 79 species from 39 families are reported as medicinal plants. The *Apocynaceae* and *Fabaceae* (six species), the *Euphorbiaceae* and *Liliaceae* (five species), the *Arecaceae* and *Verbenaceae* (four species) are the best represented families. Production systems in ornamental horticulture in Togo are very diverse in terms of speculation, access to land (variable surfaces, direct or indirect forms of tenure, acquisition methods, land use, etc.) and socio-economic profiles of farmers (men, women, young, old, people with little or no qualifications, rural to urban, etc.). The family horticultural production system, which represents over 90% of 55 horticultural farms of this study, is the main production system. It is characterized by areas of less than 0.1 hectare and farms in relative land insecurity (97% of land used belongs to the state). Throughout the system, there is a salaried labor representing 5–8% of turnover. Temporary and permanent employees are paid on weekends or at the end of the month. Farmers use gardening equipment and processing plant more or less modern including clippers, shears, pruners, and sprayers. Production units provide direct employment (more than 3 employees per unit) and directly to several hundred people. Horticulture in urban and peri-urban areas improves the living conditions of farmers (income) and the population (embellishment of streets, maintenance or creation of green area buffer) despite some negative externalities associated particularly with the use of prohibited pesticides and uncontrolled use of spaces along the roads. Its survival is threatened by many constraints, including the extension of urban housing and road building. In Togo, beneficial effects of ornamental horticulture may be more noticeable if the political authorities, private stakeholders, and the researchers work together to organize the sector. It could thus participate effectively in the formal economy and the emergence of true development plans at the municipal level.

Keywords: biodiversity, ornamental horticulture flora, socio-economics, profitability, Togo, medicinal plants, parts used

1. Introduction

Urban growth is a fundamental ingredient for global population advancement in the time where towns aggregate an increasing part of this population.

Urban growth is a fundamental ingredient in the evolution of the world's population, insofar as cities have an increasing share of this population. Urbanization represents a culmination of the economic processes in progress, both those that are part of a developmental logic and those that are the result of a breakdown of rural societies. Moreover, urbanization often causes a conflict exacerbation between the population and its environment. Urban forests bring many benefits in terms of sustainability. They help to regulate the urban climate and are a benefit source of biodiversity, but they also constitute spaces for social practices and economic resources.

Created in the early 19th century [1], the word horticulture means, according to the French language dictionary (2008), the cultivation of ornamental plants, vegetables, and fruits. It is synonymous with gardening. In the Larousse dictionary (2008), horticulture is an art of cultivating gardens. It is a branch of agriculture including the cultivation of vegetables, flowers, trees, fruit, and ornamental shrubs. It is divided into two branches:

- food or edible horticulture that includes vegetable crops in the open field, vegetable or vegetable crops, and fruit trees;
- ornamental or inedible horticulture that includes potted plant production, bedding plants and cut flowers, ornamental arboriculture, and nurseries and bulbous plant production.

Acclimatization of plants is becoming more and more a commercial business and the collecting botanists sent on mission by large horticultural firms compete with the traveling botanists of scientific institutions. These explorations are current today and the transport of plants is not without dangers: phytosanitary problems, invasive plants and more recently the question of the ownership of the genetic resource [2].

In Togo, the colonial period (1884–1960) was active in the knowledge of Togo's flora and vegetation, thanks to civilian administrators, ethnologists, foresters, agronomists, doctors, etc. [3]. However, botanical prospections really started only with the creation of the first university (University of Lomé) in 1970. Since then, several authors, including Aké Assi [4], Ern [5, 6], Brunel [7], Brunel et al. [3], Brunel [8], and Scholz and Scholz [9] have contributed to improve this knowledge. The synthesis of all these works resulted in the publication of the Flora of

Togo [3], where 2500 species are described botanically. Since then, botanical investigations are continuing to complete this list [10]. Botanical harvests of Akpagana and Guelly [11] have increased this number of 235 new species. The works of Batawila [12], Kokou [13], Kokou et al. [14], and Akpagana and Gumedzoe [15] enabled a more exhaustive census of the plant species of Togo's spontaneous flora.

In Togo's flora, 40 species are reported in note as introduced plants used for ornamental purposes in parks and gardens [3]. From 1997 to date, the work of Radji has made it possible to formally establish Togo's horticultural flora.

These various studies have allowed, according to the major plant groups, to count about 3451 species gathering the spontaneous flora and that introduced. However, although rich enough, Togolese flora is still incompletely known [13, 16, 17].

2. Aims

The usefulness of this chapter is to constitute a part of the best knowledge of Togo's flora in particularly its horticultural flora made up to more than 90% of exotic plants, while the native local flora remains poorly known. It is also an interpellation of the botanist community from local universities and research institutions in botany and ethnobotany, to deepen the knowledge of the local flora, especially its valorization in many branches like the horticultural sector which is gaining considerable momentum in the country, by proposing a greater diversity of local ornamental plants.

3. Location of the study area

Lomé city and its surroundings in south, Kara city in the north and Atakpamé and Kpalimé in center was explored for floristic inventory (**Figure 1**). No flower planting has been identified in the other cities in Togo. However, the study takes into account the landscaped areas and the private gardens in public or private administrative institutions. Each planting or landscaped areas or garden was considered as a botanical survey of 68 plantings across the country, including 65 in Lomé and its surroundings and one in each of the three cities (Atakpamé, Kara, and Kpalime).

The identification of species was made with the flora of Berhaut [18], Byrd Graf [19], Houerou and Houerou [20], Grisvard et al. [21, 22], and Hessayon [23]. Further information was collected from those of Hutchinson and Dalziel [24, 25], Brunel et al. [3], and Lebrun and Stork [26–30]. The nomenclature used is that of the mentioned authors. Data from the International Code of Botanical Nomenclature [31] and those online as the Index Nominum Genericorum (ING) [32], the International Plant Names Index (IPNI) [33], and the Harvard University Herbaria (HUH) [34] were used for the names of authors and synonyms. The classification of ornamental plants followed that of Vidalie [35] and Aké-Assi et al. [36]. The works of Radji et al. [37, 38], Aké-Assi et al. [36], and Porter et al. [39] allowed clarifying the origin countries of the plants.

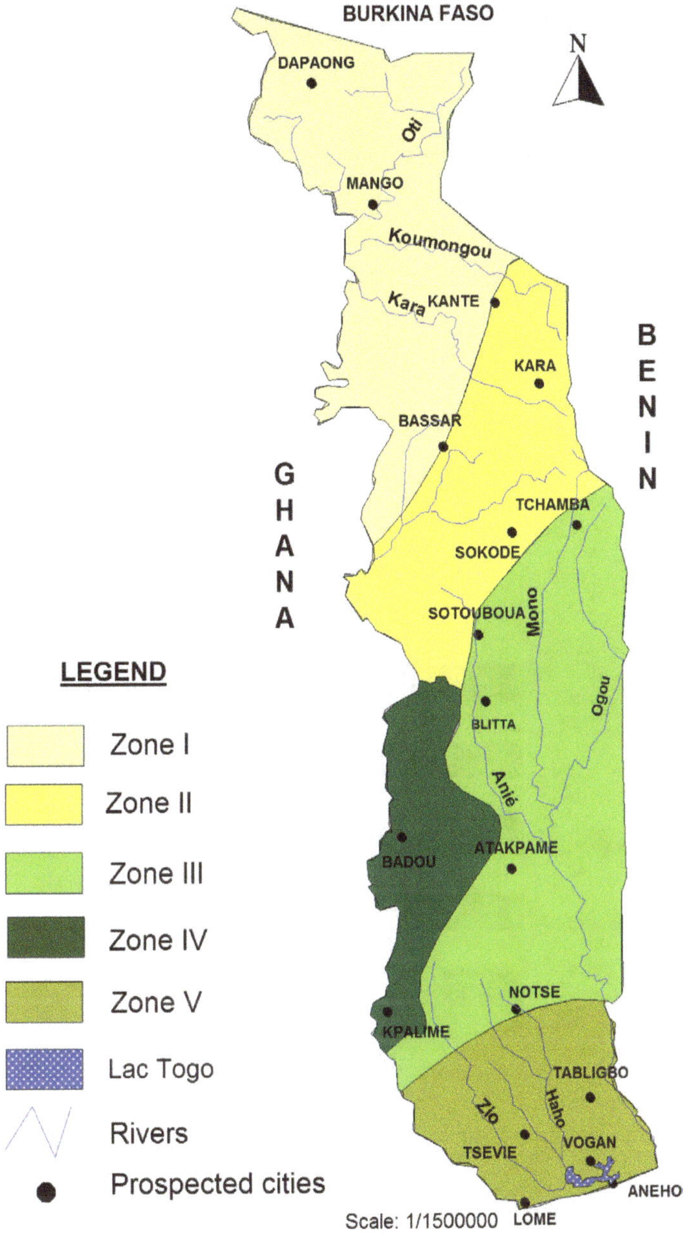

Figure 1. Togo eco-floristic subdivisions and major cities surveyed.

In each planting, an ethnobotanical survey has covered therapeutic uses of the identified species. This work was conducted with 279 respondents. The data collected were supplemented with two Traditional Practioners in Lomé. They were subsequently brought into line with those that already existed in the reference literature [36, 40, 41]. The ethno botanical information collected were recorded on sheets of raw data and then transferred to a database. They were processed and analyzed under RGui 2.7.0 statistical software [42] to obtain standardized data concerning the key constituents, therapeutic properties and uses assigned to each reported species, the parts used, and the most commonly mentioned methods of preparation. The status of rare species is indicated by an index of scarcity RI obtained from the equation of Géhu and Géhu [43], where ni indicates the number of readings in which the species i is present and N is the total number of readings.

4. Organization of Togo's ornamental firms

In 2016, the number of horticultural exploitations listed in Togo was 68 geographically distributed as; one farm in Kara, one in Atakpamé, another one in Kpalimé, and 65 in Lomé. From 1997 to 2010, the number of horticultural farms (22 in Lomé in 1997) has increased by 66.66%. Ninety-four percent (52 farms) are located in the "administrative reserves" belonging to the government area and six percent are setting on individual private property (**Figure 2**).

More than 90% of the nurseries are private family farms. Their average size is 291.47 m^2. (**Figure 3**). The gap (202.76) is very high, showing a great variability of the surface area of horticultural farms.

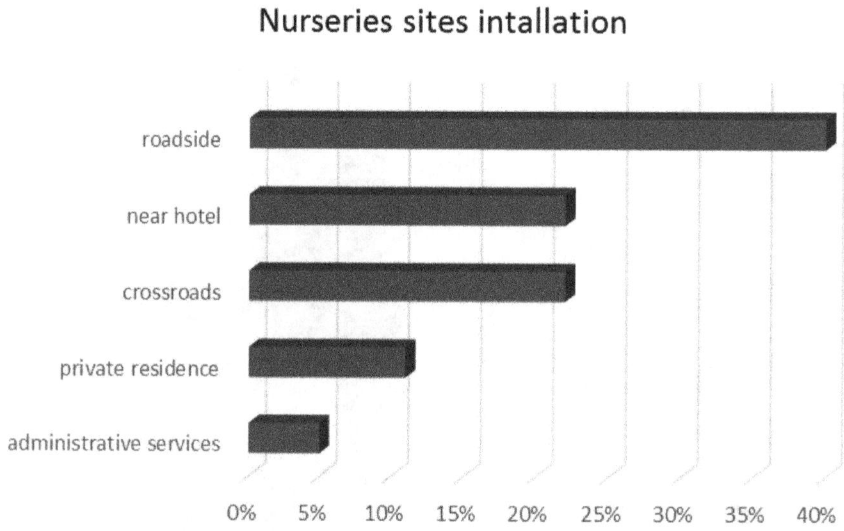

Figure 2. Distribution of nurseries according to places of installation.

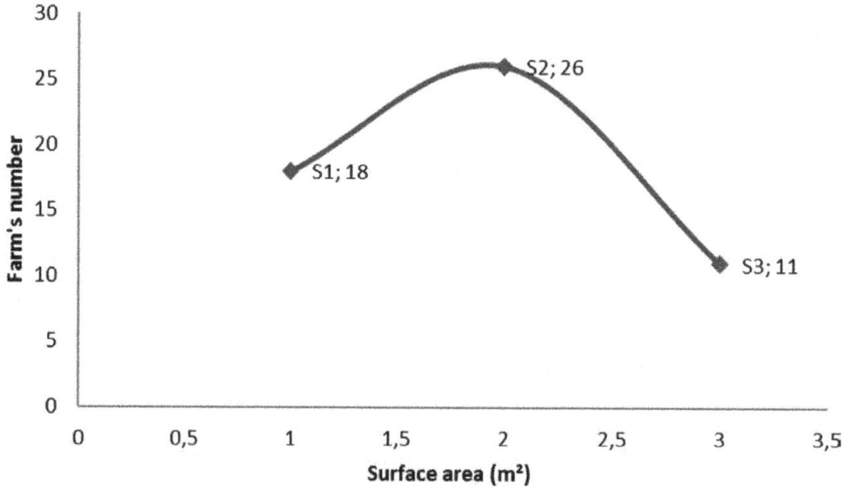

Figure 3. Distribution of farms according to dimensions (m^2). The figures indicate the number of nurseries concerned, S1 = farms having less than 150 m^2; 150 < S2 < 500 m^2; S3 = farms having a surface area above 500 m^2).

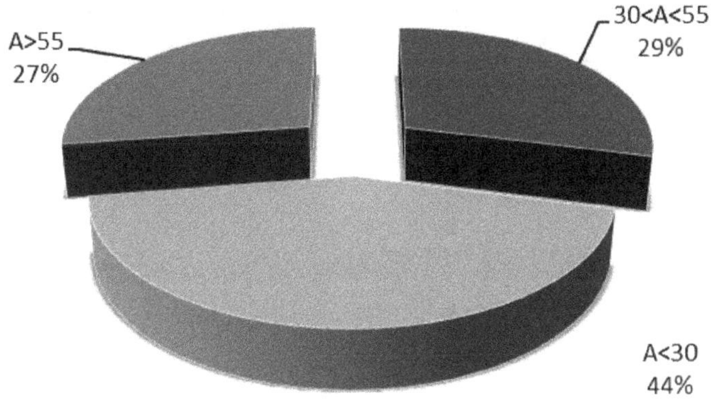

Figure 4. Age bracket of the farmers (A = age).

Otherwise, 179 people work in the ornamental horticulture sector in Togo and 58% are male. The farm lords are predominantly the youth: 44% are less than 30 year old (**Figure 4**), 29% aged between 30 and 55 years, whereas 27% are more than 55 years.

5. Togo ornamental biodiversity

5.1. Floristic richness

In Kara (north Togo) like Atakpamé (Togo center), the inventory allowed to identify 123 species distributed in 42 families and 93 genera. In Kpalimé (west), 138 species were identified and divided into 54 families and 103 genera. In Lomé (south and capital), the number of species identified is 612 which divided into 246 genera and 80 families. All the species identified outside Lomé (estimated at 380 species) are at 100% found in the list of the species identified in Lomé and its surroundings. However, nearly 500 species accounted in Lomé are not yet grown in the other cities.

In term of genus, dicotyledonous represent 59% of Togo's ornamental plants; then Monocots are 38% and Pteridophyta and Gymnospermae are less than 6%. Otherwise, all dicotyledonous belong to 49 family, 145 genera, and 315 species **Table 1**.

The average of the floristic richness is 142.91 ± 57.92 species per planting and the equation $y = 0.069x + 0.449$ with $R^2 = 0.236$ justifies the relationship between the floristic richness and the age of the nursery (**Figure 5**). This significant relationship shows that the experience acquired by the planting operator over time, especially by taking into account the demands and requirements of customers and the need to meet customer demand, is a major factor that can improve the species richness of horticultural planting.

5.2. Preferential species

This study shows that 274 species representing 44.77% of the species surveyed are the "preferential species" (RI < 80%). As listed, there are 15 most preferential species: *Aloe vera* L. var. *chinensis, Bougainvillea glabra, Catharanthus roseus* (L.) G. Don, *Cordyline terminalis* (L.) *Kunth., Dieffenbachia amoena, Dracaena arborea* (Willd.) Link., *Erythrina indica* Lamk., var. *picta* L. *Euphorbia millii* Des Moul. var. *breonii, Ficus bengamina* L., *Ficus retusa* L., *Ixora macrothyrsa,*

Fragrant plants		
Family	Taxa	Part exhaling gasoline
Annonaceae	*Cananga odorata* (Lam.) Hook. f. & Thoms.	Flowers
Agavaceae	*Polianthes tuberosa* L.	Flowers
Caesalpiniaceae	*Cynometra megalophylla* Harms	Leaves
Moraceae	*Artocarpus communis* J.R. & G. Forst	Fruits
Oleaceae	*Jasminum nitidum Skan*	Flowers
	Jasminum sambac Ait.	Flowers
Poaceae	*Cymbopogon citratus* (DC.) Stapf	Leaves
Rutaceae	*Murraya paniculata* (L.) Jacq.	Flowers

Table 1. Fragrant plants.

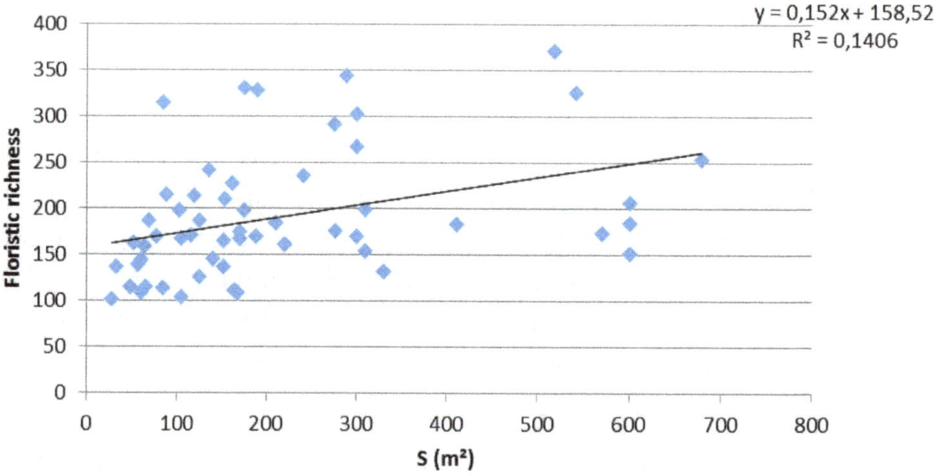

Figure 5. Relationship between floristic richness and age of nursery.

Murraya paniculata, Nerium oleander, Pithecellobium dulce (Roxb.) Thunb., and *Polyalthia longifolia.*

With RI ≥ 80%, 338 species (55.23%) are "rare" and were listed in less than 10% of the readings. Among them, 67 have only one occurrence (RI ≥ 98.18%). This is the case of *Calathea makoyana E.* Morr. & Boom, *Cryptostegia grandiflora* Br. R. ex Lindl., *Davallia bullata* Wall. Ex Hook., *Echinocereus pectinatus* Engelm., *Ficus elastica* Roxb. var. *decora, Monstera deliciosa* Liebm., *Oxalis ovate* Ruiz ex Knuth, A. *Tithonia diversifolia* A. Gray, *Vanda coerulea* Griff. ex Lindl., and *Zamia furfuracea* L.f.

6. Plants provenance of Togo's horticultural flora

Fifty-two percent of the horticultural flora species in Togo are coming from Americas against less than 20% from Africa (**Figure 6**). Outside the Americas (30%) and Asia (25%), each of

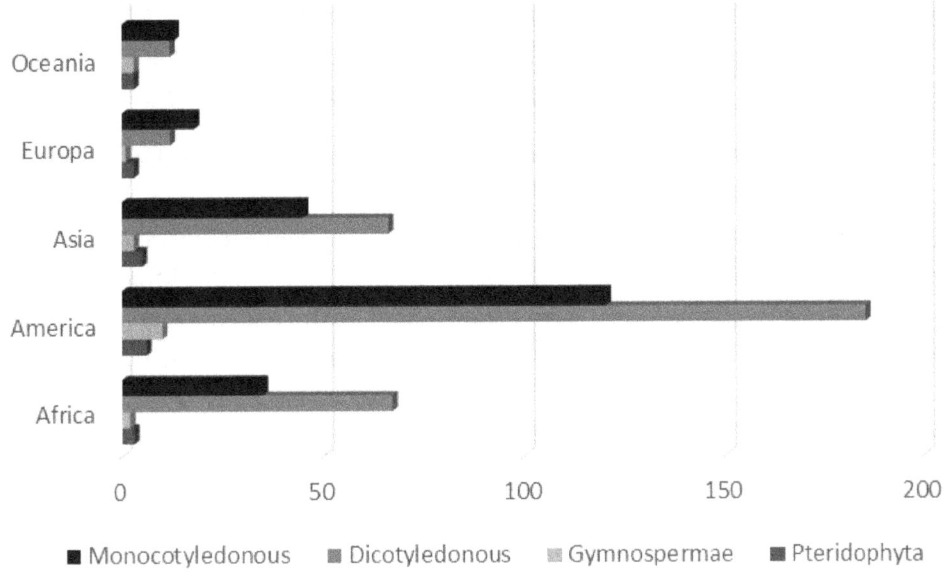

Figure 6. Species distribution according to taxonomic groups and continents.

the three other continents include 15% of Pteridophyta identified. The Gymnosperms are 53% from the Americas and only 10% are from Africa. Concerning the monocotyledons, the species originated from Africa account for only 15% of the total. This percentage is 20% for dicotyledons. Among dicotyledons, the plants originating from the Americas include only 54% of the total.

7. Species distribution according to decorative parts

In this study, 311 species grouped into 106 genera and 30 families are grown and used in Togo as ornamental decorative foliage plants. The most representative families are, respectively, *Arecaceae* with 18 genera and 22 species, *Araceae* with 14 genera and 53 species, *Acanthaceae* with 12 genera and 16 species, and the *Euphorbiaceae* with 5 genera and 31 species.

Twelve families including 34 genera and 69 species have ornamental architecture. The following species are listed as an illustration: *Polyalthia longifolia* (Sonn.) Hook. f. & Thomson (*Annonaceae*) *Ravenala madagascariensis* Gmel. J.F. (*Strelitziaceae*) and *Terminalia mantaly* H. Perrier (*Combretaceae*).

It is about 166 species from 37 families of monocotyledons and dicotyledons. This is the case of *Allamanda cathartica* L. (*Apocynaceae*), *Gardenia jasminoides* Ellis (*Rubiaceae*), *Guaiacum officinale* L. (*Zygophyllaceae*), *Hibiscus rosa-sinensis* L. (*Malvaceae*), *Mussaenda philippica* A. Rich. (*Rubiaceae*), and *Plumbago capensis* Thunb. (*Plumbaginaceae*).

Nine species are grown for the scent of their flowers, leaves, and fruits. They are from eight families (**Table 1**).

In the list of horticultural plants in Togo, three families with only one genus are ornamentals by their fruits. Families are *Arecaceae, Bignoniaceae,* and *Moraceae.* Species concerned are *Cocos nucifera* L., *Crescentia cujete* L., and *Artocarpus altilis* (Parkinson) Fosberg.

Depending on the season, the characteristics previously described (foliage, habit/look, flowering, and fruit) can be combined to give the plant its ornamental or decorative appearance. It may be the foliage and flowers (as for *Adenium obesum* (Forssk.) Roem & Schult. and *Plumeria rubra* L. from the *Apocynaceae* family; or flowers-habit/look association (as for *Guaiacum officinale* L. (*Zygophyllaceae*); or foliage and inflorescences (case of *Bougainvillea glabra* Choisy *variegata, Ligustrum ovalifolium* Hassk., and the aquatic plant *Victoria regia* Lindl.).

8. Distribution of ornamental species according to their use

The inventory gives 35 species grouped into 17 families and 26 genera used as curbs, walkways or paths plants. There are generally decorative trees and shrubs by their leaves and look and seasonally by their inflorescences such as *Khaya senegalensis* and *Delonix regia.*

Others used as outdoor garden plants are trees, shrubs, and herbs potted, installed outdoors or putted in the ground on lawns. Isolated plants, lawns plants, bedding, ornamental or protective hedges, ground cover or coating plants, and water decorative plants can also be distinguished.

Isolated plants are 49 species grouped into 34 genera belonging to 21 families. For a species to be planted alone in a garden, it must offer a spectacle of beauty either by its foliage (*Nerium oleander variegatum*) or by its look (*Araucaria excelsa* R. Br and *Cycas revoluta* Thunb.) or by its flowers (*Mussaenda philippica, Hibiscus rosa-sinensis, and Polianthes tuberosa* L.) or its fruits (*Crescentia cujete* L.).

The *Poaceae* are mostly used in Togo as lawn plants. In alphabetical order, this study was collected: *Chrysopogon aciculatus* (Retz.) Trin, *Cynodon dactylon* Pers, *Paspalum distichum* L. *Stenotaphrum secundatum* (Walt.) Kuntze. *variegatum,* and *Zoysia tenuifolia* Trin.

A part from lawns, other plants are used to cover non-grassed bare surfaces. These include creepers such as *Episcia cupreata* (Hook.) Hanst. (*Gesneriaceae*) or twining plants such as *Ipomoea quamoclit* (*Convolvulaceae*). Depending on the structure of the plant, some species are used as carpets; this is the case *Wedelia trilobata* (L.) Hitch. (*Asteraceae*) or in wall carpet as the case of *Ficus pumila* L. (*Moraceae*).

Twenty-eight species grouped into 18 genera and 15 families are bedding grown ornamental plants. Species commonly inventoried are *Acanthus mollis* L. *Barleria lupulina* Lindl. (*Acanthaceae*), *Canna generalis* L. H. Bailey (*Cannaceae*), *Lantana camara* L. (*Verbenaceae*), *Thunbergia erecta* (Benth.) T. Anders., *Turnera ulmifolia* L. (*Turneraceae*), and *Yucca aloifolia* L. (*Agavaceae*).

Ornamental hedges include 33 species of 18 genera and 14 families. Species commonly appreciated are *Clerodendrum inerme* (*Verbenaceae*) for its dense foliage, *Bougainvillea glabra* (*Nyctaginaceae*) for its purple flowers and its thorns, *Pithecellobium dulce* (*Fabaceae*) mainly for its thorns and its dense foliage when the plant is young, and *Pereskia grandiflora* (*Acanthaceae*) for its thorns and ease of its cuttings pushing.

As water ornamental plants, two families are identified: *Pontederiaceae* represented by *Eichhornia crassipes* (Mart.) Solms. and *Nymphaeaceae* with 4 genera *Euryale, Nelumbo, Nymphaea,* and *Victoria.*

9. Indoor garden plants

House plants are species that fit inside houses and apartments: balconies, verandas, hallways or inside the offices. 193 species grouped into 44 genera and 17 families are identified as house plants. The genera *Pteris* (*Adiantaceae*), *Asplenium* (*Aspleniaceae*), *Aglaonema, Alocasia, Caladium, Dieffenbachia, Monstera* and *Philodendron* (*Araceae*), *Schefflera* (*Araliaceae*), *Oxalis* (*Oxalidaceae*) and *Licuala, Kentia,* and *Livistona* (*Arecaceae*) can be mentioned.

Several species are grown in order to provide useful elements to the preparation of floral bouquets. These bouquets are placed in clay pots or glass jars for home decor. Among the species recorded and identified, we have the genera *Caesalpinia and Melia* for inflorescences *Alpinia, Anthurium, Aster,* and *Chrysanthemum, Polyanthes* and *Heliconia* for flowers, *Alpinia, Cordyline, Cycas, Dracaena, Maranta,* and *Sansevieria* for the leaves, and *Cyperus* for the stem and the leaves.

10. Ornamental plants for therapeutic purposes

Seventy-seven ornamental species belonging to 39 families are listed as medicinal plants. The most represented families in number of species are the *Apocynaceae* and *Fabaceae* represented by six species, *Euphorbiaceae* and *Liliaceae* five species, and *Arecaceae* and *Verbenaceae* with four species are the most represented families.

The ethnobotanical survey revealed that 49% of the identified species are used against diseases of digestive system and 17% for dermatoses. Visual, bone, urinary, and auditory parts of the nervous system is represented by less than 7% (**Figure 7**).

Some species are used to treat many diseases and the use of others requires a combination with non-ornamental plants.

For about 43% of the species, the leaves are the most solicited parts (**Figure 8**). They are followed by, in descending order, the association stem leaves (27%), the underground parts, the bark, the flowers, the fruits, and the entire plant.

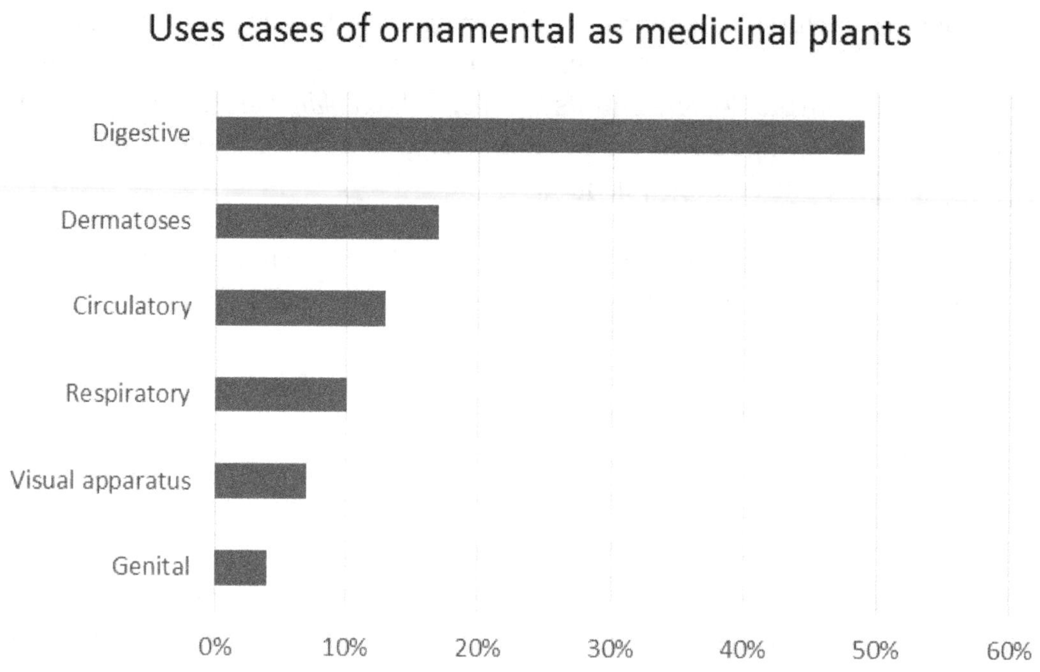

Figure 7. Uses cases of ornamental medicinal plants.

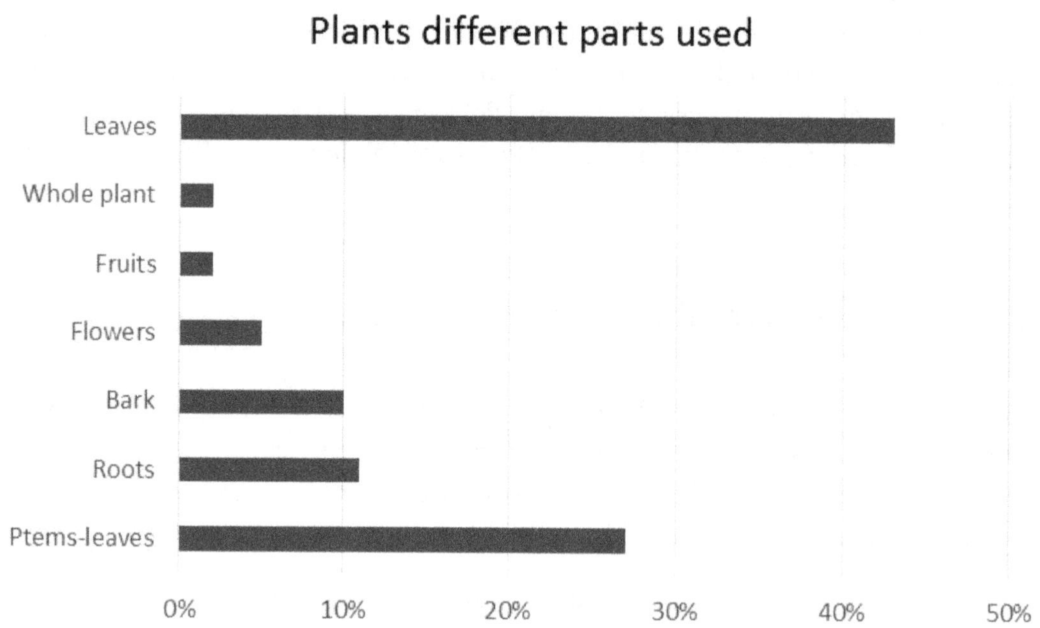

Figure 8. Distribution of plants different parts used.

Decoction is the most common method of preparation (47%). It is followed by the poultice (17%) and the bath (13%). The other methods (maceration, nature, fumigation, infusion, and powder) represent 23% (**Figure 9**). Among the modes of administration, the most used is the oral absorption (82%).

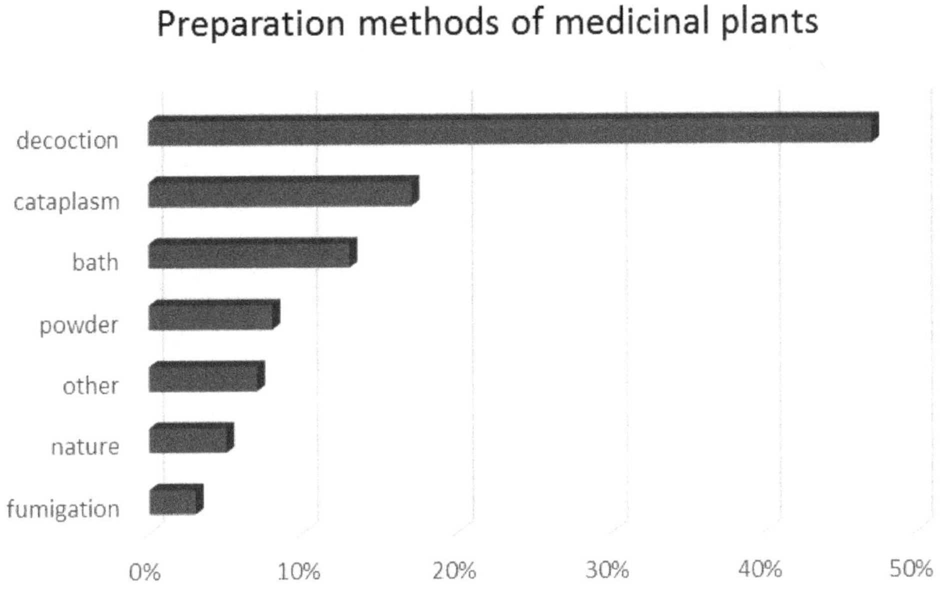

Figure 9. Spectrum of different methods of preparation of medicinal plants.

11. Other uses identified for ornamental plants in Togo

Although the first reason for their cultivation is ornamentation and beauty, many ornamental plants are used for other purposes. In Togo, the present study identified: plants for psycho-magic, feeding, traditional or industrial cosmetics (**Table 2**), and toxic plants (**Table 3**).

For food plants, fruits and leaves (*Eugenia malaccensis* L. *Gomphrena globosa* L.) are the most used while wood is very useful as timber (*Gmelina arborea* Roxb.) in buildings (*Acacia auriculiformis* L.) and as firewood (*Cassia siamea* Lam.).

For human consumption, 15 plant species was identified, 3 others are reported for traditional cosmetics and 11 in the agro-food system, industrial cosmetics, and wood industry.

12. Discussions

The farms surveyed are less than 0.1 hectare in size. This constraint prevents the development of the sector, its modernization and competition, and making its participation in the formal economy invisible [44, 45]. The equipment is basic and the use of inputs, especially fertilizers, is minimal. Squatting (illegal occupation) of land is the dominant tenor.

Although entrepreneurial horticultural production systems constitute only 5.45% of horticultural holdings, they hold almost the entire public market as they have completed the administrative formalities toward the municipal administration and paid the taxes. They are private farms characterized by the use of permanent wage labor, the use of modern tools (lawn mower, hedge shears, and motorized sprayer). However, employees have no specialty in horticulture. The main destination of production is local consumption. The ornamental horticultural sector in Togo is therefore not registered for export.

Scientific name	Organs consumed	Products obtained
Food		
Artocarpus communis J.R. & G. Forst.	Fruit	
Caladium bicolor (Ait.) Vent.	Leaves and bulb	
Chrysophyllum albidum G. Don	Fruit	
Citrus maxima; C. sinensis	Fruit (pulp and juice) and flowers	Beverages, jams, flavoring, and beekeeping
Cocos nucifera L.	Fruit	
Cycas revoluta Thunb.	Marrow	
Elaeis guineensis Jacq.	Seed, sap	Fermented beverages, alcohol, and vegetable oils
Eugenia malaccensis L.	Fruit	
Gmelina arborea Roxb.	Leaves	Food package
Gomphrena globosa L.	Leaves	
Mangifera indica L.	Fruit (nature, dried or cooked)	Jams, marmalades, jellies, compotes, and alcohol
Pithecellobium dulce (Roxb.) Benth.	Fruit (aril)	
Punica granatum L.	Fruit (pulp)	Jams
Samanea saman (Willd.) Merril.	Clove	
Terminalia cattapa L.	Almond	
Cosmetic		
Bixa orellana L.	Seed	Dyestuff
Elaeis guineensis Jacq.	Fruit	Palm kernel oil
Lawsonia inermis L.	Leaves and flowers	Dyestuff, tincture, perfume, and toothpicks
Industrial		
Acacia auriculiformis L.	Wood	Cabinetmaking
Azadirachta indica A. Juss.	Wood, seeds, and gum	Softwood lumber, firewood and charcoal, oil, tincture, lubricants, disinfectants, cosmetics, and insecticides
Cananga odorata (Lam.) Hook. f. & Thoms.	Flowers	Perfume
Cassia siamea Lam.	Wood	Cabinetmaking, firewood, and charcoal
Catharanthus roseus L.	Leaves, flowers, and fruits	Drugs
Citrus maxima; C. sinensis	Wood and branches	Woodwork, turning, and marquetry

Scientific name	Organs consumed	Products obtained
Cocos nucifera L.	Fruit	Vegetable oil consumption
Ealeis guineensis Jacq.	Seed, fruit, and stipe	Palm kernel cake, construction, bridges, straw, fencing, and brush
Gmelina arborea Roxb.	Wood	Frames, poles, woodmaking, sculpture, crates, plywood, Firewood, and for matches
Lantana camara L.	Leaves	Sandpaper to polish wood
Mangifera indica L.	Green or dried fruit, wood	Fuel, construction, tincture, and black ink
Psycho-magic		
Aloe vera	Leaves	Negative influences and household accidents protection
Bambusa vulgaris	Trunk, leaves, and stems powder	Protection against negative energies, bad luck, brings luck, and fortune
Croton zambesicus Müll Arg.	Whole plant	Protection and hunting evil spirits
Thevetia neriifolia Juss.	Fruit	Ordeals and divinations

Table 2. Other uses identified for ornamental plants.

Scientific names	Part used	Main chemical composition	Effects
Acalypha hispida Burm.f.	Whole plant	Steroids, saponoides, polyphenols, tannins, and cyanogenic derived	
Adenium obesum (Forssk.) Roem. & Schult.	Latex and root	Cardiotoxic heterosides	Dangerous for eyes and cause of violent diarrhea
Allamanda cathartica L.	Whole plant	Cardiotoxic heterosides	
Alocasia macrorhiza Schott.	Whole plant	Alkaloid, oxalate de calcium	
Aralia balfouriana Bailey	Whole plant	Saponin, alkaloid	
Asclepias curassavica L.	Whole plant	Cardiotoxic heterosides	
Caesalpinia pulcherrima (L.)Sw.	Racines and seeds	Cyanhydric acide	
Caladium bicolor (Ait.) Vent	Whole plant	Calcium oxalate	
Calotropis procera (Aiton) W.T. Aiton	Whole plant and latex	Cardiotoxic	
Crescentia cujete L.	Pulpe, fruit, seed	Cyanhydric acid and acids oleic acid	
Cycas circinalis L.	Whole plant	Methylazoxyméthanol, and alkylating agents	

Scientific names	Part used	Main chemical composition	Effects
Cycas revoluta Thunb.	Whole plant	Carcinogenic alkaloid and neurotropic agent	
Duranta repens L.	Fruit	Saponin (sterols)	
Erythrina indica Lam.	Whole plant	Saponin (sterols), alkaloids, and acid cyanhydrique	
Euphorbia millii Des Moul.	Latex	Euphorbone, résine, caoutchouc, substance non identifiée crystallized into needless	
Euphorbia turicalli L.	Latex	Euphol, tirucallol, tinyatoxine, triterpénoïdes (euphorbinol, cycloeuphorbinol, euphoron, 31-nortriterpen cycloeuphordénol), and macrocyclic diterpen (tirucalicin)	Very caustic and toxic, makes blind, and irritating to the stomach
Jatropha multifida L.	Leaves and seeds	Saponosides, tannins, sterols, and alkaloids	
Lantana camara L.	Leaves and flowers	Toxic terpenoid	
Pedilanthus tithymaloides (L.) Poit.	Seeds	Caustic latex	
Rhoeo spathacea (Sw.) Stern	Whole plant	Anthocyan βD-glucan	
Setcreasea purpuracea Boan	Sap		
Thevetia neriifolia Juss. ex Steud	Whole plant	Cardiotoxic heterosides	
Thevetia peruviana (Pers.) Merr.	Whole plant	Cardiotoxic heterosides	

Table 3. Ornamental plants with toxic constituents.

The cultivation of cut flowers is fairly technical and requires technical training. Large companies operating in Africa belong to European groups that market in the European community and participate in major international exhibitions. Currently, the best way for the African horticulturist to integrate this commercial network is to outsource, as recommended by IQRHH [46] to Quebec horticulturists. Indeed, it is a question of settling near big companies to benefit from advice, technical aids and outlets without having to risk a depreciation of the goods during the transport, and the search of a final clientele. Organizing in a formal grouping (ADF, 2007) is also a solution. Such a grouping is a success in various countries exporting other products (coffee, cocoa, etc.). Government assistance through chambers of commerce and industry to the organization of the sector is a necessity and a non-negligible source of foreign exchange for producers but also for the state. Everything argues in favor of this sector in Togo which has land, labor, and the climate allowing the development (reproduction and harvesting) of crops throughout the year. Technicity remains to be formed.

Like foliage and flowers, the colorful fruits are used to beautify the gardens. Fruit size, shape, and appearance contribute to the recognition of taxa. Of a generally bright color (yellow, orange, red, green-yellow, etc.), the fruits are easily enhanced in the greenish foliage.

Age, size, shape, and ecological requirements can provide information on the use of plants. Among these criteria, the ecological requirements are decisive for the place of a harmonious development of the plant. According to these requirements, three main categories are identified for the place of use of ornamental plants:

- plants of shade or alignment along avenues and roads;
- outdoor garden plants: planted in isolated, massive, ornamental or safe hedges, ground cover, and water plants;
- indoor garden plants that can be installed on the balcony or in apartments.

Moreover, the use of ornamental plants is related to the type of housing and the standard of living of the populations. We can thus distinguish three types of populations [40]:

- the wealthy population is found in the "common courtyards", rental dwellings of which the inhabitants do not own. In these dwellings, the inhabitants are placed in front of their room, the ornamental plants in the cement pots. These are usually shrubs or herbs. The few trees found there are planted in the middle of the common courtyard and serve as species of shade and rest.
- The population with an average level of living resides in more spacious dwellings with an interior space where they can install pots with indoor plants or a cemented terrace, where they also have pots with green plants, variegated foliage or able to bloom. In these types of concessions, the external space is cemented and does not allow the installation of an outdoor garden requiring resources for its maintenance.
- The high-income population is found in residential neighborhoods or in private and individual dwellings. Outside the pots of indoor plants or terraces/balconies, an outside garden is implanted. It is often composed of lawns, flower beds, hedges, and odoriferous plants and requires a labor often permanent for its maintenance.

The use of ornamental plants, in relation to the living environment, informs about the standard of living and cultures. Whatever the standard of living of the population, the current trend is to have a plant in its place of residence [38] and many people are convinced that contact with trees and other [47, 48]. For the population still in rental house or not very well, this is expressed by the presence at the portal of a pot containing a plant symbol, usually an herbaceous. This is the case of *Setcreasea purpurea*, Aloe vera, *Rhoeo spathacea*, *Euphorbia millii* or a shrub (*Schefflera arboricola*, *Codiaeum variegatum*, etc.). For the average or well-to-do population, it is a garden of varying size with a flowerbed and a flower-lined terrace. The type of pot and its contents are also indicators of the standard of living of the owner of the concession.

The work of Waliczek et al., (2005) supports the idea that gardening is a hobby that can enhance the level of satisfaction with life and improve general health [49].

There are plants inspiring all sorts of beliefs [50, 51]; some are known as beneficial or lucky (*Croton zambesicus*), and others evil or mischievous (case of Cactaceae). The presence or absence of a type of plant in the life environment can have a positive or negative influence on the activities of the inhabitants of this framework. This fact is called superstition. Indeed, superstition is the belief in the manifestation of mysterious forces linked to acts, objects or phenomena [52]. Biley [50] describes different types of trees that can be planted in the perimeter of a hospital where patients, visitors, and employees will not only benefit from the esthetic appeal of trees but also mythical powers. Vendors and users of ornamental plants report that plants with abundant flowering or pleasantly perfumed flowers are lucky plants. This is the case of *Mussaenda philippica, Cananga odorata,* and *Murraya paniculata. Setcreasea purpurea* is also listed in this category. They attribute to Draceana fragrans massangeana the role of removing the quarrels from the homes, thus ensuring by their presence in the concession a certain tranquility and security to the couples [40]; to *Euphorbia tirucalli* the role of moving creeping animals away and to *Croton zambesicus,* that of keeping evil spirits away. The color of the flowers also has an interpretation. White symbolizes peace, reconciliation, reunion, the red expresses a sincere love, and the pink is in relation with the friendship. On the other hand, plants whose organs exhale an unpleasant odor are termed dishwashing plants and sellers avoid cultivating them on their site [38].

Despite the extensions of the Togolese cities, the social stratification is not pronounced. All types of housing can be found in all neighborhoods. Apart from residential cities, spatial segregation [40] is not characteristic of cities in Togo.

Plants are still the first reservoir of new drugs. Approximately 75% of drugs are of plant origin and each year new drugs are developed [53]. African countries have a long tradition of medicine and traditional medicinal plant-based know-how. Approximately 80% of the populations in developing countries use traditional medicine for primary health care [54–58], either through cultural tradition, or due to the lack of other alternatives, such as the difficulty of accessing conventional care or the higher cost of conventional medicines [59]. The majority of the medicinal species in this study are used 49% to treat diseases of the digestive system and 17% against skin diseases. This is in line with the work of Mehdioui and Kahouadji [60] who find 50% and 15%, respectively, for the same conditions.

The diversity of parts of the plants from which natural medicines are extracted is surprising. In addition to leaves and flowers, sap (Aloe vera), bark (*Khaya senegalensis*), seeds, fruits, wood (Gaiacum officinale), walnut (Cocos nucifera), stem, resin, straw, tuber, bulb, and roots ([61, 62]). These organs are used raw, dried or extracted in "sodabi" and local alcoholic beverage [38]. In this study, leaves (43%) are mostly solicited and this confirms the work of Mehdioui and Kahouadji [60], Poffenberger et al., (1992) in Ouattara [63], Zihiri (1991), and Adjanohoun and Aké Assi [64]. If the value of 30% found by Mehdioui and Kahouadji [60] seems lower than that of 43% of the present work, the other authors found percentages ranging from 50.90% for Vangah (1986) to 64.49% for Zirihi [65]. Poffenberger et al. (1992) quoted by Ouattara [63] estimates that harvesting 50% of the leaves of a tree would not significantly affect its survival.

In addition, the most used method of preparation in this study is the decoction (47%). This result is very close to the 42.30% established by N'Guessan et al. [66] and equals the 47% found by Mehdioui and Kahouadji [60]. On the other hand, it is very much higher compared to the 32.94% reported by Adjanohoun and Aké Assi [64]. As for the mode of administration of the drugs, the buccal absorption is solicited to 82%. If this mode is similarly reported in other studies, its value is far superior to that of N'Guessan et al. [66] which indicate 48.97%, to that of 32.35% established by Ouattara [63] or of 27.06% indicated by Adjanohoun and Aké Assi [64].

Ornamental horticulture that incorporates the cultivation of medicinal plants could reduce the pressure on the medicinal plant species most used in traditional pharmacopeia. In the case of rare, threatened or overexploited plants for commercialization, cultivation is the only way to obtain the necessary plant quantities without further compromising the survival of these species [67].

13. Conclusion

More than 600 species divided into 20 Pteridophyta, 17 Gymnosperms, and 575 Angiosperms are the constituent elements of Togo ornamental flora taxonomically predominated by dicotyledonous with *Rubiaceae* and *Annonaceae* as preferential families. Among the monocotyledonous, the strongly represented families are *Araceae* and *Liliaceae*. These species are diverse across continents and over 82% are exotic to Africa. They are classified on the one hand, according to the ornamental organs, in particular the leaves, the flowers, the pace or the port, and the fruits and, on the other hand, depending on the place of use as an interior garden (apartment and balconies), outdoor garden (lawns, massifs, siding, and hedges), and alignment plants along streets and avenues. Depending on the presence or not of these plants and their diversity in a dwelling, a social stratification can be envisaged. Of the species listed, 77 clustered in 39 botanical families are used as medicinal plants. The most representative families in terms of specific richness are *Apocynaceae, Fabaceae, Euphorbiaceae, Liliaceae, Arecaceae,* and *Verbenaceae.* This study revealed that 49% of the listed species are used against diseases of the digestive system and 17% for dermatoses.

1. Apart from decorative and medicinal use, horticultural species are used for other purposes especially in traditional therapeutics. Despite advances in therapeutics, there is room for new medications for poorly tolerated or accustomed active substances, for new or emerging microbial strains or for disarming against a number of mainly tropical diseases. The exploration of the resources of the vegetable and horticultural world remains current. The wild destruction of forests deprives humankind of a source of material essential for the discovery of new molecules necessary for the development of future medicines. Is crop production in ornamental horticulture not a panacea to this destruction and to the *in situ* or *ex situ* conservation of overexploited species and put in a situation of threats or even disappearance?

Ornamental horticulture in Togo is characterized by the weakness of the areas exploited and the inorganization of the sector. It makes it possible to satisfy in part the ornamental plants needs of the capital, to participate in the generation of direct and indirect income, to partially reduce unemployment and to improve the environmental landscape of cities. The mode of acquisition and exploitation of the spaces is precarious as soon as the cities of the country and especially Lomé (the capital) are under construction especially in the construction of road infrastructures. This work, as is currently the case for operators located on the Boulevard du Mono (Lomé—Benin Border) axis, force operators installed on public roads to forcible removal, plunging the sector into precarious situations. The maintenance of a balance between urbanization and ornamental horticulture may exist provided that the authorities concerned are aware of its importance in improving the quality of urban life, in supplying cities with ornamental species and diversity, the elimination of unemployment and participation in the national economy. It is therefore important that all the actors on the ground have an integrated and concerted approach not only to improve the performance of production systems but also to take horticultural holdings into account in urban planning schemes.

As part of the policy of embellishing cities, flowers can play a very important role. It is therefore necessary to take action to enable the nascent horticultural sector to play its full part in this policy and in the economy of the country. The training of a skilled workforce would be an asset for its development.

Author details

Radji Raoufou* and Kokou Kouami

*Address all correspondence to: pradji@hotmail.com

Laboratory of Forest Research, Faculty of Science, University of Lomé, Lomé, Togo

References

[1] Durnerin A. Les grandes expéditions botaniques du XVIIe au XIXe siècle et le transport des plantes. In: Le voyage des plantes: Les jardins, acteurs culturels de la biodiversité. Deuxième Cahier du Conseil national des parcs et jardins; 2008. pp. 10-17

[2] Bellec F, Girault de Coursac P. La généreuse et tragique expédition Lapérouse. Ouest-France; 1985

[3] Brunel JF, Hiepko P, Scholz H. Flore analytique du Togo. Eschborn: GTZ éd; 1984. 750p

[4] Aké Assi L. Présence d'un Piper d'Amérique du Sud sur les pentes de la montagne Klouto (Togo). Mitteilungen der Botanischen Staatssammlung München. 1971;**10**:169

[5] Ern H. Die Vegetation Togos. Gliederrung, Gefährdung, Erhaltung. Willdenowia. 1979;**9**: 295-312

[6] Ern H. Proposals for the installation and organisation of a botanical garden in Togo (Tropical West Africa). Proc. Gen. Meet. & Conf. IABG, Canberra. 1982;**8**:38

[7] Brunel JF. Contribution à l'étude de quelques Phyllanthoideae africains et à la taxonomie du genre Phyllanthus L. (Euphorbiaceae). Thèse 3e cycle, ULP, Strasbourg; 1975. 206 p

[8] Brunel JF. Sur le genre Phyllanthus L. et quelques genres voisins de la Tribu des Phyllanthoidae Dumort. (Euphorbiaceae) en Afrique intertropicale et Madagascar. Thèse ès-Sci, Univ. Strasbourg; 1987. 768 p

[9] Scholz H, Scholz U. Flore descriptive des Cypéracées et Graminées du Togo. Phanerog. Monogr. 1983;**15**:360

[10] Akpagana K, Bouchet Ph. Nouvelles espèces d'Angiospermes pour la flore du Togo. Acta botanica Gallica. 1994;**141**(3):367-372

[11] Akpagana K, Guelly AK. Nouvelles espèces d'Angiospermes pour la flore du Togo. Acta botanica Gallica. 1994;**141**(6/7):781-787

[12] Batawila. Recherches sur les formations dégradées et jachères de la plaine côtière du sud Togo. Mém. DEA. Biol Dév. Univ. Lomé; 1997. 65 p

[13] Kokou K. Les mosaïques forestières au sud du Togo: biodiversité, dynamique et activités humaines. Th. Doct. Univ. Montpellier II; 1998. 140 p

[14] Kokou K, Caballé G, Akpagana K, Batawila K. Les îlots forestiers au sud du Togo: dynamique et relations avec les végétations périphériques. Rev. Ecol. (Terre vie). 1999;**54**:301-313

[15] Akpagana K, Gumedzoe MYD. La flore. In: Rapport national sur la diversité biologique. PNAE Togo; 2001. 102 p

[16] Akpagana K. Espèces nouvelles pour la flore du Togo. Ann. Univ. Bénin, sect. sci. 1992;**10**:25-32

[17] Radji AR. Contribution à l'étude de la diversité floristique du Togo : la flore horticole de la ville de Lomé et ses aspects économiques. Mém. DEA Biol. Vég. App., Univ. Bénin, Togo; 1997. 52 p

[18] Berhaut J. Flore illustrée du Sénégal, Dicotylédones, Tomes I, II, IV, V, VI & IX. Sainte-Rufine, France: Presses de l'Imprimerie Maisonneuve; 1971-1988

[19] Byrd Graf A. Tropica: Color Cyclopedia of Exotic Plants and Trees, 2e éd. N.J., USA: Roehrs Company East Rutherford; 1981. 1136p

[20] Houerou V, Houerou J-Y. L'arbre du voyageur. Abidjan: Les Nouvelles Editions Africaines; 1987. 186p

[21] Grisvard P, Chaudun V, Chouard P, Guillaumin A, Schneiter P. Le Bon jardinier Encyclopédie horticole. Paris: La Maison Rustique; 1990. 152e éd. Tome I. 883p

[22] Grisvard P, Chaudun V, Chouard P, Guillaumin A, Schneiter P. Le Bon jardinier Encyclopédie horticole. Paris: La Maison Rustique; 1990. 152e éd. Tome II. 783p

[23] Hessayon DG. The New House Plant Expert. Herts: PBI Publications; 1992. 256p

[24] Hutchinson J, Dalziel JM. In: Keay RWJ, Hepper FN, editors. Flora of West Tropical Africa. 2nd ed. Vol. I-III. London; 1954-1972

[25] Hutchinson J, Dalziel JM. In: Keay RWJ, editor. Flora of West Tropical Africa. 2nd ed. Vol. 1. London: Crown Agents for Overseas; 1954

[26] Lebrun J-P, Stork A. Tropical African Flowering Plants. Ecology and Distribution. vol. 1 Annonaceae-Balanitaceae. Conservatoire et Jardin botaniques de la Ville de Genève. hors-série 9; 2003. 793p

[27] Lebrun J-P, Stork A. Tropical African Flowering plants. Ecology and distribution. vol. 2 Euphorbiaceae-Dichapetalaceae. Conservatoire et Jardin botaniques de la Ville de Genève. hors-série 9a; 2006. 306p

[28] Lebrun J-P, Stork A. Tropical African Flowering Plants. Ecology and Distribution. vol. 3 Mimosaceae-Fabaceae. Conservatoire et Jardin botaniques de la Ville de Genève. hors-série; 2008. 325p

[29] Lebrun J-P, Stork A. Tropical African Flowering Plants. Ecology and Distribution. vol. 4 Fabaceae (Desmodium-Zornia). Conservatoire et Jardin botaniques de la Ville de Genève. hors-série; 2008. 291p

[30] Lebrun J-P, Stork A. African flowering plants database (online). Available from: http://www.ville-ge.ch/musinfo/bd/cjb/africa/recherche.php [Consulted on: July 10, 2011]

[31] International Code of Botanical Nomenclature (online). Available from: http://www.bgbm.fu-berlin.de/iapt/nomenclature/code/SaintLouis/0000St.Luistitle.htm [Consulted on: July 9, 2012]

[32] ING: Index Nominum Genericorum (online). Available from: http://botany.si.edu/ing/ [Consulted on: July 10, 2011]

[33] IPNI: The International Plant Name Index (online). Available from: http://www.ipni.org/ [Consulted on: July 10, 2012]

[34] HUH: Harvard University Herbarium databases (online). Available from: http://kiki.huh.harvard.edu/databases/botanist_index.html [Consulted on: July 2012]

[35] Vidalie H. Les productions florales. Paris: Lavoisier TEC & DOC; 1998

[36] Aké-Assi E, Adou Yao CY, Ipou Ipou J, Neuba DF, Aké Assi L, Traoré D. Représentations des plantes ornementales pour les populations d'Abidjan et San Pedro, en Côte d'Ivoire. In: van der Burgt X, van der Maesen J, Onana J-M, editors. Systématique et Conservation des Plantes Africaines. Royal Botanic Gardens: Kew; 2010. pp. 289-296

[37] Radji R, Bouchet P, Akpagana K. La flore horticole du Togo. Journal de Botanique de la Société Botanique de France. 1998;**8**:87-94

[38] Radji AR, Kokou K, Akpagana K. Étude diagnostique de la flore ornementale du Togo. International Journal of Biological and Chemical Sciences. 2010;**4**(2):491-508

[39] Porter L, Bongers F, Kouamé FN, Hawthorne WD. Biodiversity of west African Forest. In: An Ecological Atlas of Woody Plant Species. CABI Publishing; 2004

[40] Aké Assi AE. Contribution à l'étude des plantes ornementales cultivées dans les régions d'Abidjan et de San-Pedro, en Côte d'Ivoire. Volume 1—Textes et iconographies. Volume 2—Catalogue illustré de 241 espèces et variétés horticoles. Th. Doct. 3è cycle, Univ. Cocody (RCI); 2002. 242p

[41] Florence J, Chevillotte H, Ollier C, Meyer J-Y. Base de données botaniques Nadeaud de l'Herbier de la Polynésie française (PAP). 2007. Available from: http://www.herbier-tahiti.pf

[42] Ihaka R, Gentleman RR. A language for data analysis and graphics. Journal of Computational and Graphical Statistics. 1996;**5**(3):299-314

[43] Géhu JM, Géhu J. Essai d'objection de l'évaluation biologique des milieux naturels. Exemples littoraux. In: Géhu JM, editor, Séminaire de Phytosociologie Appliquée. Amicale Francophone de Phytosociologie. Metz; 1980. pp. 75-94

[44] Nevena D, Oderab MM, Reardonc T, Wang H. Kenyan supermarkets, emerging middle-class horticultural farmers, and employment impacts on the rural poor. World Development. 2009;**37**(11):1802-1811

[45] Niang I, Stinghlamber G. Étude pour un programme de développement du secteur horticole au Sénégal. Rapport final. Dakar: Cabinet JEXCO; 2001. 71p + annexes

[46] IQRHH. Diagnostic sur la main d'œuvre en horticulture ornementale au Québec. Secteur commercialisation et services. Rapport final; 2003. 91p

[47] Watson G. Overview of the Canadian horticulture industry. In: Introduction to the Floriculture Crop Profile Series. Centre pour la lutte antiparasitaire d'AAC (non publié); 2006

[48] Waylen K. Botanic Gardens: Using Biodiversity to Improve Human Well-Being. Botanic Gardens Conservation International. 2006. Available from: http://www.bgci.org/wellbeing/report/ [Consulted on: December 30, 2011]

[49] Brethour C, Watson G, Sparling B, Bucknell D, Moore T. Revue de la littérature sur les bienfaits des produits de l'horticulture ornementale sur la santé et l'environnement. In: Rapport Final. George Morris Centre; 2007. 76p

[50] Biley FC. Utilizing the Mythical and Folkloric Power of Trees in the Modern Hospital Environment. Complementary Therapies in Nursing & Midwifery. 2001;**7**:207-210

[51] Barloy JJ. Le monde végétal: Les plantes dans l'évolution de la vie et des civilisations. Ed. Hachette Jeunesse, Paris; 1984. 73 p

[52] Anonymous. Dictionnaire universel de poche. Hachette Livre et Librairie Générale Française; 1993. 759 p

[53] Fouchet JG, Marquet A, Hambuckers A. Les plantes médicinales: de la plante au médicament. Observatoire du Monde des Plantes, Liège; 2000. 18 p

[54] Jiofack T, Ayissi I, Fokunang C, Guedje N, Kemeuze V. Ethnobotany and phytomedicine of the upper Nyong Valley forest in Cameroon. African Journal of Pharmacy and Pharmacology. 2009;**3**(4):144-150

[55] Betti JL, Mebere Yemefa SR. Contribution à la connaissance des produits forestiers non ligneux du parc de Kalamaloué, Extrême nord Cameroun : les plantes alimentaires. International Journal of Biological and Chemical Sciences. 2011;**5**(1):291-303

[56] CIB-UNESCO (Comité international de bioéthique—Organisation des Nations-Unies pour l'Education, la science et la culture). Avant-projet de rapport sur la médecine traditionnelle et ses implications éthiques. SHS/EST/CIB-17/10/CONF.501/3. Paris; 2010. 19p

[57] Dibong SD, Mpondo Mpondo E, Ngoye A, Betti JL. Ethnobotanique et phtytomédecine des plantes médicinales vendues sur les marchés de Douala, Cameroun. Journal of Applied Biosciences. 2011;**37**:2496-2507

[58] Ngono Ngane RA, Koanga Mogtomo ML, Tchinda Tabou A, Magnifouet Nana H, Motso Chieffo PR, Mballa Bounou Z, Ebelle Etame RM, Ndifor F, Biyiti L, Amvam Zollo PH. Ethnobotanical survey of some camerounian plants used for treatment of viral diseases. African Journal of Plant Science. 2011;**5**(1):15-21

[59] Okafor J, Ham R. Identification, utilisation et conservation des plantes médicinales dans le sud-est du Nigéria. Thèmes de la biodiversité africaine. 1999;**3**:1-8

[60] Mehdioui R, Kahouadji A. Etude ethnobotanique auprès de la population riveraine de la forêt d'Amsittène: cas de la Commune d'Imi n'Tlit (Province d'Essaouira). Bulletin de l'Institut Scientifique, Rabat, Section Sciences de la Vie. 2007;**29**:11-20

[61] Chevallier A. Encyclopédie des plantes médicinales. Montréal: Sélection du Reader's Digest (Canada); 1996

[62] Dobelis IN. Magic and Medicine of Plants. NewYork: The Reader's Digest Association Inc. Pleasantville; 1986. pp. 53-67

[63] Ouattara D. Contribution à l'inventaire des plantes médicinales significatives utilisées dans la région de Divo (sud forestier de la Côte-d'Ivoire) et à la diagnose du poivrier de Guinée: *Xylopia aethiopica* (Dunal) A. Rich. (Annonaceae). Doctorate Dissertation of Thèse de Doctorat de l'Université de Cocody-Abidjan (Côte-d'Ivoire), UFR Biosciences, Laboratoire de Botanique; 2006. 184p

[64] Adjanohoun JE, Aké Assi L. Contribution au recensement des plantes médicinales de Côte-d'Ivoire. Université d'Abidjan, Centre National de Floristique (C.N.F.). 1979; 358 p

[65] Zirihi GN. Contribution au recensement, à l'identification et à la connaissance de quelques espèces végétales utilisées dans la médecine traditionnelle et la pharmacopée chez les Bété du Département d'Issia, Côte d'Ivoire. Thèse de Doctorat de 3ème Cycle, Université d'Abidjan, F.A.S.T.; 1991. 150 pp

[66] N'Guessan K, Kadja B, Zirihi G, Traoré D, Aké-Assi L. Screening phytochimique de quelques plantes médicinales ivoiriennes utilisées en pays Krobou (Agboville, Côte-d'Ivoire). SciNat. 2009;**6**(1):1-15

[67] WHO, IUCN, and WWF. Principes directeurs pour la conservation des plantes médicinales. Suisse: Gland; 1993. 35p

3

Towards an Integrative Taxonomy of the Genus *Alstroemeria* (Alstroemeri-aceae) in Chile

Victor L. Finot, Carlos M. Baeza, Eduardo Ruiz,
Oscar Toro and Pedro Carrasco

Abstract

The genus *Alstroemeria* encompasses approximately 80 species endemic to South America, with 2 centers of diversity (Chile and Brazil). In Chile, *Alstroemeria* represents one of the most diverse genera of vascular monocotyledons, comprising more than 50 recognized or accepted taxa (36 species, 11 subspecies and 10 varieties) from which ca. 82% are endemic to the Mediterranean zone of central Chile, one of the world's diversity hotspots. The taxonomy of the genus is very difficult due to the great variability of the vegetative and floral traits. Moreover, a number of taxa have been recently described and several nomenclatural changes have been proposed. In order to elucidate the taxonomy of some Chilean complexes of *Alstroemeria*, an integrative approach including morphology, colorimetry, cytogenetic, multivariate statistical analyses of morphological variation and DNA-molecular studies have been conducted. In this chapter, we review the literature concerning these approaches; a checklist of the species growing in Chile is provided including all published names, references to the original protologues, accepted names, synonyms and the biogeographic status (endemic or native) of the accepted taxa; maps illustrating the diversity of the genus in South America and its distribution in Chile were constructed.

Keywords: checklist, Chilean hotspot, cytogenetic, endemism, geographical distribution, South America, taxonomy

1. Introduction

The family Alstroemeriaceae Dumort. *nom. Cons.*, belongs to the monocotyledon angiosperm clade (Subclass: Liliopsida, Order: Liliales) [1, 2]. It is distributed in Central and South America [3], and comprises about 200 species distributed in the genera *Alstroemeria* L., *Bomarea* Mirb., *Leontochir* Phil. and *Schickendantzia* Pax [4]. Some authors recognized *Taltalia* Ehr. Bayer as an independent genus [5, 6] while others included *Taltalia* and *Schickendantzia* in *Alstroemeria* [7]. Genera have been classified in two subfamilies: (a) Subfamily: Luzuriagoideae with *Drymophila* R. Br. and *Luzuriaga* Ruiz & Pav. and (b) Subfamily: Alstroemerioideae with *Alstroemeria* and *Bomarea* (incl. *Leontochir*) [3, 8] or treated Luzuriagoideae as a separate family Luzuriagaceae [3]. *Alstroemeria* comprises about 80 species endemic to South America (southern South America and Eastern Brazil); *Bomarea* includes 120 species in Central and South America; *Luzuriaga* comprises 4 species (3 species in Chile, 1 in New Zealand) and *Drymophila* 1 species from Australia and Tasmania [8]. Phylogenetic studies using morphology and DNA sequences (*rps16, rbcL*) recognized *Alstroemeria* as monophyletic and certainly different from *Leontochir* and *Bomarea*; moreover, three subclades of *Alstroemeria* have been recognized that correspond to northern Chile, central Chile and Brazil [9]. *Alstroemeria* was established by Linnaeus in 1762 in honor to the Swedish botanist Claus von Alströmer [10]. It has its boreal distribution in Venezuela (3°N) and its austral limit in the Patagonia of Chile and Argentina, with two main distribution centers in the continent: Brazil (and adjacent areas of Paraguay and Argentina) and Chile (and the adjacent countries Peru, Bolivia and Argentina) [11–15]. In Chile, *Alstroemeria* represents 1 of the most diverse genera of vascular monocotyledons, comprising more than 50 recognized or accepted taxa (36 species, 11 subspecies and 10 varieties) from which about ca. 82% are endemic to the Mediterranean zone of central Chile [14]. Due to the beauty of their flowers, Chilean species of *Alstroemeria*, locally known as "astromelias or lirios del campo" are appreciated all over the world as ornamental plants [16–18]. Many hybrids and cultivars have been developed in several countries, such as, The Netherlands, England, United States and Japan [14]. The taxonomy of the genus is very complex due to the great variability both in vegetative and floral characters [19]. Moreover, several taxa have been recently described (e.g., *A. werdermannii* var. *flavicans* [20], *A. philippi* var. *albicans* [14], *A. philippi* subsp. *adrianae* [21], *A. hookeri* subsp. *sansebastiana* [22], *A. marticorenae* [19], *A. traudliae* [23]) or nomenclatural changes have been made, affecting the rank status of many taxa. Due to its geographical isolation, Chile contains a unique flora which includes an extraordinary number of endemic plants. Furthermore, the area between the Regions of Atacama and Biobío comprises about 60% of the vascular species of the Chilean flora, with nearly 50% endemic to Chile [24]. In this area, identified as a hotspot of biodiversity [25], lives most of the Chilean species of *Alstroemeria*. This area harbors most of the Chilean population, and is characterized by a strong disturbance triggered mainly by agriculture, industry and forestry. In this chapter, we reviewed the taxonomic literature to make an updated checklist of the species growing in Chile and their synonyms. In addition, we construct distribution maps based on the literature as well as on the database of the Herbarium of the University of Concepción (CONC). Recent studies integrating different source of evidence, such as morphometry, cytogenetic, colorimetry and molecular data for better taxonomic species delimitation are discussed.

2. Methods

The database of the Herbarium of the University of Concepción (CONC) was used to construct preliminary lists of species and the geographic distribution of each taxon. This database contains the following fields: (1) Taxon name; (2) Collector's name; (3) Collector's number; (4) Latitude; (5) Longitude; (6) Elevation; (7) Administrative region; (8) Locality; (9) Collection date (month); (10) Collection date (Year); (11) Herbarium and (12) Herbarium number. A total of 714 specimens were included in the CONC-DB. In addition to historical specimens, plants were collected and photographed in the field and kept in CONC. A checklist is provided including all published combinations, references to original publication, accepted names and biogeographic status (endemic or native). Maps illustrating the diversity of the genus in South America and distribution in Chile were constructed based on data taken from the CONC-DB and literature [14, 15, 23], using the software DIVA-GIS 7.5.0.

3. Taxonomy and distribution of the genus *Alstroemeria* in Chile

3.1. General morphology

Alstroemeria comprises mostly perennial species. In 1998, Bayer [5] established the monotypic genus *Taltalia* to separate the annual *A. graminea* Phil., endemic to northern Chile, from *Alstroemeria*. Perennial species have cylindrical rhizomes, from which two kinds of roots born: thin roots and thick roots which contain starch (**Figure 1A**); *A. ligtu* (locally known as "liuto") was used by indigenous people (Mapuches or Araucanos) to produce starch from the thick roots. According to Molina [26], the farmers made from the roots of this plant "a white, light, nutritious and so healthy flour that they usually gave it to the sick persons…" [26]. Aerial stems are erect or decumbent. The leaves are often resupinated, that is, twisted from the petiole or the leaf blade so the lower surface becomes functionally the upper surface; sometimes the leaves form basal rosettes; leaf blades thin or thick, sometimes with papillae; the blade varies in shape from linear to elliptic or ovate; fertile stems usually have reduced leaves but sterile stems have well developed leaves (**Figure 1B**). The flowers are slightly zygomorphic (**Figure 1C-D**), with six free tepals in two verticils; the three outer tepals are similar in shape and color; the inner three tepals are differentiated in two upper inner tepals and one lower inner tepal. Upper inner tepals with colored lines (nectar guides) on a lighter background (**Figure 1D**). Stamens 3: the ovary is inferior, 3-carpellate, 3-loculate. The fruit is a six-ribbed loculicidal capsule (**Figure 1C**) with explosive dehiscence, with numerous globose seeds (**Figure 2**).

3.2. Colorimetric studies in Chilean *Alstroemeria*

In *Alstroemeria*, the flowers varies in color from white to yellow, pink, red, purple and violet according to the species [11, 14]; color is regulated by several pigments including anthocyanin-like 6-hydroxydelphinidine 3-rutinoside, 6-hydroxycyanidin 3-rutinoside, delphinidin 3-malonyl-glucoside among others, carotenoids and flavonoids [27]. In some groups of plants, as occurs in *Alstroemeria*, it is possible that the taxonomic characters traditionally used do not have sufficient discriminant power to differentiate very close species or varieties within a species complex. In such cases, it may be useful to have characteristics that pose a new perspective on the problem. It has been shown that the color of the corolla, objectively measured, had high taxonomic value when the traditional characters were less informative to distinguish cryptic taxa [28]. In *Alstroemeria*, the color of the flower has often been used in keys and descriptions [11, 14, 15, 22, 23], however, most of the time, the described color corresponds to a subjective perception of the same by the human eye. The color of the flowers is taxonomically significant in *Alstroemeria* [14, 29, 30]; the

Figure 1. Morphology of the roots, rhizomes, leaves, flowers and fruits of *Alstroemeria*. (A) Roots and rhizomes of *A. x chrysantha*; (B) sterile leaves in fertile stems and basal rosettes of *A. magenta*; (C) fruits (capsules) of *A. hookeri* subsp. *recumbens*; (D) flower of *A. x chrysantha*. Photos A and D by V.L. Finot; photos B and C by C.M. Baeza.

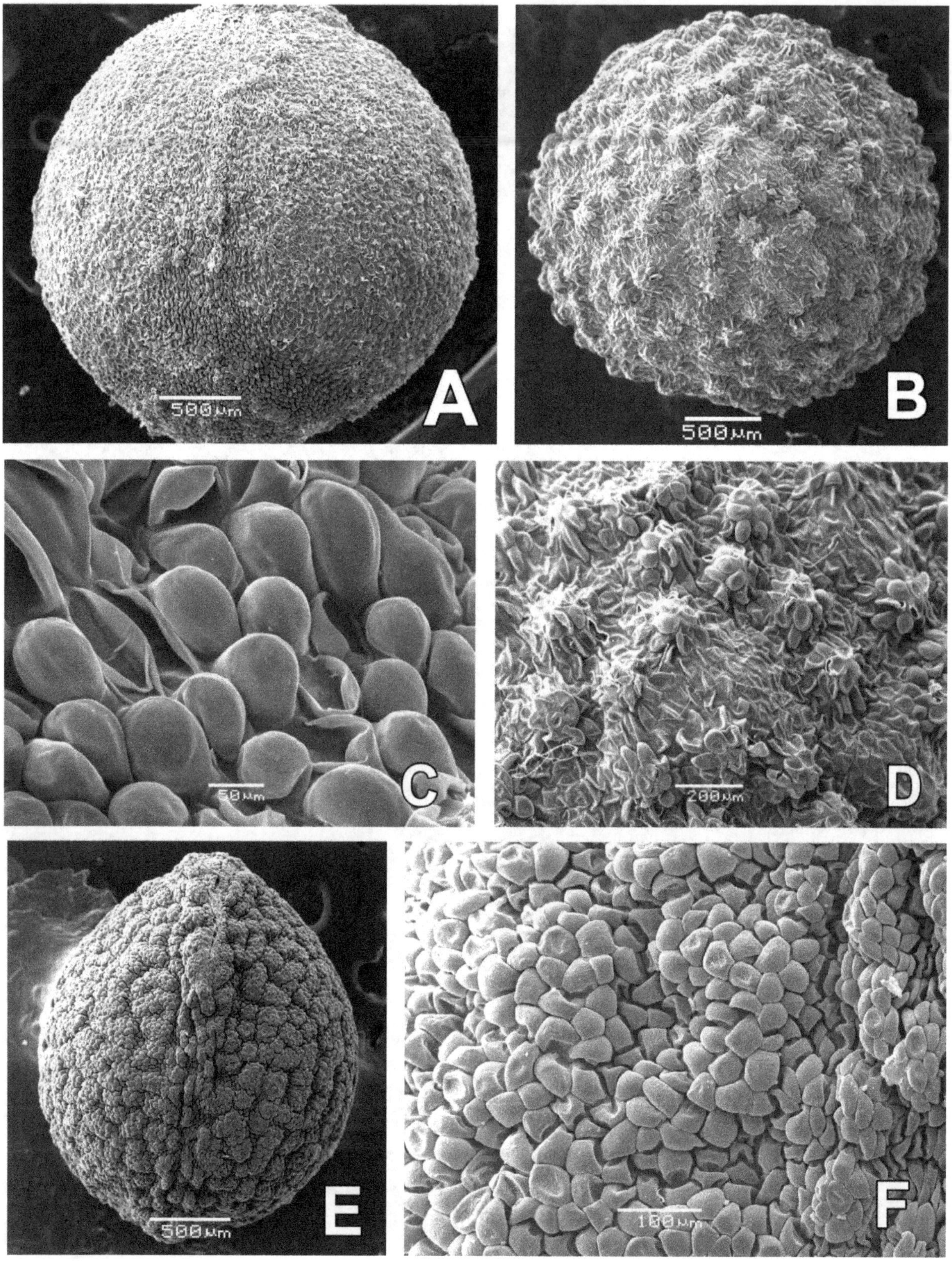

Figure 2. Seeds of *Alstroemeria* as seen with scanning electron microscope (SEM). (A) *Alstroemeria presliana* subsp. *australis*. (B) *Alstroemeria magenta*. (C) *Alstroemeria ligtu* subsp. *simsii*. (D). *Alstroemeria magenta*, surface details. (E) and (F) *Alstroemeria x chrysantha*.

outer tepals and the lower inner tepals are similar in shape and color but the upper internal tepals are different, showing unique patterns of maculae (actually nectar guides) that are species specific; however, some species showed variability in the background color [30], because it depends on several ecological factors such as temperature and pH [31]. In order to describe objectively the color (mainly the background color) of the different tepals (external, upper internal and lower internal), the CIElab system [32] has been used in some species complexes of Chilean *Alstroemeria* [29, 30] (**Figure 3**). The CIE (Commission Internationale de l'Eclairage) color system uses three coordinates to locate a given color in the color space. The spectrophotometer registers the reflected wavelengths as numerical values (spectral curve) from which the coordinates that place a given color in the color space are calculated. The color expressed in the CIELab scale uses the Cartesian coordinates L*, a* and b*. L* expresses the luminosity, a* denotes the green/red value and b* the blue/yellow value. The degree of luminosity of L* determines that a color appears lighter or darker and is expressed in a scale from 0 (black or total absorption) to 100 (white). The axis a* moves from negative values (green) to positive values (red) while axis b* moves from blue (negative values) to yellow (positive values). The color expressed in the CIELCh scale uses polar coordinates (L*, C*, h°), derived from the CIELab scale. C* denotes chroma (saturation, intensity) and h° denotes hue, expressed as angular measures. Chroma is the distance of the color from the axes a* and b* of L*, calculated as $(a^{*2} + b^{*2})^{1/2}$ and represents the color saturation; the hue, h° is calculated as arctg (b*/a*). In *A. magnifica* complex, the colorimetric study of the flower helped to elucidate the taxonomic position of *A. pulchra* var. *maxima*. This taxon, originally described by Philippi in 1864 [33], was transferred by Bayer in 1987 to *A. magnifica* with the subspecific rank (*A. magnifica* subsp. *maxima*) [11].

The colorimetric differences between *A. magnifica* and *A. pulchra* as shown in the reflectance spectra were due mainly to the parameters a* and b* indicating that *A. magnifica* have tepals comparatively more intense violet than those of *A. pulchra* var. *maxima*. Our results suggest that the color of the flowers can be used as a new taxonomic character in *Alstroemeria* and that var. *maxima* probably belongs to *A. pulchra* as originally proposed and not to *A. magnifica* [29]. Colorimetric studies were carried out also in *A. presliana* [30]. This species comprises two subspecies: subsp. *presliana* and subsp. *australis*. *Alstroemeria presliana* subsp. *presliana* grows in Chile (Regions of Maule and Biobío) and Argentina (Neuqén) [34]; subsp. *australis* in endemic to Chile (Regions of Biobío and Araucanía) [11, 14].

Although the color of the flowers is one of the most important characters to distinguish the subspecies [11], there is a huge variability in color in the flowers of both subspecies. Differences in the spectral reflectance curves were detected between 440 and 540 nm and between 660 and 700 nm in the outer and lower inner tepals. Upper internal tepals differ mainly between 640 and 700 nm. The color measured in the CIELab space is related to the content of anthocyanins so that the flowers containing delphinidin-3-glucosides take on a more blue hue than those containing exclusively cyanidin-3-glucosides [35]. The presence of delphinidin-3-glucosides detected in subsp. *presliana* but not in subsp. *australis* [27] could explain the bluer hue observed in subsp. *presliana* in comparison with subsp. *australis* and the difference observed in the parameter b*, which takes negative values in subsp. *presliana* and positive values in subsp. *australis* both in outer and lower inner tepals. On the other hand, in the upper inner tepals, b* was positive (yellow), reaching higher values in subsp. australis and therefore denoting a more intense yellow color in subsp. *australis* than in subsp. *presliana* [30].

Figure 3. Flowers of some species of Chilean *Alstroemeria*. (A) *Alstroemeria pulchra*, Maule region, Talca, Río Claro (Baeza 4393). (B) *Alstroemeria ligtu* subsp. *simsii*, Maule region, Talca, Río Claro (Baeza 4395). (C) *Alstroemeria magnifica* var. *sierrae*, Coquimbo region, Caleta Hornos (Baeza 4375). (D) *Alstroemeria x chrysantha*, Coquimbo region, Huanaqueros (Baeza 4376). Photos A and B by C. Baeza; C and D by V. Finot.

3.3. Geographical distribution

In Chile, there are 38 species of *Alstroemeria* and 16 infraspecific taxa (8 subspecies and 8 varieties) and 1 nothospecies (*A. x chrysantha*) [23]. Nevertheless, more than 116 species have been described, most of which are considered synonyms or are names of uncertain application because there is no original material (types) in herbaria (see Checklist below). The description of such high number of taxa can be explained by the extent of morphological variation, especially of the flowers, which harbors most of the characters useful to taxonomy, and because microevolutionary processes are still active so that species are not yet completely separated. Chile shares only few taxa with its neighboring countries (Argentina, Bolivia and Peru). In Peru, five species have been mentioned [14, 36, 37], one of which is also present in Chile: *A. violacea*. In Bolivia, three species are found, one of which growth in Chile: *A. aurea* ([38], under *A. aurantiaca*). In Argentina, there are 10 species [34], 5 shared with Chile: *A. andina* var. *venustula*, *A. aurea*, *A. patagonica*, *A. presliana* subsp. *presliana*, *A. pseudospathulata*. Thus, more than 88% of the genus is represented by taxa endemic to Chile (**Figure 4**). In Chile, *Alstroemeria* spreads from 20°S (Tarapacá Region) to 53°S (Magallanes Region) [14, 23]. Most taxa have a very restricted distribution in Chile (**Table 1**). The vast majority of the species are distributed in north (Tarapacá-Coquimbo) and central (Valparaíso-Biobío) Chile (**Figure 5**); only six species growth in southern Chile (Araucanía-Magallanes). The most boreal taxa are *A. lutea* and *A. violacea* that reach the Region of Tarapacá (20°S) in northern Chile. *Alstroemeria lutea* is restricted to the coast of the Tarapacá Region (Iquique) whereas *A. violacea* extends southern to 28°S in the Atacama Region; this species if known also from Peru (Arequipa) [14, 23, 37]. The regions with the largest number of taxa are Atacama (14 taxa), Coquimbo (26), Valparaíso (19) and Metropolitan (14) (**Figure 6**). The number of taxa decreases abruptly southern the Maule Region where 12 taxa are found; in Los Ríos and Los Lagos, only one

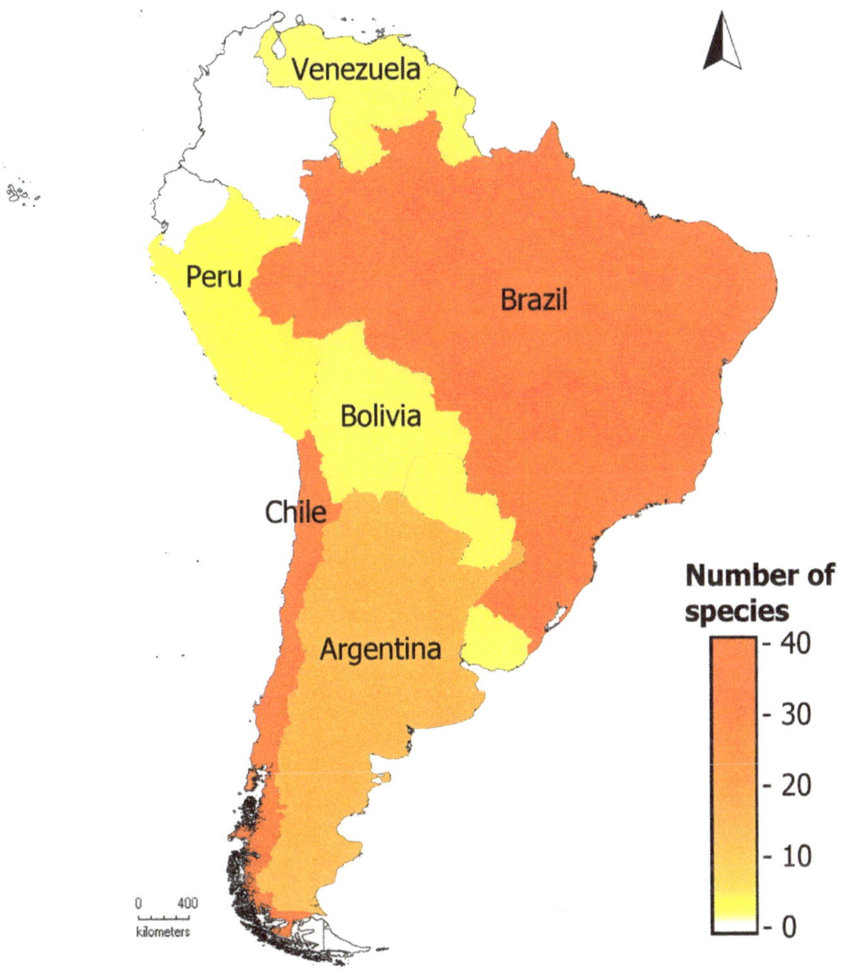

Figure 4. Species diversity of *Alstroemeria* in South America.

No.	Taxa	AYP	TAR	ANT	ATA	COQ	VAL	RME	LBO	MAU	NUB	BIO	ARA	LRI	LLA	AYS	MAG
1	*A. achirae*									x							
2a	*A. andina* var. *andina*				x	x											
2b	*A. andina* var. *venustula*					x											
3a	*A. angustifolia* var. *angustifolia*						x	x									
3b	*A. angustifolia* var. *velutina*					x	x										
4	*A. aurea*								x	x	x	x	x	x	x	x	x
5	*A. citrina*					x	x										
6	*A. crispata*				x	x											
7	*A. cummingiana*					x	x										
8	*A. diluta*						x	x	x	x							
9	*A. exerens*							x	x	x							
10	*A. garaventae*						x	x									
11	*A. graminea*			x	x												
12a	*A. hookeri* subsp. *hookeri*									x	x	x					
12b	*A. hookeri* subsp. *maculata*					x											
12c	*A. hookeri* subsp. *recumbens*					x	x										
12d	*A. hookeri* subsp. *sansebastiana*											x					
13	*A. kingii*				x												
14	*A. leporina*				x	x											
15a	*A. ligtu* subsp. *ligtu*									x	x	x					
15b	*A. ligtu* subsp. *simsii*						x	x	x	x							
15c	*A. ligtu* subsp. *splendens*								x	x							
16	*A. lutea*		x														

No.	Taxa	AYP	TAR	ANT	ATA	COQ	VAL	RME	LBO	MAU	NUB	BIO	ARA	LRI	LLA	AYS	MAG
17	*A. magenta*					x	x										
18a	*A. magnifica* var. *magnifica*					x											
18b	*A. magnifica* var. *sierrae*					x											
18c	*A. magnifica* var. *tofoensis*					x											
19	*A. marticorenae*						x										
20	*A. mollensis*					x											
21	*A. pallida*						x	x									
22	*A. parvula*						x	x									
23	*A. patagonica*															x	x
24	*A. pelegrina*					x	x										
25a	*A. philippi* subsp. *adrianae*				x	x											
25b	*A. philippii* var. *albicans*				x												
25c	*A. philippii* var. *philippii*				x	x											
26	*A. polyphylla*				x												
27a	*A. presliana* subsp. *australis*												x				
27b	*A. presliana* subsp. *presliana*									x	x	x					
28	*A. pseudospathulata*									x							
29a	*A. pulchra* subsp. *lavandulacea*											x	x				
29b	*A. pulchra* subsp. *pulchra*					x	x	x	x	x							
29c	*A. pulchra* var. *maxima*					x	x	x									
30	*A. revoluta*						x	x	x	x	x	x	x				
31a	*A. schizanthoides* var. *alba*					x											
31b	*A. schizanthoides* var. *schizanthoides*				x	x											

No.	Taxa	AYP	TAR	ANT	ATA	COQ	VAL	RME	LBO	MAU	NUB	BIO	ARA	LRI	LLA	AYS	MAG
32	*A. spathulata*					x	x	x									
33	*A. traudliae*					x											
34	*A. umbellata*							x	x								
35	*A. versicolor*							x	x	x	x	x	x				
36	*A. violacea*		x	x	x												
37a	*A. werdermannii* subsp. *flavicans*				x	x											
37b	*A. werdermannii* subsp. *werdermannii*				x	x											
38	*A. x chrysantha*				x	x	x										
39	*A. zoellneri*						x	x									

AYP = Arica-Parinacota Region; TAR = Tarapacá Region; ANT = Antofagasta Region; ATA = Atacama Region; COQ = Coquimbo Region; VAL = Valparaíso Region; RME = Metropolitan Region; LBO = O'Higgins Region; MAU = Maule Region; NUB = Ñuble Region; BIO = Biobío Region; ARA = Araucanía Region; LRI = Los Ríos Region; LLA = Los Lagos Region; AYS = Aysén Region; MAG = Magallanes and Antártica Chilena Region.

Table 1. Presence (x) of the accepted species of ***Alstroemeria*** in the administrative political regions of Chile.

species has been collected (*A. aurea*) and *A. patagonica* is found in Aysén and Magallanes being the most austral species of the genus *Alstroemeria* in the world. The latter species grow from 46°30′S to 52°45′S [14] and also in Argentina (Neuquén to Tierra del Fuego) [14, 34]. *Alstroemeria aurea* is the species with the widest distribution in Chile (this species spreads over

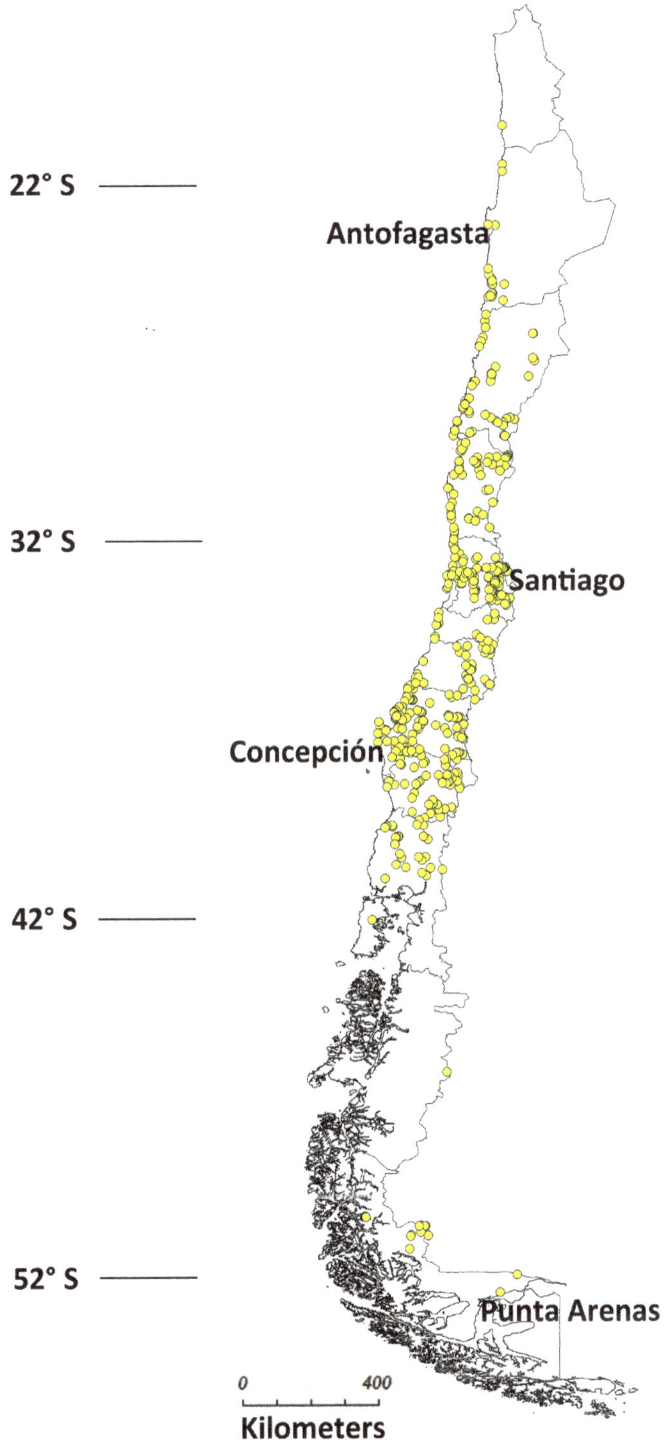

Figure 5. Distribution of the genus *Alstroemeria* in Chile. Each point represents at least one collection housed in the herbarium of the University of Concepción, Chile (CONC).

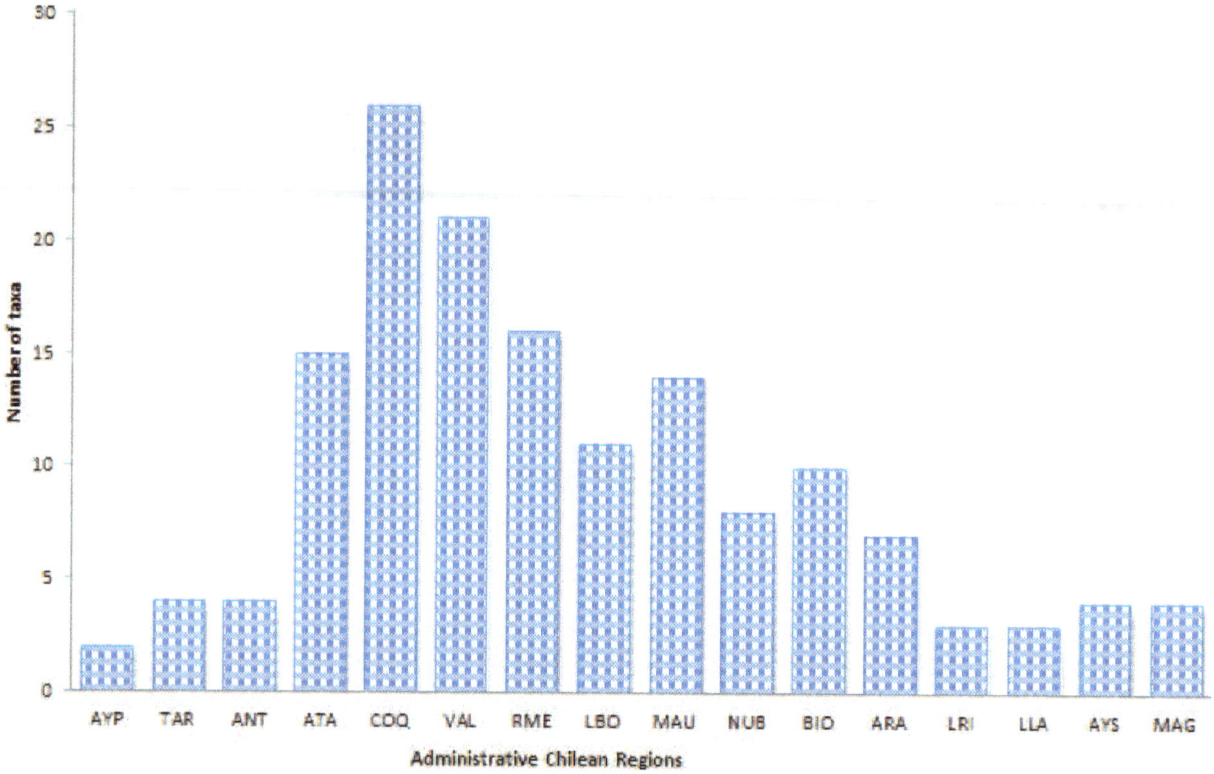

Figure 6. Number of taxa in the 16 administrative regions of Chile. For regions names see **Table 1**.

10 regions, from the O'Higgins Region, 34°12′S to Torres del Paine National Park, Magallanes Region, 51°21′S). *Alstroemeria revoluta*, the second widely distributed species, spreads from Valparaíso (La Campana National Park, 32°57′S) to Araucanía Region (Traiguén-Galvarino, 38°16′S) and *A. versicolor* ranges from the Metropolitan Region (Rio Clarillo National Reserve, 33°40′S) to Araucanía (Malleco, Renaico, 37°48′S). With these exceptions, most species show very narrow distribution, some of them being confined to a single region, such as *A. lutea* (Tarapacá Region), *A. kingii*, *A. philippi* var. *albicans* and *A. polyphylla* (Atacama Region), *A. andina* var. *venustula*, *A. hookeri* subsp. *maculata*, *A. magnifica*, *A. schizanthoides* var. *alba* and *A. mollensis*, *A. traudliae* (Coquimbo Region), *A. marticorenae* (Valparaíso Region), *A. achirae* (Maule Region), *A. hookeri* subsp. *sansebastiana* (Biobío Region), *A. presliana* subsp. *australis* (Araucanía Region, Nahuelbuta National Park). For the latitudinal distribution of each species, see reference ([14], **Figure 5**). Altitudinally, the genus *Alstroemeria* spreads from the sea level to nearly 4000 m.a.s.l. although most species are found below 2000 m.a.s.l. *Alstroemeria andina*, *A. crispata*, *A. exerens*, *A. pallida*, *A. parvula*, *A. spathulata* and *A. umbellata* can be found above 3000 m of elevation.

4. Cytogenetic studies in Chilean *Alstroemeria*

Cytogenetic studies in *Alstroemeria* have proved to be useful in delimiting species, since each studied taxon has a unique karyotype. These studies have contributed to the delimitation of the different taxa, as well as to the understanding of the chromosomal processes that determine the divergence among them [39]. Recent studies at the infraspecific level, in taxonomic complexes of the genus, have also been shown to be useful in the recognition of these taxa, either due to differences in the chromosomal architecture or in the asymmetry indexes of the chromosomes [40, 41]. Strasburger [42] was the first researcher to perform chromosome studies in *Alstroemeria* and until 1989 the number of cytological published papers involved no more than 10 different species [43]. In the last 25 years, a wide variety of cytogenetic studies have been carried out in the genus, including physical location of repetitive DNA sequences in *A. aurea* [16, 44, 45], meiosis and mitosis [46], karyology [47–49], variation and size of the genome [50], fluorescent *in situ* hybridization [39, 51, 52] and cytotaxonomy [40, 41, 52–58]. In 15 geographically isolated populations of five species of *Alstroemeria* (*A. aurea, A. hookeri, A. ligtu, A. pelegrina* and *A. presliana*) collected in Chile, karyotypes and variation of RAPD markers have been investigated. Tandemly repeated DNA sequences—5S and 18/25S rDNA genes and the sequence A001-1 were used to characterize karyotypes by fluorescence in situ hybridization (FISH). Ten somatic metaphases per population were used for measurement of chromosome length. Differences in RAPD marker bands were used for characterization of populations, creating a similarity index. FISH with all three DNA probes shows a high degree of polymorphism among and sometimes also within accessions of *A. aurea, A. hookeri* and *A. ligtu*. The number of chromosome pairs showing 5S rDNA signals is more different for the investigated species *A. aurea, A. hookeri, A. ligtu, A. pelegrina* and *A. presliana* with 5, 7, 5, 3 and 7, respectively, than the number of 18/25S rDNA signals in this succession with 7, 7, 6, 5 and 7 chromosome pairs, showing a high evolutionary dynamics within the genus. Furthermore, among the four populations of *A. hookeri*, accession 4181 was different in arm length of chromosome 3. RAPD markers (index of similarity) also showed a greater genetic distance of accession 4181 from the other three accessions of *A. hookeri* [39].

The study of the chromosomes in *Alstroemeria* has already helped to clarify a number of taxonomic issues within the genus. For example, study of karyotypes in the *A. hookeri* complex permitted change a subspecies to the species rank (*A. cummingiana*), the recognition of a new subspecies (*A. hookeri* subsp. *sansebastiana*) and description of a new species (*A. marticorenae*) [19, 22, 40, 41]. Similar situation occurred in the *A. presliana* complex, where after completing a comparative karyotypic study in 11 populations, it was suggested that *A. presliana* subsp. *australis*, endemic to the cordillera of Nahuelbuta, should be raised to species rank [18].

A number of cytological studies have been completed in the *Alstroemeria ligtu* complex. Buitendijik and Ramanna [16] and Buitendjik et al. [50] found variation in the DNA content and polymorphism of C bands in the chromosomes of subsp. *ligtu*, subsp. *simsii* and subsp. *splendens*. Zhou et al. [51], utilizing FISH, completed the characterization of the genomic DNA of eight highly repetitive sequences in subsp. *ligtu* and *simsii*, showing detailed karyotypes with localization of specific DNA sequences. DAPI staining and acetic orcein, completed a comparative karyotype study of five populations of subsp. *ligtu* from the Region of Biobío and one population of subsp. *simsii* from the Region of Valparaiso [39]. The six populations studied revealed an asymmetric karyotype with $2n = 2x = 16$ chromosomes. The populations of subsp. *ligtu* have a haploid formula of four metacentric chromosomes (chromosomes 1 and 2 with microsatellites), one submetacentric with a microsatellite and three telocentric with microsatellites. The population of subsp. *simsii* is characterized by having five metacentric chromosomes (chromosome 2 with a microsatellite and 6 with a secondary constriction) and three telocentric chromosomes with satellites. Baeza et al. [39] analyzed four populations of subsp. *ligtu*, defining localization on the chromosomes of the ribosomal genes 5S and 18-45S. Low polymorphic hybridization sites were detected in the populations, and only chromosome 1 presented a polymorphic site of 5S and 18/25S rDNA in the proximal and distal positions, respectively. Three subspecies are recognized within *A. ligtu* complex: subsp. *ligtu*, subsp. *splendens* and subsp. *simsii*. Fourteen populations were collected throughout its distributional range. Chromosome number, karyotype formulae, karyotypes, ideograms, intrachromosomal asymmetry index M_{CA}, and interchromosomal asymmetry index CV_{CL} were calculated [57, 58]. All studied populations showed 2n = 2x = 16 chromosomes. Subspecies *ligtu* and *simsii* are clearly differentiated from each other in M_{CA} and together from subsp. *splendens* with CV_{CL}. Intrachromosomal asymmetry index revealed two population groups within subsp. *splendens*. These populations also differ in karyotype formulae, habitat, soil type and distribution. We concluded that a fourth subspecies should be described from populations located in the lower part of the cordillera de los Andes in the Region of Maule. Populations of higher elevations correspond to those already described as subsp. *splendens* [57]. A comparative karyotype study was carried out among four populations of *A. diluta* subsp. *diluta* and three populations of *A. diluta* subsp. *chrysantha*. The seven populations presented an asymmetric karyotype, with 2n = 2x = 16 chromosomes, and with same karyotype formulae: 3 m + 1sm + 1st + 3 t. The architecture of the karyotype between the subspecies is the same. The scatter plot among M_{CA} versus CV_{CL} shows different groupings between populations of the two subspecies, and the total chromosomes length (TCL) is highest in the populations of subsp. *chrysantha*. According to the results obtained, the populations growing in Valparaíso Region should be considered belong to subsp. *diluta* [58]. We analyzed the karyotypes of 10 populations of *A. magnifica* complex along its natural distribution. All the populations showed an asymmetric karyotype, with 2*n* = 16 chromosomes but with 3 different karyotype formulae. *Alstroemeria magnifica* var. *magnifica* and *A. magnifica* var. *sierrae* presented the same karyotype formula, and *A. magnifica* var. *magenta* and *A. magnifica* var. *tofoensis* each had a different formula. The scatter plot among CV_{CL} versus M_{CA} shows different groupings between populations of the four varieties. Based on these results it is possible to consider raising *Alstroemeria magnifica* var. *magenta* to species rank and *A. magnifica* var. *tofoensis* to subspecies; *A. magnifica* var. *magnifica* and *A. magnifica* var. *sierrae* should each remain as varieties. Nevertheless, these taxonomic changes should be considered tentative, as additional sources of evidence become available.

5. Molecular studies in *Alstroemeria*

During recent years, an increasing accessibility to molecular data and the development of a vast range of bioinformatics analysis has favored the successful implementation of genetic tools in the identification and conservation of biological diversity [59]. Dominant molecular markers based on random fragment alleles e.g. Inter Simple Sequence Repeat (ISSR), Random Amplified Polymorphic DNA (RAPD) and Amplified Fragment Length Polymorphism (AFLP) have been used for characterizing genetic diversity in *Alstroemeria* [60–62], becoming the marker of choice for the identification of cultivar varieties with ornamental value [63–65] and conducting population genetic analyses [66, 67]. The use of DNA sequences has been more related to the construction of phylogenetic hypotheses and establishment of biogeographic patterns [8, 9]. Interestingly, near 30% of the Chilean species of *Alstroemeria* form species complexes, comprising from two to four infraspecific taxa each. This pattern is likely explained by adaptation to a wide range of environmental heterogeneity present in Chile [68], which is possibly driving processes of microevolutionary divergence [14]. Given the complexity of interpreting the integrity within and among these species complexes, we started several initiatives for applied genetic studies with the purpose of disentangling the discernibility of intraspecific patterns of divergence, especially in groups highly regarded for their ornamental and conservation value.

5.1. Molecular markers in assessing genetic diversity for conservation in *Alstroemeria*

A priority goal in conservation is to evaluate levels of apportionment of genetic diversity in targeted species, given the association between population genetic diversity and their potential for local adaptation and evolutionary resilience. Genetic variability is the result of the dynamics of gene flow, for which a homogeneous distribution of allelic frequencies is expected under high levels of gene flow among populations [69]. Interestingly, this situation is rarely found in nature, since the strong effect that geographic isolation and selection represents for local populations of plants. As a result, it is not surprising that peripheral populations tend to increase gene differentiation and population structure levels; hence, contributing to the local isolation that eventually could result in different isolated species (**Figure 7**) [69].

Such patterns of isolation and divergence are no exception in *Alstroemeria*, for which high levels of structuration are documented. For example, *Alstroemeria hookeri* represents a species complex that comprises four subspecies, two of them (subspecies *recumbens* and *maculata*) distributed in North-Central of Chile and two (subspecies *hookeri* and *sansebastiana*) in southern Chile. Based on ISSR (Inter Simple Sequence Repeat) markers, high levels of population structure were found among southern subspecies (**Figure 8**, **Table 2**); also concomitant with previous findings found with allozymes markers [67]. Similarly, high levels of within population diversity was found in *A. presliana* complex using AFLP markers (**Table 2**), exhibiting significant levels of among population variability and moderate levels of genetic population structure (**Table 2**). This complex comprises of two varieties (var. *presliana* and var. *australis*), both separately distributed across Coastal and Andean mountain ranges in Chile (**Figure 9**). The results from both complexes showed the existence of two heterogeneous genetic groups with no evident spatial congruence suggesting genetic differentiation among varieties or subspecies. Interestingly, several populations are individually differentiated in their genetic profiles, despite of occurring closely enough with other neighbored populations to sustain substantial levels of gene flow (**Figure 9**).

Among explanations of the observed patterns of genetic diversity found in *Alstroemeria*, strategies of reproduction and dispersal become plausible enough to be considered. *Alstroemeria* species have a restricted capacity of seed and pollen dispersal [66], which in combination with their vegetative reproduction by rhizomes [14, 67], could contribute to maintaining restricted levels of gene flow and sustaining high levels of genetic structure

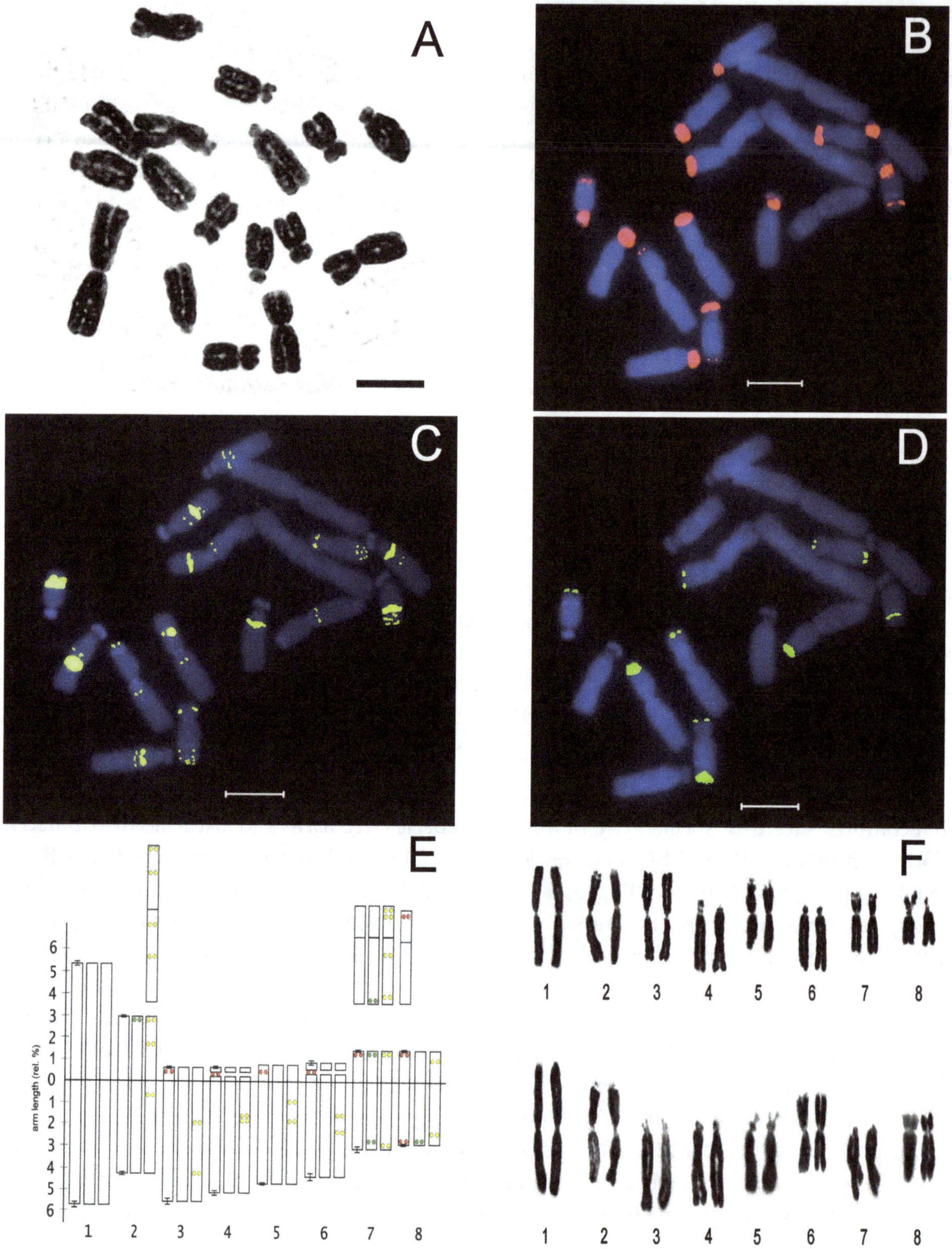

Figure 7. Cytogenetic studies in Chilean *Alstroemeria*. (A) Mitotic metaphase of *Alstroemeria hookeri* Ssp. *hookeri*. (B) Mitotic metaphase of *Alstroemeria hookeri* ssp. *hookeri* using 5S genes. (C) Mitotic metaphase of *Alstroemeria hookeri* ssp. *hookeri* using 18-25S genes. (D) Mitotic metaphase of *Alstroemeria hookeri* ssp. *hookeri* using A001 genes. (E) Ideogramm of *Alstroemeria hookeri* ssp. *hookeri* showing genes 5S, 18-25S, and A001 (FISH). (F) Karyotypes of *Alstroemeria presliana*: above, *A. presliana* ssp. *presliana*; below, *A. presliana* ssp. *australis*.

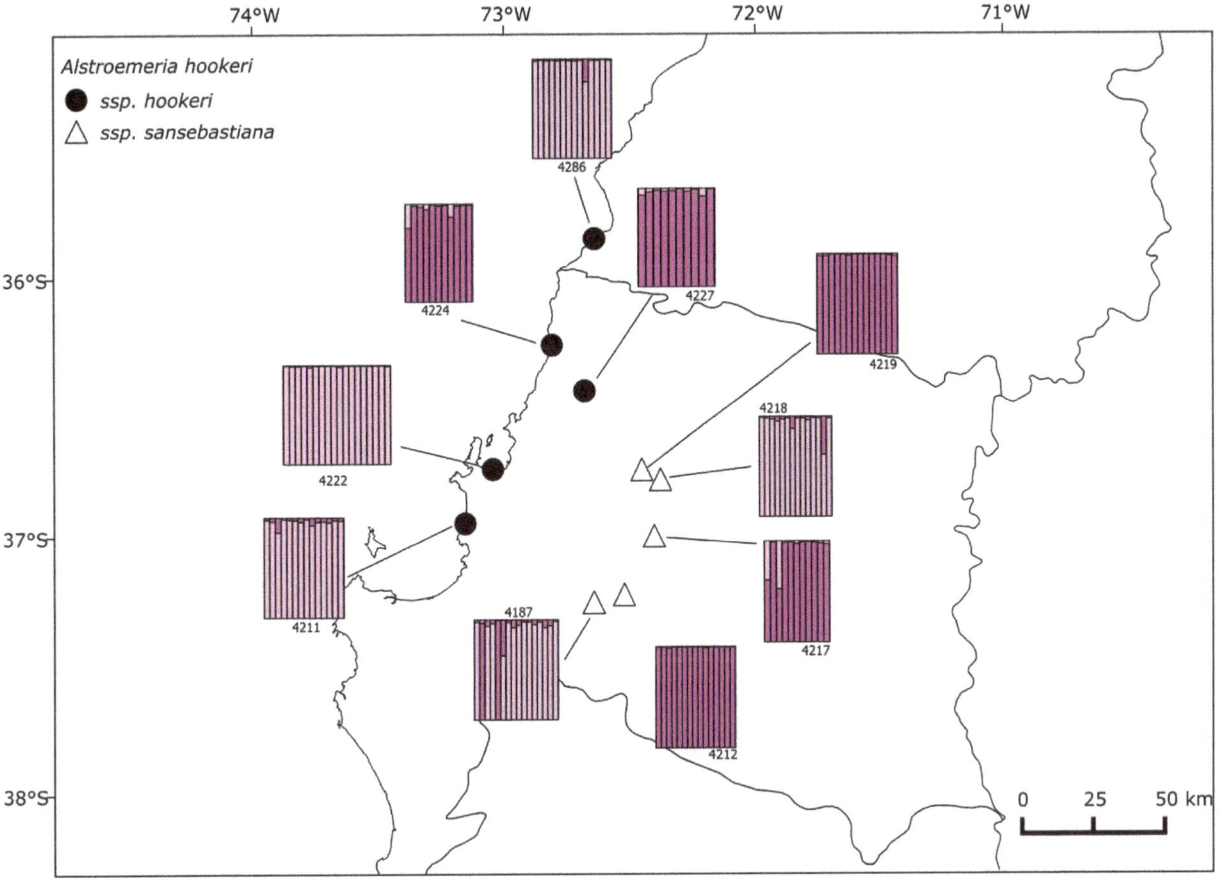

Figure 8. Population structure inferred with ISSR in the *A. hookeri* complex. Bar plots colors represent levels of genetic membership (k = 2) in each individual per sampled population, as inferred under Bayesian admixture inference criterion with the program STRUCTURE [70].

among populations [67]. The sum of these factors implies that local populations could be subject to strong geographic and ecological isolation, which would explain the diversity of infraspecific taxa found in this and other species complexes [67]. In general, moderate to high levels of among population genetic diversity were detected in the studied *Alstroemeria* species complexes. From a conservation perspective, this pattern suggests that protection

Species	Subspecies	He	Fst	AMOVA (%)		Marker/source
				Within pop.	Among pop.	
A. hookeri	*hookeri*	0.052	0.582	41.71	58.28	Allozymes/Ruiz et al. [67]
	hookeri	0.248	0.415	58.47	41.53	ISSR/unpublished
	sansebastiana	0.246	0.36	63.99	36.01	ISSR/unpublished
A. presliana	*presliana*	0.200	0.171	82.91	17.0	AFLP/unpublished
	australis	0.198	0.179	82.10	17.9	AFLP/unpublished

Table 2. Genetic diversity values obtained with allozymes and DNA markers (fragments analyses), for *A. hookeri* and *A. presliana* complexes.

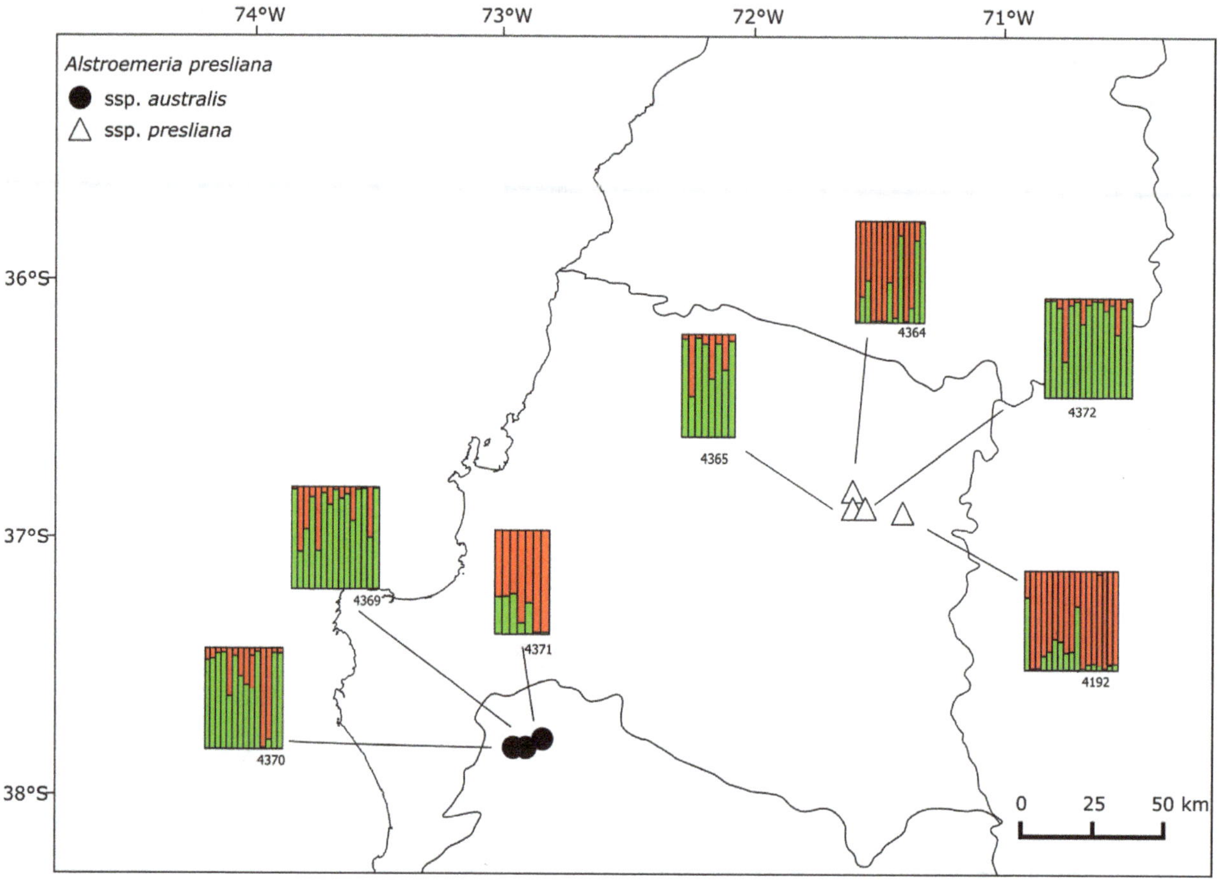

Figure 9. Population structure inferred with AFLP in the *A. presliana* Complex. Bar plots colors represent levels of genetic membership (k = 2) in each individual per sampled population, as inferred under Bayesian admixture inference criterion with the program STRUCTURE [70].

initiatives should consider as many populations as possible, in order to preserve the largest proportion of total species genetic diversity.

5.2. Molecular phylogenetic studies in *Alstroemeria*

Phylogenetic studies provide a theoretical framework to understand the relationships among populations and species. It is desirable that taxa must represent monophyletic lineages, thus reflecting the genetic, evolutionary and biogeographical integrity of lineages and species. Under such premise, several phylogenetic studies in *Alstroemeria* have been conducted integrating a diverse array of molecular, morphological and cytological data [8, 9]. Chacón et al. [8] conducted the most recent and comprehensive phylogenetic studies in genus *Alstroemeria*, based on DNA sequences and cytological data. From taxonomic and evolutionary perspective, three are the most relevant results: (1) Samples belonging to the same species were retrieved as monophyletic; (2) a biogeographic break exists between Brazilian and Chilean species groups and (3) a relatively recent divergence has occurred with the most species of the genus, being diverged during the last 8 millions of years. Interestingly, some of these results have been confirmed from previous initiatives, especially those reflecting the monophyly of the Brazilian species group with alternative molecular markers (i.e., AFLP) [61].

Despite the promising of these results, species from the Chilean group were mostly under-represented, making difficult to obtain relevant evidence of local patterns of diversification, particularly to those depicting evolutionary trends or taxonomic integrity in species complexes. Nonetheless, while some progress has been achieved scrutinizing chloroplast sequences (*rpl*32-*trn*l), discordant results challenge the hypothesis integrity previously stated in several of these groups. While *Alstroemeria hookeri* and *A. presliana* complexes are retrieved as monophyletic clades, other groups like *A. magnifica* and *A. ligtu* are retrieved as paraphyletic (**Figure 10**).

5.3. Molecular markers in eliciting taxonomic status in *Alstroemeria* species complex

The delimitation of species is a fundamental step for conducting natural and applied sciences, as they represent the main study unit for most areas of research (global evaluations of biodiversity, assessment and initiatives for biological conservation, etc.) [72]. In this sense, molecular approaches in taxonomy have been used under the assumption that observed

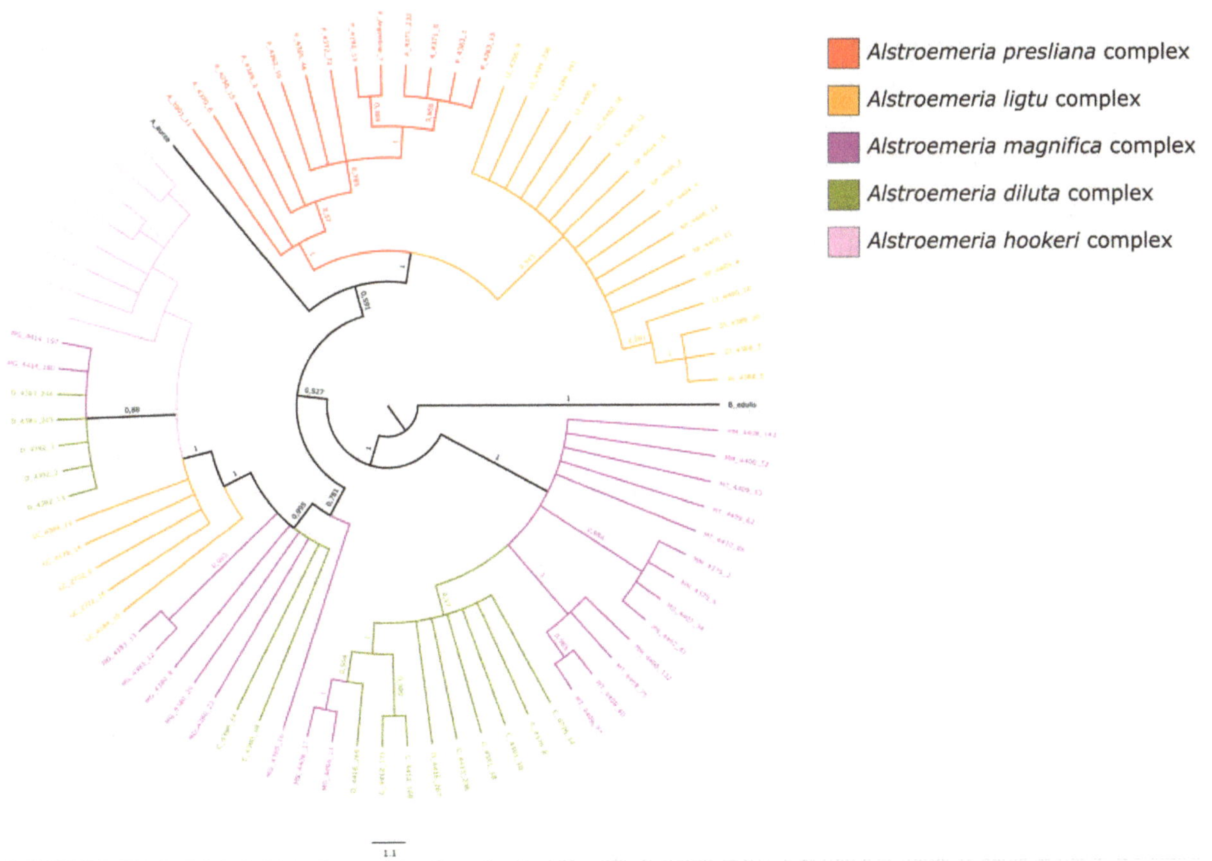

Figure 10. Maximum clade credibility tree (MCCT) inferred with the combination of chloroplast regions (trnL-S, rpl32-trnL, petA) for five species complexes of *Alstroemeria*, calculated with Bayesian inference criterion inferred with Mr. Bayes 3.2 [71]. Each tip represents an individual sampled per population and labels on branches depict posterior probabilities for each clade.

divergent patterns of genetic variation are the direct result of breaks in gene flow, leading to phenotypic and genotypic differences that sustain isolated and differentiated species and populations [69]. The simultaneous use of multiple molecular markers and criteria of delimitation has improved the taxonomic work, particularly helping to contextualize the role of microevolutionary processes in the species generation. As previously stated, species of *Alstroemeria* share attributes that could heavily influence patterns of micro evolutionary isolation and divergence, such as restricted seed and pollen dispersal and vegetative reproduction by rhizomes. Therefore, the wide distribution of taxonomic complexes in areas with contrasting topography and climatic conditions implies the existence of restrictions for gene flow, where substantial effects of ecogeographic isolation and divergence are expected in local diversification patterns [67]. Hence, micro evolutionary processes are currently underway and active [14], probably producing decoupled or unnoticeable patterns of divergence. Traditional taxonomic treatment in *Alstroemeria* has heavily relied on the interpretation of floral diversity and vegetative attributes, which has resulted in an important number of recognized taxa and species complexes described [14]. Nonetheless, because active of micro-evolutionary divergence may not ensure congruence among the diverse phenotypic and genotypic characters, it is likely that an under or overrepresentation of taxa is currently occurring in *Alstroemeria*. Therefore, given that molecular data could reflect patterns of divergence more accordingly to the dynamic of local gene flow, an interesting approach is to evaluate taxonomic boundaries integrating both molecular and phenotypic data as potential taxonomic characters. Recent molecular and phenotypic integrated studies conducted in *Alstroemeria* complexes resulted beneficial when multiple sources of evidence are placed to solve questions about the integrity or the validity of previous taxonomic treatments. For example, when morphometric, cytogenetic and molecular data were employed in *A. hookeri* complex taxa [41, 67], all analyzed characters were partially consistent with the recognition of the new subspecies *Alstroemeria hookeri* subsp. *sansebastiana* [22], and supported the hypothesis of Muñoz & Moreira [14] of elevating *Alstroemeria hookeri* ssp. *cummingiana* to species level. Subsequent investigations were also conducted in other three complexes (*A. ligtu*, *A. magnifica* and *A. presliana*), eliciting similar evidence with significant taxonomic impact. In the *A. magnifica* complex, evidence from morphology, colorimetry and cytology support the change of the taxonomic status of *A. magnifica* var. *magenta*. Preliminary analyses based on chloroplast sequences (*rpl*32-*trn*L) also supported this observation, validating the separation of var. *magenta* from the other *A. magnifica* varieties (**Figure 11**). In the *A. ligtu* complex, a new entity was discovered based on cytogenetic data, and its taxonomic status was redefined [57, 58]. The molecular data, based on chloroplast DNA (*rpl*32-*trn*L region) support the separation of Coastal populations of *A. ligtu* subsp. *ligtu* from populations of the inland distribution range (**Figure 12**). Finally, in *A. presliana*, same chloroplast markers also confirm the lack of structure observed with AFLP data; nonetheless, both of them seem not concordant with previous studies conducted with phenotypic data [18, 30]. It is likely that different sources of divergence are shaping idiosyncratic processes of differentiation among species complexes of *Alstroemeria*, suggesting that a case by case evaluation might be required before reaching a consensus for a more genus-wide taxonomic perspective.

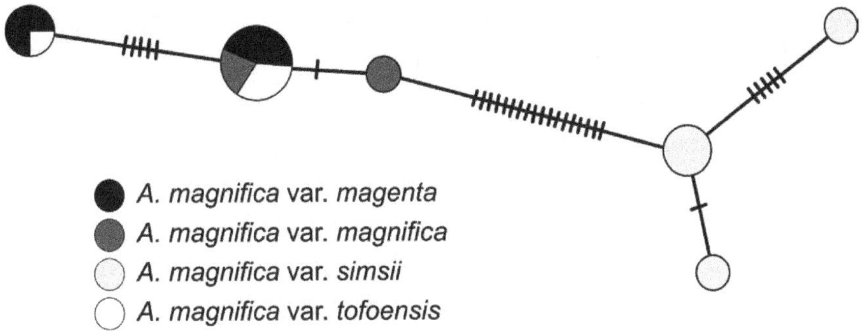

Figure 11. Haplotype network inferred with trnL-F chloroplast spacer for individuals sampled from population of *A. magnifica* species complex. Network was constructed under parsimony criteria with TCS [73], as implemented in PopART [74].

5.4. Perspectives and future work

Previous studies have demonstrated that a consensus about the integrity of the taxa of *Alstroemeria* is far from being reached, as different patterns of differentiation may difficult to be elicited separately. In this sense, molecular markers have provided a natural framework to contextualize their evolutionary process, reconciling discordance observed from different character sources. Nonetheless, despite of their utility, molecular markers are not exempt of limitations that should be addressed in subsequent studies. One of the main limitations to reach a robust taxonomic hypothesis is the recurrent difficulty to obtain consistent molecular markers adaptable enough for interspecific and intraspecific analyses. These difficulties arise from the extraordinary large and complex genomic architecture of *Alstroemeria*, which is likely comprised of a large proportion of repetitive DNA (18–34 pg.) [17]. Our experience suggests that most nuclear markers tend to recurrently fail to retrieve single and readable copies through recurrent Sanger sequencing techniques, especially for the Internal Transcribed Spacer or ITS. Similarly, fragment analyses also exhibit levels of difficulties for consistent scoring, since the effect that

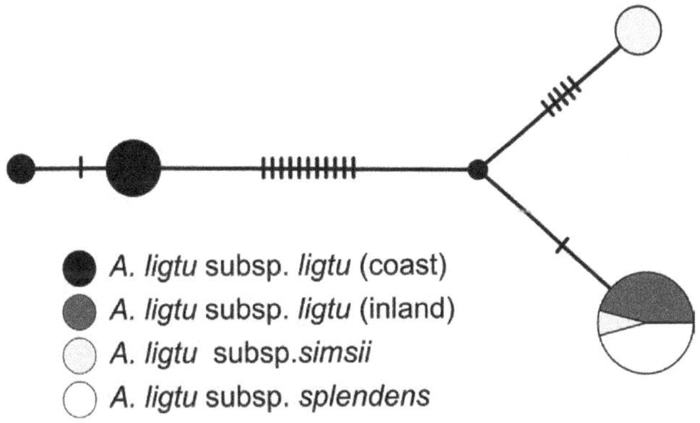

Figure 12. Haplotype network inferred with trnL-F chloroplast spacer for individuals sampled from population of *A. ligtu* species complex. Network was constructed under parsimony criteria with TCS [73], as implemented in PopART.

repetitive DNA has in the proportion of cut sites with restriction enzymes [17]. Since a more widespread consensus exist about the necessity of integrating different sources of molecular evidence and methodologies in species delimitation analyses [75], further work is required in the design of reliable and stable molecular markers for the study of natural species of *Alstroemeria*.

With the onset of new and more accessible technologies for genome sequencing (Next Generation Sequencing or NGS), new possibilities have opened for the generation of more representative analyses of genetic diversity [76, 77]. Unfortunately, such techniques have been not widely implemented in *Alstroemeria*, except important breakthroughs like the sequencing and the annotation of the chloroplast genome in *A. aurea* [78]. The generation of single nuclear polymorphisms (SNP) in non-model organisms has been the approach of choice for high-throughput genome sequencing, adding improved genome representation and resolution for inter and intraspecific levels relationships [79]. The implementation of SNPs might result in a significant improvement in the estimation of genetic diversity and species limits in *Alstroemeria*, as SNPs represent codominant markers capable of providing a higher statistical power and an easier species comparability considering the available genomic resources compared to AFLP [80]. For taxonomic purposes, SNP could greatly improve the use of DNA barcodes to identify species through the use of specific DNA regions, especially when traditional approaches of taxonomy fail [81]. Obviously, the use of NGS and SNP techniques in *Alstroemeria* requires adjustments to overcome the limitations imposed by genome size and complexity, for which recent alternatives have been shown from the study of other equally complex organisms [82]. As such, the perspective of solving the taxonomic problems with molecular techniques in *Alstroemeria* remains promising, yet keeping in perspective its own limitations and challenges to reach the require tools to finally approach the inherent dynamics of macro and micro evolutionary patterns in this group.

6. Concluding remarks

Genetic divergence and population structure estimated with AFLPs and ISSR, have demonstrated the importance of molecular markers for conservation purposes in *Alstroemeria*. Integrative use of molecular data with other source of evidence (morphology, cytology and morphometry) give a best interpretation of lineage divergence with better argumentation for taxonomic delimitation in species complexes of *Alstroemeria*. Due to the high proportion of species complex in genus *Alstroemeria* in Chile, is necessary to carry out phylogenetic studies including the most infraspecifc taxa and more representative sampling, in addition with a major representation of the genome in the analyses. More efforts are needed in producing more stable molecular markers, in order to further implement integrative analyses. In this sense, it is likely that NGS will play a pivotal role helping to overcome present limitations of molecular work in *Alstroemeria*.

Acknowledgements

We gratefully acknowledge the Curators and Director of the Herbarium CONC for providing access to the studied specimens. This work was supported by the projects Fondecyt 1130349 "Morphological, molecular and cytogenetic overview of some species complexes of the genus *Alstroemeria* in Chile" and VRID-ENLACE 217.111.063-1.0 "Integrative taxonomy in the genus *Alstroemeria* L. (Alstroemeriaceae): The use of morphological, colorimetric, cytological and molecular approaches on several Chilean complexes". We acknowledge Nicolás Villalobos for technical support.

A. Appendix

Checklist of the Chilean species of Alstroemeria

Accepted names were written in **bold**, synonyms in *italics*. Names written in normal fonts corresponds to those whose acceptance or status has not yet been clarified. (**E**) = Endemic to Chile. (**N**) = Native.

1. **Alstroemeria achirae** Muñoz-Schick & Brink, Gayana, Bot. 57(1): 56. 2000 (**E**)
2. *Alstroemeria albiflora* C. Presl, Reliq. Haenk. 1(2): 121. 1827 = **Alstroemeria pallida** Graham
3. *Alstroemeria amoena* Salisb., Prodr. Stirp. Chap. Allerton 248. 1796 = *Alstroemeria pelegrina* L.
4. **Alstroemeria andina** Phil., Linnaea 29: 69. 1858.
    a. subsp. *venustula* (Phil.) Ehr. Bayer, Mitt. Bot. Staatssamml. München 24: 74. 1987 = **Alstroemeria andina** Phil. var. **venustula** (Phil.) Muñoz-Schick, Not. Mens. Mus. Nac. Hist. Nat. 352: 22. 2003.
    b. var. **andina** (**E**)
    c. var. **venustula** (Phil.) Muñoz-Schick, Not. Mens. Mus. Nac. Hist. Nat. 352: 22. 2003. (**N**)
5. **Alstroemeria angustifolia** Herb., Amaryllidaceae 96. 1837.
    a. subsp. *velutina* Ehr. Bayer, Mitt. Bot. Staatssamml. München 24: 165. 1987.
    b. var. *acuminata* Herb., Amaryllidaceae 97. 1837 = **Alstroemeria angustifolia** var. **angustifolia**
    c. var. **angustifolia** (**E**)
    d. var. *conferta* Herb., Amaryllidaceae 97. 1837 = **Alstroemeria angustifolia** var. **angustifolia**
    e. var. *intermedia* Herb., Amaryllidaceae 97. 1837 = **Alstroemeria angustifolia** var. **angustifolia**

f. var. *solliana* Herb., Amaryllidaceae 97. 1837 = **Alstroemeria angustifolia** var. **angustifolia**

g. var. **velutina** (Ehr. Bayer) Muñoz-Schick, Mitt. Bot. Staatssamml. München 24: 165. 1987. (**E**)

6. Alstroemeria aulica Ravenna, Onira 4(10): 41. 2000.

7. *Alstroemeria aurantiaca* D. Don, Brit. Fl. Gard., ser. 2, t. 205. 1835 = *Alstroemeria aurea* Graham

8. *Alstroemeria araucana* Phil., Anales Univ. Chile 43: 547. 1873 = *Alstroemeria aurea* Graham

9. *Alstroemeria aurea* Graham, Edinburgh Philos. J. 181. 1833. (**N**)

   a. var. *valparadisiaca* Herb., Amaryllidaceae 98. 1837 = **Alstroemeria ligtu** L. subsp. **simsii** (Spreng.) Ehr. Bayer

10. *Alstroemeri aurea* Meyen, Reise Erde 1: 311. 1834, hom. Illeg. = **Alstroemeria ligtu** L. subsp. **simsii** (Spreng.) Ehr. Bayer

11. *Alstroemeria bicolor* Hook., Exot. Fl. 2(9): t. 95. 1824 = **Alstroemeria pulchra** Sims subsp. **pulchra**

12. Alstroemeria bilabiata Ravenna, Phytologia 64(4): 282. 1988.

13. Alstroemeria cantillanica Ravenna, Phytologia 64(4): 285. 1988.

14. *Alstroemeria chilensis* Lem., Fl. Serres 1(5): 98. 1845 = **Alstroemeria ligtu** L. subsp. **ligtu**

15. *Alstroemeria chillanensis* Grau & Ehr. Bayer, Mitt. Bot. Staatssamml. München 18: 220. 1982 = **Alstroemeria presliana** Herb. subsp. **presliana**

16. *Alstroemeria chiloensis* Phil., Linnaea 29: 71. 1858 = *Alstroemeria aurea* Graham

17. *Alstroemeria ciliata* Poepp., Fragm. Syn. Pl. 6. 1833 = **Alstroemeria ligtu** L. subsp. **simsii** (Spreng.) Ehr. Bayer

18. **Alstroemeria citrina** Phil., Linnaea 22: 264. 1864 (**E**)

19. *Alstroemeria concolor* Steud., Berberid. Amer. Austr. 53. 1857

20. **Alstroemeria x chrysantha** (Ehr. Bayer) J.M. Watson & A.R. Flores, Fl. Silvtr. Chile 1: 90. 2015.(**E**)

21. **Alstroemeria crispata** Phil., Linnaea 29: 70. 1858 (**E**)

22. *Alstroemeria crocea* Phil., Linnaea 33: 262. 1864, non Ruiz & Pav. 1802 = **Alstroemeria pseudospathulata** Ehr. Bayer

23. **Alstroemeria cummingiana** Herb., Amaryllidaceae 96. 183 (**E**)

24. Alstroemeria decora Ravenna, Onira 4(10): 42. 2000.

25. *Alstroemeria dentata* Klotzsch ex Kunth, Enum. Pl. 5: 780. 1850, nom. Nud. = **Alstroemeria ligtu** L. subsp. **simsii** (Spreng.) Ehr. Bayer

26. *Alstroemeria diazii* Phil., Linnaea 33: 261. 1864 = **Alstroemeria exerens** Meyen

27. *Alstroemeria diazii* auct. Non Phil., Fl. Patagonica 2: 162. 1969 = **Alstroemeria presliana** Herb. subsp. **presliana**

28. **Alstroemeria diluta** Ehr. Bayer, Ehr. Bayer, Mitt. Bot. Staatssamml. München 24: 178. 1987

- **a.** subsp. *chrysantha* Ehr. Bayer, Mitt. Bot. Staatssamml. München 24: 178. 1987 (**E**) = **Alstroemeria x chrysantha** (Ehr. Bayer) J.M. Watson & A.R. Flores
- **b.** subsp. **diluta** (**E**)

29. Alstroemeria discolor Ravenna, Onira 4(10): 44. 2000.

30. *Alstroemeria epulauquensis* Ravenna, Phytologia 64(4): 283. 1988 = **Alstroemeria presliana** Herb. subsp. **presliana**

31. **Alstroemeria exerens** Meyen, Reise Erde 1: 34. 1834 (**E**)

32. *Alstroemeria exserens* Meyen, Reise Erde 1: 34. 1834 **=** **Alstroemeria exerens** Meyen

33. *Alstroemeria flava* Phil., Linnaea 33: 263. 1864 = **Alstroemeria ligtu** subsp. **ligtu**

34. *Alstroemeria flos-martini* Ker Gawl., Bot. Reg. 9: t. 731. 1823 = **Alstroemeria pulchra** Sims subsp. **pulchra**

35. **Alstroemeria garaventae** Ehr. Bayer, Gattung Alstroemeria in Chile 60. 1987 (**E**)

36. *Alstroemeria gayana* Phil., Linnaea 29: 71. 1857 = **Alstroemeria magnifica** Herb. var. **magnifica**

37. **Alstroemeria graminea** Phil., Anales Univ. Chile 93: 161. 1896 (**E**)

38. *Alstroemeria haemantha* Ruiz & Pav., Fl. Peruv. 3: 60. 1802 **=** **Alstroemeria ligtu** L.

- **a.** var. *haemantha* = **Alstroemeria ligtu** L.subsp. **ligtu**
- **b.** var. *pilosa* Herb., Amaryllidaceae 100. 1837 **=** **Alstroemeria ligtu** L. subsp. **simsii** (Spreng.) Ehr. Bayer
- **c.** var. *simsiana* Herb., Amaryllidaceae 99. 1837 = **Alstroemeria ligtu** L. subsp. **simsii** (Spreng.) Ehr. Bayer

39. *Alstroemeria haemantha* auct. Non Ruiz & Pav. = **Alstroemeria ligtu** L. subsp. **simsii** (Spreng.) Ehr. Bayer

40. *Alstroemeria herbertiana* M. Roem., Fam. Nat. Syn. Monogr. 250. 1847 = **Alstroemeria revoluta** Ruiz & Pav.

41. *Alstroemeria hirtella* Phil., Linnaea 29: 70. 1858, hom. Illeg. **=** **Alstroemeria leporina** Ehr. Bayer & Grau

42. **Alstroemeria hookeri** Lodd., Bot. Cab. 13: t. 1272. 1827

- **a.** subsp. *cummingiana* (Herb.) Ehr. Bayer, Mitt. Bot. Staatssamml. München 24: 157. 1987 = **Alstroemeria cummingiana** Herb.
- **b.** subsp. **hookeri** (**E**)

c. subsp. **maculata** Ehr. Bayer, Mitt. Bot. Staatssamml. München 24: 153. 1987 (**E**)

d. subsp. **recumbens** (Herb.) Ehr. Bayer, Mitt. Bot. Staatssamml. München 24: 149. 1987 (**E**)

e. subsp. **sansebastiana** C.M. Baeza & E. Ruiz, Gayana, Bot. 68(2): 313. 2011 (**E**)

43. *Alstroemeria hookeriana* Schult., Syst. Veg. 7(1): 733. 1829 = **Alstroemeria hookeri** Lodd. subsp. **hookeri**

44. Alstroemeria huemulina Ravenna, Phytologia 64(4): 285. 1988

45. *Alstroemeria inconspicua* Phil., Anales Univ. Chile 43: 546. 1783 = **Alstroemeria revoluta** Ruiz & Pav.

46. Alstroemeria jocunda Ravenna, Phytologia 64(4): 284. 1988

47. **Alstroemeria kingii** Phil., Anales Univ. Chile 43: 548. 1873 (**E**)

48. *Alstroemeria lacrima-solis* Ravenna, Phytologia 64(4): 283. 1988 = **Alstroemeria pseudospathulata** Ehr. Bayer

49. **Alstroemeria leporina** Ehr. Bayer & Grau, Mitt. Bot. Staatssamml. München 18: 222. 1982 (**E**)

50. **Alstroemeria ligtu** L., Pl. Alströmeria 10. 1762.

a. Fm*a. pulchra* (Sims) Voss, Bot. Mag. 50: t. 2421. 1823[1822] = **Alstroemeria pulchra** Sims subsp. **pulchra**

b. subsp. *incarnata* Ehr. Bayer, Mitt. Bot. Staatssamml. München 24: 131. 1987, nom. Illeg. = **Alstroemeria ligtu** L. subsp. **splendens** Muñoz-Schick

c. subsp. **ligtu** (**E**)

d. subsp. **simsii** (Spreng.) Ehr. Bayer, Mitt. Bot. Staatssamml. München 24: 122. 1987 (**E**)

e. subsp. **splendens** Muñoz-Schick, Notic. Mens. Mus. Nac. Hist. Nat. 352: 22. 2003 (**E**)

f. var. *andina* Phil., Linnaea 33: 261. 1864 = **Alstroemeria pallida** Graham

g. var. *pulchra* (Sims) Baker, Handb. Amaryllideae: 139. 1888 = **Alstroemeria pulchra** Sims subsp. **pulchra**

51. *Alstroemeria ligtu* auct. Non L., Trab. Inst. Bot. Farmacol. 33: 26. 1915 = *Alstroemeria aurea* Graham

52. *Alstroemeria lothiana* Utinet. J. Jardins Jahrgang 1841: 348. = **Alstroemeria ligtu** subsp. **ligtu**

53. **Alstroemeria lutea** Muñoz-Schick, Gayana, Bot. 57(1): 55. 2000 (**E**)

54. *Alstroemeria macreana* Herb., Amaryllidaceae 90. 1837 = **Alstroemeria ligtu** L. subsp. **simsii** (Spreng.) Ehr. Bayer

55. **Alstroemeria magenta** Ehr. Bayer, Mitt. Bot. Staatssamml. München 24: 191. 1987 = *Alstroemeria magnifica* Herb. var. *magenta* (Ehr. Bayer) Muñoz-Schick

56. **Alstroemeria magnifica** Herb., Bot. Reg. 29: 64. 1843

   a. subsp. *gayana* (Phil.) Ehr. Bayer, Gatt. Alstroemeria Chile 252. 1987[1986] = **Alstroemeria magnifica** Herb. var. **magnifica**

   b. var. **magnifica** (E)

   c. var. *magenta* (Ehr. Bayer) Muñoz-Schick, Notic. Mens. Mus. Nac. Hist. Nat. 352: 22. 2003 = **Alstroemeria magenta** Ehr. Bayer

   d. var. **sierrae** (Muñoz) Muñoz-Schick, Notic. Mens. Mus. Nac. Hist. Nat. 352: 22. 2003 (E)

   e. var. **tofoensis** Muñoz-Schick, Notic. Mens. Mus. Nac. Hist. Nat. 352: 22. 2003 (E)

57. **Alstroemeria magenta** Ehr. Bayer, Mitt. Bot. Staatssamml. München 24: 191. 1987 (E)

58. Alstroemeria magna Ravenna, Phytologia 64(4): 284. 1988

59. **Alstroemeria marticorenae** Negritto & C.M. Baeza, Syst. Bot. 40(1): 70. 2015 (E)

60. *Alstroemeria meyeniana* Schauer, Nova Acta Acad. Caes. Leop.-Carol. German. Nat. Cur. Suppl. 1: 440. 1843 = **Alstroemeria versicolor** Ruiz & Pav.

61. **Alstroemeria mollensis** Muñoz-Schick & Brinck, Gayana, Bot. 60(2): 102. 2003 (E)

62. Alstroemeria monantha Ravenna, Phytologia 64(4): 286. 1988

63. *Alstroemeria mutabilis* Kunze, Enum. Pl. 5: 777. 1850 = *Alstroemeria aurea* Graham

64. *Alstroemeria nana* Rendle, J. Bot. 42: 325. 1904 = **Alstroemeria patagonica** Phil.

65. Alstroemeria nidularis Ravenna, Phytologia 64(4): 282. 1988

66. *Alstroemeria nivalis* Phil., Linnaea 29: 69. 1858, hom. Illeg. = **Alstroemeria pallida** Graham

67. *Alstroemeria nubigena* Phil., Linnaea 29: 69. 1858 = **Alstroemeria pallida** Graham

68. *Alstroemeria oxyphylla* Ravenna, Phytologia 64(4): 286. 1988, nom. Superfl. = **Alstroemeria pseudospathulata** Ehr. Bayer

69. *Alstroemeria pallens* Phil., Linnaea 33: 265. 1864 = **Alstroemeria exerens** Meyen

70. **Alstroemeria pallida** Graham, Edinburgh New Philos. J. 344. 1829 (E)

71. **Alstroemeria parvula** Phil., Linnaea 33: 261. 1864 (E)

72. **Alstroemeria patagonica** Phil., Anales Univ. Chile 84: 160. 1894 (N)

   a. fma. *Biflora* Ravenna, Sellowia 19: 34. 1967 = **Alstroemeria patagonica** Phil.

73. *Alstroemeria paupercula* Phil., Fl. Atacam. 51. = **Alstroemeria violacea** Phil.

74. *Alstroemeria pelegrina* L., Sp. Pl. (ed. 2) 1: 461. 1762 (E)

   a. var. *albescens* Herb. Amaryllidaceae 91. 1837. = *Alstroemeria pelegrina* L.

75. **Alstroemeria philippii** Baker, Handb. Amaryllidae 140. 1888

a. subsp. **adrianae** J.M. Watson & A.R. Flores, Herbertia 63: 102. 2009[2010] (**E**)

b. subsp. **philippii (E)**

c. var. **albicans** Muñoz-Schick, Alstroemerias Chile 41. 2003 (**E**)

d. var. **philippii** (E)

76. Alstroemeria poetica Ravenna, Phytologia 64(4): 285. 1988

77. Alstroemeria polpaicana Ravenna, Phytologia 64(4): 283. 1988

78. **Alstroemeria polyphylla** Phil., Anales Univ. Chile 93: 160. 1896 (**E**)

79. **Alstroemeria presliana** Herb., Enum. Pl. 5: 773. 1850

a. subsp. **presliana** (**N**)

b. subsp. **australis** Ehr. Bayer, Gatt. Alstroemeria Chile 122. 1987 (**E**)

80. **Alstroemeria pseudospathulata** Ehr. Bayer, Mitt. Bot. Staatssamml. München 24: 49. 1987 (**N**)

81. *Alstroemeria pulchella* auct. Non L.f. = **Alstroemeria ligtu** subsp. **simsii** (Spreng.) Ehr. Bayer

82. *Alstroemeria pulchella* L.f. var. *pilosa* Lindl., Edwards's Bot. Reg. 17: t. 1410. 1831 = **Alstroemeria ligtu** subsp. **simsii** (Spreng.) Ehr. Bayer

83. **Alstroemeria pulchra** Sims, Bot. Mag. 50, t. 2421. 1823[1822]

a. subsp. **lavandulacea** Ehr. Bayer, Mitt. Bot. Staatssamml. München 24: 187. 1987 (**E**)

b. subsp. **maxima** (Phil.) Ehr. Bayer, Mitt. Bot. Staatssamml. München 24: 202 t.14, 18. 1987 (**E**)

c. subsp. **pulchra** (**E**)

d. var. **maxima** Phil., Linnaea 33: 266. 1864–65 (**E**)

84. *Alstroemeria pygmaea* auct. Non Herb., Svenska Exped. Magell. 3(5): 205. 1901 = **Alstroemeria patagonica** Phil.

85. *Alstroemeria quillotensis* Herb., Amaryllidaceae 97, t.2., f.2, 1837 = **Alstroemeria ligtu** L. subsp. **simsii** (Spreng.) Ehr. Bayer

86. *Alstroemeria recumbens* Herb., Amaryllidaceae 97, t.3. 1837 = **Alstroemeria hookeri** Lodd. subsp. **recumbens** (Herb.) Ehr. Bayer

87. *Alstroemeria reflexa* M. Roem., Fam. Nat. Syn. Monogr. 4: 252. 1847 = **Alstroemeria revoluta** Ruiz & Pav.

88. **Alstroemeria revoluta** Ruiz & Pav., Fl. Peruv. 3: 59. 1802 (**E**)

89. *Alstroemeria rosea* Hook., Exot. Fl. 3(27): t.181. 1825 = **Alstroemeria hookeri** Lodd. subsp. **hookeri**

90. *Alstroemeria rosea* Phil., Sert. Mendoc. Alt. 43. 1871 = **Alstroemeria pallida** Graham

91. Alstroemeria sabulosa Ravenna, Phytologia 64(4): 281. 1988

92. **Alstroemeria schizanthoides** Grau, Ehr. Bayer, Mitt. Bot. Staatssamml. München 18: 213. 1982

a. var. **alba** Muñoz-Schick, Notic. Mens. Mus. Nac. Hist. Nat. 352: 22. 2003 (**E**)

b. var. **schizanthoides** (E)

93. *Alstroemeria sericantha* Schauer, Nov. Actorum Acad. Caes. Nat. Cut. 19(Suppl. 1): 441. 1843 = **Alstroemeria umbellata** Meyen

94. *Alstroemeria sierrae* Muñoz, Fl. Silvestr. Chile: 64. 1966 = **Alstroemeria magnifica** Herb. var. **sierrae** (Muñoz) Muñoz-Schick

95. *Alstroemeria simsii* Spreng., Syst. Veg. [Sprengel] 2: 80. 1825 = **Alstroemeria ligtu** L. subsp. **simsii** (Spreng.) Ehr. Bayer

96. *Alstroemeria sotoana* Phil., Anales Univ. Chile 93: 159. 1896 = **Alstroemeria versicolor** Ruiz & Pav.

97. **Alstroemeria spathulata** C. Presl, Reliq. Haenk. 1(2): 122, t.22, f.2., 1827 (**E**)

98. *Alstroemeria spathulata* auct. Non C. Presl, Fl. Patag. 2: 162. 1969 = **Alstroemeria pseudospathulata** Ehr. Bayer

99. Alstroemeria spectabilis Ravenna, Phytologia 64(4): 284. 1988

100. *Alstroemeria stenopetala* Phil., Anales Univ. Chile 43: 547. 1873 = *Alstroemeria aurea* Graham

101. Alstroemeria timida Ravenna, Phytologia 64(4): 281. 1988

102. *Alstroemeria tigrina* Phil., Linnaea 29: 68. 1857 = **Alstroemeria versicolor** Ruiz & Pav.

103. **Alstroemeria traudliae** J.M. Watson & A.R. Flores, Fl. Sylvestr. Chil. 1: 118. 2015 **(E)**

104. *Alstroemeria tricolor* Hook., Exot. Fl. 1(5): t. 65. 1823 = **Alstroemeria pulchra** Sims subsp. **pulchra**

105. **Alstroemeria umbellata** Meyen, Reise Erde 1: 356. 1835 (**E**)

106. Alstroemeria venusta Ravenna, Phytologia 64(4): 282. 1988

107. *Alstroemeria venustula* Phil., Linnaea 33: 260. 1864 = **Alstroemeria andina** Phil. var. **venustula** (Phil.) Muñoz-Schick

108. **Alstroemeria versicolor** Ruiz & Pav., Fl. Peruv. 3: 59. 1802 (**E**)

109. **Alstroemeria violacea** Phil., Fl. Atacam. 51. 1860 (**N**)

110. *Alstroemeria violacea* Knight & Perry, nom. Nud. = **Alstroemeria violacea Phil.**

111. **Alstroemeria werdermannii** Ehr. Bayer, Gatt. Alstroemeria Chile 87. 1986[1987]

a. var. **werdermannii (E)**

b. var. *flavicans* Muñoz-Schick, Gayana, Bot. 57(1): 57, f.3. 2000 = **A. werdermannii** Ehr. Bayer subsp. **flavicans** (Muñoz-Schick) J.M. Watson & A.R. Flores **(E)**

c. subsp. **flavicans** (Muñoz-Schick) J.M. Watson & A.R. Flores, Fl. Silvestr. Chile 1: 122. 2015.

112. *Alstroemeria xanthina* Phil., Anales Univ. Chile 43: 546. 1873 = **Alstroemeria versicolor** Ruiz & Pav.

113. Alstroemeria yaelae Ravenna, Phytologia 64(4): 282. 1988

114. **Alstroemeria zoellneri** Ehr. Bayer, Mitt. Bot. Staatssamml. München 24: 245, t. 143–144. 1987(**E**)

115. *Taltalia graminea* (Phil.) Ehr. Bayer, Sendtnera 5: 7, f. 1–4. 1998 = **Alstroemeria graminea** Phil.

Author details

Victor L. Finot[1*], Carlos M. Baeza[2], Eduardo Ruiz[2], Oscar Toro[2] and Pedro Carrasco[2]

*Address all correspondence to: vifinot@udec.cl

1 Department of Animal Production, Faculty of Agronomy, University of Concepción, Chile

2 Department of Botany, Faculty of Natural Sciences and Oceanography, University of Concepción, Chile

References

[1] The Angiosperm Phylogeny Group. An update of the angiosperm phylogeny group classification for the orders and families of flowering plants: APG III. Botanical Journal of the Linnean Society. 2009;**161**:105-121

[2] The Angiosperm Phylogeny Group. An update of the angiosperm phylogeny group classification for the orders and families of flowering plants: APG IV. Botanical Journal of the Linnean Society. 2016;**181**:1-20

[3] Stevens PF. Angiosperm Phylogeny Website. Version 12, July 2012. [Internet]. 2001. Available from http://www.mobot.org/MOBOT/reasearch/APweb [Accessed: Aug 9, 2017]

[4] Watson L, Dallwitz MJ. The Families of Flowering Plants: Descriptions, Illustrations, Identification, and Information Retrieval. [Internet]. 1992. Available from: http://www.delta-intkey.com/angio/www/alstroem.htm [Accessed: July 28, 2017]

[5] Bayer E. *Taltalia,* eine neue Gattung in der Familie der Alstroemeriaceae. Sendtnera. 1998;**5**:5-14

[6] Sanso AM, Xifreda C. Generic delimitation between *Alstroemeria* and *Bomarea* (Alstroemeriaceae). Annals of Botany. 2001;**88**:1057-1069

[7] Camargo de Assis M. Neotropical Alstroemeriaceae. In: Milliken W, Klitgard W, Baracat A, editors. Neotropikey-Interactive Key and Information Resources for Flowering Plants of the Neotropics. [Internet]. 2009. Available from: http://www.kew.org/science/tropamerica/neotropikey/families/Alstroemeriaceae.htm [Accessed: July 28, 2017]

[8] Chacón J, Camargo de Assis M, Meerow AW, Renner S. From East Gondwana to Central America: Historical biogeography of the Alstroemeriaceae. Journal of Biogeography. 2012a;**39**:1806-1818

[9] Aagesen L, Sanso M. The phylogeny of the Alstroemeriaceae, based on morphology, rps16 Intron, and rbcL sequence data. Systematic Botany. 2003;**28**:47-69

[10] Linnaeus C. Planta *Alströmeria*. Johannes Petrus Falck, Vestrogothus. Upsala. 1762:**8**

[11] Bayer E. Die Gattung *Alstroemeria* in Chile. Mitteilungen der Botanischen Staatssammlung München. 1987;**24**:1-362

[12] Aker S, Healy W. The phytogeography of the genus *Alstroemeria*. Herbertia. 1990;**46**:76-87

[13] Meerow AW, Tombolato AFC, Meyer F. Two new species of *Alstroemeria* (Alstroemeriaceae) from Brazil. Brittonia. 1999;**51**:439-444

[14] Muñoz-Schick M, Moreira A. Alstroemerias de Chile. Diversidad, Distribución y Conservación. Taller La Era: 2003. 140 p

[15] Hofreiter A, Rodríguez EF. The Alstroemeriaceae of Peru and neighbouring areas. Revista Peruana de Biología. 2006;**13**:5-69

[16] Buitendijk J, Ramanna M. Giemsa C-banded karyotypes of eight species of *Alstroemeria* L. and some of their hybrids. Annals of Botany. 1996;**78**:449-457

[17] Buitendijk J, Boon E, Ramanna M. Nuclear DNA content in twelve species of *Alstroemeria* L. and some of their hybrids. Annals of Botany. 1997;**79**:343-353

[18] Baeza CM, Finot VL, Ruiz E. Comparative karyotype analysis of populations in the *Alstroemeria presliana* Herbert (Alstroemeriaceae) complex in Chile. Genetics and Molecular Biology. 2015;**38**:199-204

[19] Negritto MA, Baeza CM, Ruiz E, Novoa P. *Alstroemeria marticorenae* (Alstroemeriaceae), a new species from Central Chile. Systematic Botany. 2015;**40**:69-74

[20] Muñoz-Schick M. Novedades en la familia Alstroemeriaceae. Gayana Botanica. 2000; **57**:55-59

[21] Watson JM, Flores AR. A rare new subspecies and change in rank in *Alstroemeria* L. (Alstroemeriaceae) for taxa narrowly endemic to Chile's Atacama region. Herbertia. 2010;**63**:101-117

[22] Baeza CM, Ruiz E. *Alstroemeria hookeri* Lodd. Subsp. *sansebastiana* C. Baeza & E. Ruiz, new for the Chilean flora. Gayana Botanica. 2011;**68**:313-315

[23] Hoffmann A, Watson J, Flores A. Flores silvestres de Chile. Cuando el desierto florece. Vol. 1. Monocotiledóneas y otros taxones. Fundación Claudio Gay. 2015. 263 pp

[24] Arroyo MTK, Marquet P, Marticorena C, Simonetti J, Cavieres L, Squeo F, Rozzi R, Massardo F. El hotspot chileno, prioridad mundial para la conservación. In: Comisión Nacional del Medio Ambiente, editors. Biodiversidad de Chile: Patrimonio y Desafíos. Ocho Libros Editores Ltd., Santiago de Chile. 2006. pp. 94-97

[25] Myers N, Mittermeier RA, Mittermeier CG, De Fonseca GAB, Kent J. Biodiversity hotspots for conservation priorities. Nature. 2000;**403**:853-858

[26] Molina JA. Compendio de la Historia Geográfica Natural y Civil del Reyno de Chile, Primera Parte. Madrid: Imprenta de Sancha; 1788 418 pp

[27] Norbaek R, Christensen LP, Bojesen G, Brandt K. Anthocyanins in Chilean species of *Alstroemeria*. Phytochemistry. 1996;**42**:97-100

[28] Abdelaziz M, Lorite J, Muñoz-Pajares AJ, Herrador MB, Perfectti M, Gómez JM. Using complementary techniques to distinguish cryptic species: A new *Erysimum* (Brassicaceae) species from North Africa. American Journal of Botany. 2011;**98**:1049-1060

[29] Finot VL, Baeza CM, Melín P, Novoa P, Campos J, Ruiz E, Fuentes G, Albornoz P, Carrasco P. ¿Es *Alstroemeria pulchra* Sims var. *maxima* Phil. (Alstroemeriaceae) un miembro del complejo *A. magnifica* Herbert?: Estudio colorimétrico de los tépalos externos e internos de la flor. Gayana Botanica. 2015;**72**:101-113

[30] Finot VL, Baeza CM, Ruiz E, Carrasco P, Melín P. Análisis colorimétrico y morfométrico de la flor de *Alstroemeria presliana* (Alstroemeriaceae). Journal of the Botanical Research Institute of Texas. 2016;**10**:89-108

[31] He Q, Shen Y, Wang M, Yang R, Azhu S, Wang L, Xu Y, Wu R. Natural variation in petal color in *Lycoris longituba* revealed by anthicyanin components. PLoS One. 2011;**6**:e22098. DOI: 10.1371/journal.pone.0022098

[32] CIE (Commision Internationale de L'éclairage). Recommendations on Uniform Color Spaces: Color-difference Equations, Psychometric Color Terms. Paris: CIE Publ. 15, supl. 2; 1978

[33] Philippi RA. Plantarum novarum chilensium. Linnaea. 1864;**33**:260-261

[34] Sanso AM. El género *Alstroemeria* (Alstroemeriaceae) en Argentina. Darwin. 1996;**34**: 349-382

[35] Norbaek R, Christensen LP, Brandt K. An HPLC investigation of flower color and breeding of anthocyanins in species and hybrids of *Alstroemeria*. Plant Breeding. 1998;**117**:63-67

[36] Macbride JF. Flora of Peru. Publication: Field Museum of Natural History. 1936;**13**:631-690

[37] Missouri Botanical Garden. Tropicos.org. [Internet]. Available from: www.tropicos.org [Accessed: Aug 3, 2017]

[38] Jorgensen PM, Nee MH, Beck SG. Catálogo de las plantas vasculares de Bolivia. Monographs in Systematic Botany from the Missouri Botanical Garden. 2014;**127**:1-1744 [Internet]. Available from: www.tropicos.org/projectwebportal.aspx Accessed: Aug 3, 2017

[39] Baeza C, Schrader O, Budahn H. Characterization of geographically isolated accessions in five *Alstroemeria* L. species (Chile) using FISH of tandemly repeated DNA sequences and RAPD analysis. Plant Systematics and Evolution. 2007a;**269**:1-14

[40] Cajas D, Baeza M, Ruiz E, Negritto M. Análisis citogenético en poblaciones de *Alstroemeria hookeri* Lodd. ssp. *hookeri* (Alstroemeriaceae) en la Región del Biobío, Chile. Gayana Botanica. 2009;**66**:117-126

[41] Baeza C, Ruiz E, Negritto M. Comparative karyotypic analysis in the *Alstroemeria hookeri* Lodd. (Alstroemeriaceae) complex sensu Bayer (1987). Genetics and Molecular Biology. 2010a;**33**:119-124

[42] Strasburger E. Über den Teilungsvorgang der Zellkerne und das Vehaltnis der Kernteilung zur Zellteilung. Archiv für Mikrobiologie und Anatomie. 1882;**21**:476-596

[43] Tsuchiya T, Hang A. Cytogenetics in the genus *Alstroemeria*. Herbertia. 1989;**45**:163-170

[44] De Jeu M, Lasschuit J, Kuipers A, Kamstra S, Visser R. Characterization and localization of repetitive DNA sequences in the ornamental *Alstroemeria aurea* Graham. Theoretical and Applied Genetics. 1997;**94**:982-990

[45] Kamstra S, Kuipers A, De Jeu M, Ramanna M, Jacobsen E. Physical localization of repetitive DNA sequences in *Alstroemeria*: Karyotyping of two species with species-specific and ribosomal DNA. Genome. 1997;**40**:652-658

[46] Stephens J, Tsuchiya T, Hughes H. Chromosome studies in *Alstroemeria pelegrina* L. International Journal of Plant Sciences. 1993;**154**:565-571

[47] Sanso A, Hunziker J. Karyological studies in *Alstroemeria* and *Bomarea* (Alstroemeriaceae). Hereditas. 1998;**129**:67-74

[48] Sanso A. Chromosome studies in Andean taxa of *Alstroemeria* (Alstroemeriaceae). Botanical Journal of the Linnean Society. 2002;**138**:451-459

[49] Jara-Seguel P, Palma-Rojas C, von Brand E. Karyotype and C-bands in the annual Inca lily *Alstroemeria graminea*. Belgian Journal of Botany 2004;**137**:199-204

[50] Buitendijk J, Peters A, Jan-Quené R, Ramanna M. Genome size variation and C-band polymorphism in *Alstroemeria aurea, A. ligtu* and *A. magnifica* (Alstroemeriaceae). Plant Systematics and Evolution. 1998;**212**:87-106

[51] Zhou S, De Jeu M, Visser R, Kuipers A. Characterization of distant *Alstroemeria* hybrids: Applications of highly repetitive DNA sequences from *A. ligtu* ssp. *ligtu*. Annals of Applied Biology. 2003;**142**:277-283

[52] Baeza CM, Schrader O, Ruiz E, Negritto M. Análisis comparativo del cariotipo en poblaciones de *Alstroemeria aurea* R. Graham (Alstroemeriaceae) de Chile. Gayana Botanica. 2007;**64**:31-37

[53] Baeza CM, Schrader O, Ruiz E, Negritto M. *Alstroemeria presliana* herb. (Alstroemeriaceae) in Chile from a cytogenetic perspective. Chilean Journal of Agricultural Research. 2008;**68**:328-333

[54] Baeza CM, Ruiz E, Novoa P. The karyotype of *Alstroemeria diluta* Ehr. Bayer Subsp. *chrysantha* (Alstroemeriaceae). Chilean Journal of Agricultural Research. 2010b;**70**:667-669

[55] Baeza CM, Espejo J, Ruiz E. El cariotipo de *Alstroemeria versicolor* Ruiz et Pav. (Alstroemeriacerae). Gayana Botanica. 2011a;**68**:327-329

[56] Baeza CM, Rojas G, Ruiz E. El Cariotipo fundamental de *Alstroemeria patagonica* Phil. (Alstroemeriaceae). Boletín de la Sociedad Argentina de Botánica. 2011;**46**:313-315

[57] Baeza CM, Finot VL, Ruiz E, Carrasco P, Novoa P, Stuessy T, González A. Comparative karyotypic analysis and cytotaxonomy in the *Alstroemeria ligtu* L. (Alstroemeriaceae) complex of Chile. Brazilian Journal of Botany. 2016;**39**:305-313

[58] Baeza CM, Finot VL, Ruiz E, Carrasco P, Novoa P, Rosas M. Análisis citotaxonómico del complejo *Alstroemeria diluta* E. Bayer (Alstroemeriaceae) de Chile. Gayana Botanica. 2016;**73**:221-227

[59] Allendorf FW, Luikart GH. Conservation and the Genetic of Populations. 1st ed. Blackwell Publishing; Malden, USA. 2007 652 p

[60] Han T-H, Van ECK, HJ DJMJ, Jacobsen E. Optimization of AFLP fingerprinting of organisms with a large genome size: A study on *Alstroemeria* spp. Theoretical and Applied Genetics. 1999;**98**:465-471

[61] Han T-H, De Jeu M, Van Eck H, Jacobsen E. Genetic diversity of Chilean and Brazilian *Alstroemeria* species assessed by AFLP analysis. Heredity. 2000;**84**:564-569

[62] Anastassopoulos E, Keil M. Assessment of natural and induced genetic variation in *Alstroemeria* using random amplified polymorphic DNA (RAPD) markers. Euphytica. 1996;**90**:235-244

[63] De Jeu M, Lasschuit J, Chevalier F, Visser RGF. Hybrid detection in *Alstroemeria* by use of species specific repetitive probes. Acta Horticulturae. 1995;(420):62-64

[64] Dubouzet JG, Murata N, Shinoda K. RAPD analysis of genetic relationships among *Alstroemeria* L. cultivars. Scientia Horticulturae. 1997;**68**:181-189

[65] Han T-H, van Eck H, de Jeu M, Jacobsen E. The construction of a linkage map of *Alstroemeria aurea* by AFLP markers. Euphytica 2002;**128**:153-164

[66] Souto CP, Premoli AC. Genetic divergence among natural populations of *Alstroemeeria aurea* D. Don: A dominant clonal herb of the understory in subalpine *Nothofagus* forests. Botanical Bulletin of Academia Sinica. 2003;**44**:329-336

[67] Ruiz E, Balboa K, Negritto MA, Baeza CM, Briceño V. Variabilidad genética y morfológica y estructuración poblacional en *Alstroemeria hookeri* subsp. *hookeri* (Alstroemeriaceae), endémica de Chile. Revista Chilena de Historia Natural. 2010;**83**:605-616

[68] Di Castri F. An ecological overview of the five regions of the world with mediterranean climate. In Groves, RH, F. Di Castri, editors. Biogeography of Mediterranean Invasions. Cambridge, United Kingdom: Cambridge University Press; 1990. p. 3-15

[69] Grant V. The Evolutionary Process. A Critical Study of Evolutionary Theory. 2nd ed. New York, USA: Columbia University Press; 1991 487 p

[70] Pritchard JK, Stephens M, Donnelly P. Inference of population structure using multilocus genotype data. Genetics. 2000;**155**:945-959

[71] Ronquist F, Teslenko M, van der Mark P, Ayres DL, Darling A, Höhna S, Larget B, Liu L, Suchard MA, Huelsenbeck JP. MrBayes 3.2: Efficient Bayesian phylogenetic inference and model choice across a large model space. Systematic Biology 2012;**61**:539-542

[72] Sites Jr JW, Marshall JC. Operational criteria for delimiting species. Annual Review of Ecology, Evolution, and Systematics. 2004;**35**:199-227

[73] Clements M, Snell Q, Walker P, Posada D, Crandall K. TCS: Estimating gene genealogies. Parallel and Distributed Processing Symposium, International Proceedings. Vol. 2. 2002. p. 184. [Internet]. Available from: http://ieeexplore.ieee.org/document/1016585/. Accessed: Nov 22, 2017

[74] Leigh JW, Bryant D. PopART: Full-feature software for haplotype network construction. Methods in Ecology and Evolution. 2015;**6**:1110-1116

[75] Carstens BC, Pelletier TA, Reid NM, Satler JD. How to fail at species delimitation. Molecular Ecology. 2013;**22**:4369-4383

[76] Metzker ML. Sequencing technologies-the next generation. Nature Reviews Genetics. 2010;**11**:31-46

[77] Davey JW, Hohenlohe PA, Etter PD, Boone JQ, Catchen JM, Blaxter ML. Genome-wide genetic marker discovery and genotyping using next-generation sequencing. Nature Reviews Genetics. 2011;**12**:499-510

[78] Kim JS, Kim J-H. Comparative genome analysis and pylogenetic relationship of order Liliales insight from the complete plastid genome sequences of two lilies (*Lilium longiflorum* and *Alstroemeria aurea*). Plos One. 2013;**8**:e68180

[79] Andrews KR, Good JM, Miller MR, Luikart G, Hohenhole PA. Harnessing the power of RADseq for ecological and evolutionary genomics. Nature Reviews Genetics. 2016;**17**: 81-92

[80] Allendorf FW, Luikart G, Aitken SN. Conservation and the Genetics of Populations. 2nd ed. Massachusets, USA: Wiley-Blackwell Publisher; 2013 642 p

[81] Lahaye R, Van der Bank M, Bogarin D, Warner J, Pupulin F, Gigot G, Maurin O, Duthoit S, Barraclough T, Savolainen V. DNA barcoding the floras of biodiversity hotspots. Proceedings of the National Academy of Sciences. 2008;**105**:2923-2928

[82] Pan J, Wang B, Pei Z-Y, et al. Optimization of the genotyping-by-sequencing strategy for population genomic analysis in confiers. Molecular Ecology Resouces. 2015;**15**:711-722

4

Ecosystem Services Provided by the Little Things that Run the World

Olga Maria Correia Chitas Ameixa,
António Onofre Soares, Amadeu M.V.M. Soares and
Ana I. Lillebø

Abstract

Highest extinction risk and consequently biodiversity loss are predicted to occur in invertebrates, specifically insects, and these declines are expected to cascade onto ecosystem functioning and human well-being. Although this knowledge is intrinsically present in more traditional communities, in more urban environments, mapping ecosystem services can be an important tool to raise people's awareness on the importance of preserving insect diversity. After an extensive revision of the available literature, we used a rule-based approach to assess the provisioning, regulating and maintenance, and cultural services delivered by insects. We followed the Common International Classification of Ecosystem Services (CICES) and identified several potential indicators that may help underpin the mapping and valuation of the services delivered by insects. From our search, we extracted a total of 73 indicators, divided as 17 Provisional indicators, 27 Regulation and Maintenance indicators, and 29 Cultural indicators. We concluded that insects are providers of services in the three major 'Sections' of ecosystem services defined by CICES. Despite the lack of recognition of provisioning and cultural services, the indicators provided may help to raise awareness on the importance of the little things the run the world, in order to preserve traditional and technological uses of insects and their services.

Keywords: ecological function, ecosystem services, insects, natural capital, services indicators

1. Introduction

The highest extinction risk and consequently the greatest biodiversity loss are predicted to occur in invertebrates [1, 2], specifically insects, and these declines are expected to cascade onto ecosystem functioning and human well-being [2]. For this reason, mapping ecosystem services provided by insects can be a way to raise people's awareness on the importance of preserving insect diversity.

From 1 million species described, only 5000 are considered harmful to crops, livestock, or humans [3], the so-called dis-services [4]. However, it also widely acknowledge the benefits that insects might directly or indirectly bring to humans, namely through pollination, pest control, raw materials, or as food, known as ecosystem services (ES). Moreover, there are clear evidences of the growing importance of insects as feed and food resources, especially in the context of the global population growth and to overcome the global scarcity of proteins [5] or even in the context of new drugs discovery for by pharmaceutical industries, e.g., [6, 7, 8], with insects having a prominent role in the provisioning services by ecosystems. Insects produce a huge number of defensive secretions against predators, and these molecules are potentially medicinal drugs, e.g., [7, 9]. In countries such as China, about 77 species from 14 families and belonging to 8 orders have been used in traditional medicine to treat tumors and cancer [10].

Due to the growing interest regarding this class as key mediators of ES, i.e., there is a growing demand for ES mediated by insects; there is also the need to assess and map the supply of ES that is mediated by insects and provided in the habitats they populate. The demand for ecosystem services is socio-economically driven [11], while the supply is underpinned by ecosystems process and functions [12]. In this context, this work proposes to deliver a set of potential global indicators for ES mediated by insects that can serve as a foundation for mapping and assessment of ES at local/regional, national, or even transnational levels. The use of indicators has been recognized as a useful communication tool that facilitates the simplification of human-environmental systems high complexity, e.g., [13]; they can be selected for mapping and assessment of ES [14] and to support specific management purposes involving ES trade-offs [15]. This work may also be seen as a contribution to the EU Biodiversity Strategy for 2020, Action 5—Improve knowledge of ecosystems and their services in the EU, which is linked to other global initiatives to protect biodiversity, namely International Platform of Biodiversity and Ecosystem Services (IPBES) and to the UN Natural Capital Accounting. In this sense, we performed an extensive review of the available literature to assess the supply side of ecosystem services, which results directly or indirectly from insect-mediated processes and functions. The ES classification followed the Common International Classification of Ecosystem Services (CICES) that considers three major categories: provisioning, regulating and maintenance, and cultural services in order to avoid natural capital double counting [16]. CICES has been applied to habitats, e.g., [17], to social-ecological systems, e.g., [18], or with the purpose of supporting the mapping and assessment of ES at EU Member States (Action 5 of the Biodiversity Strategy), e.g., [14]. Results from this rule-based review are presented and discussed according to the lowest possible CICES category.

2. Methodology

The ES provided by insects was categorized according to the CICES classification [16], which the latest full version can be downloaded at http://cices.eu/. In this review, we applied the CICES rationale to the world's most abundant and diverse group of organisms and acknowledge their role as major agents of the ES provided in the ecosystems where they live. Following the same principle as the MAES working group [14], we provide a set of potential indicators for each identified service, which might help underpin the mapping, valuation, and management of the delivered services.

The services were categorized to the maximum possible resolution following CICES hierarchical structure whenever possible. If sufficient data were not available, the next higher level of resolution was chosen.

We used a rule-based approach to map indicators for ES. We did this by first reviewing and summarizing existing scientific and nonscientific literature related to ES provided by insects. Our aim was to extract from the literature and by using CICES rational, to classify potential indicators that can be used to assess and map these services in different ecosystems and across different geographic regions.

The literature search was performed in specific scientific libraries (e.g., Scopus and Web of Knowledge), using the keywords (single and in combination): ES, insects, ecological function, pollination, seed predation, biological control, decomposition, seed dispersal, food, feed, materials, medicines, society, and ecosystems. However, there were many gaps for several ES categories in these referenced libraries. For this reason, we enlarged our search to include books and nonresearch literature. This approach was especially important to access the cultural services, for which available scientific literature is scarce. In fact, evidences seem to show that inclusion of nonresearch literature can be useful to validate the results of a research-based literature search [19]. In our case, we definitely would have biased our conclusions if we had excluded nonresearch-based literature, even if this only makes up ~25% of all searched literature, mainly because cultural services are rarely considered in ES assessments [20]. Actually, cultural services do not represent purely ecological phenomena but rather are the outcome of complex and dynamic relationships between ecosystems and humans [21], and for this reason, we needed to adapt our bibliographic review in order to include cultural services and produce valid potential indicators. In total, we used 90 papers indexed in Scopus, 21 authored and 1 edited books, 16 chapters, 13 reports, 7 proceedings, 8 papers not indexed in Scopus, 7 electronic articles, 3 thesis, and 8 webpages.

3. Results

In this section, a short description of each of the headings corresponds to CICES hierarchical structure identified for ES mediated by insects, while the corresponding indicators for mapping and assessment are presented in **Table 1**.

CICES hierarchical structure	Relevant examples	Potential indicator(s)
1. Provisional		
1.1. Nutrition		
1.1.1. Biomass		
1.1.1.1. Reared and wild animals and their outputs for direct consumption	Reared insects for Human nutrition	Number and distribution of insects farms of edible species
	Wild insects collected for direct Human nutrition	Honey consumption from domesticated and wild bees
		Number, distribution and biomass of species consumed by humans
1.1.1.2. Wild animals for indirect consumption	Wild insects predated by animals (fish and game birds) that are used in Human nutrition	Number and distribution of insects predated by animals (fish and game birds) used in Human nutrition
	Insect species that can be incorporated in food production systems.	Insect's number and biomass used in food production for Human nutrition
1.2. Products		
1.2.1. Biomass		
1.2.1.1. Insects products for direct use or processing	Products made with insect materials	Amount and distribution of bee products (e.g., honey and beeswax production), and other insect products by category (e.g., silk, pigments, lac production).
	Medicinal uses of insects and their derived products for humans	Amount and distribution by category of medical/ pharmaceutical active substances)
	Insects used by animals as medicines	Number and distribution of animals which rely on insects as medicines
2. Regulation and Maintenance		
2.1. Mediation of waste, toxics and other nuisances	Bio-remediation Decomposition	Number and distribution of species which clean/ decompose wastes
2.2. Mediation of flows		
2.2.1. Mass flows		
2.2.1.1. Mass stabilization and control of erosion rates	Contribution for soil erosion mitigation by influence on the soil structure due their activities	Number and distribution of species which contribute to soil erosion mitigation
2.3. Maintenance of physical, chemical, biological conditions		
2.3.1. Lifecycle maintenance, habitat and gene pool protection		

CICES hierarchical structure	Relevant examples	Potential indicator(s)
2.3.1.1. Pollination by insects	Pollinators of wild plants and cash crops	Number and distribution of pollinator species; Number and distribution of plant species which depend on insects to reproduce and that are used for human consumption Number and distribution of beehive's renting for pollination
2.3.1.2. Seed dispersal by insects	Dispersal of seeds by insects	Number and distribution of species that promote seed dispersal Number and distribution of plant species which depend on insects for seed dispersal and that are used for human consumption
2.3.1.3. Maintaining nursery populations and habitats	Wild insect species which serve as food for other predators e.g., fish, birds, spiders, lizards, bats, etc.	Number and distribution of insect species that are predated by insectivorous species. Number and distribution of insectivorous species that are used for human consumption
2.4. Maintenance of physical, chemical, biological conditions		
2.4.1. Pest and disease control		
2.4.1.1. Pest control	Biological control of different types of pests	Number and distribution of biocontrol agents Effective biological control reported cases
2.5. Maintenance of physical, chemical, biological conditions		
2.5.1. Soil formation and composition	Termites, ants and dung beetles	Number and distribution of insect species involved in pedogenese and soil bioturbation
2.5.2. Decomposition and fixing processes	Decomposition of dead organic material Herbivory Nutrient cycling	Number and distribution of insect species that decompose dead organic material, contributing to nutrient cycling and soil fertility and which can influence the rate of decomposition (herbivory).
3. Cultural		
3.1. Physical and intellectual interactions with insects		
3.1.1. Physical and experiential interactions		
3.1.1.1. Experiential use of insects and insect dependent wildlife	Observation of insects and insect dependent wildlife. Activities related with wildlife which rely on insects as a critical nutritional resource, e.g., hunting, fishing	Number and distribution of Citizen Science projects related with insects; Number of internet groups related with insects, which share pictures and other information. Number and distribution of insectivorous game and game species
3.1.2. Intellectual and representative interactions		

CICES hierarchical structure	Relevant examples	Potential indicator(s)
3.1.2.1. Scientific	Amount of scientific study areas in which insects are subject matter of research	Number of scientific studies using insects as subjects of research. Studies listed in Database Platforms, e.g., ISI Web of Knowledge. Number of insect biological model systems Number and distribution of species used in forensic studies
	Bioindication	Number and distribution of insect species which are bioindicators of ecological condition
3.1.2.2. Educational	Use of insects in schools and educational programs	Activities using insects in kindergartens and elementary schools Number of websites devoted to insect information Number of citizen science projects using insects Number of insect zoos Number of insectariums and butterfly displays
3.1.2.3. Heritage and cultural	Insect fossils	Number and distribution of paleontological data
3.1.2.4. Entertainment	Movies, books, TV series, magazines, social internet groups	Ex-situ viewing/experience of natural world through different media. Videos, documentaries on insects. Mass media indicators: TV and magazines, Web sites, social networking sites; Movies
3.1.2.5. Esthetic	Inspiration for painters, photographers and other artists interested in insects	Different types of artifacts in which insects are used Number of visits to artistic exhibits
3.2. Spiritual, symbolic and other interactions with insects 3.2.1. Spiritual and/or emblematic (sacred and/or religious)		
3.2.1.1. Symbolic	Emblematic insects (e.g., U.S., state insects or state agricultural insects)	Number and/or distribution of symbolic insect species
3.2.1.2. Sacred and/or religious	Scarab cult, cicadas as symbols of rebirth	Number of sacred species; insect species that sustain supernatural believes
3.2.1.3. Other cultural outputs (Existence and bequest)	Entomological organizations, insect collections	Number and distribution of entomological organizations, insect collections, zoos, museums, and butterfly gardens, colleges and universities where entomology can be studied. ONG's involved in sustainable insect collection by local populations.

Table 1. Potential indicators for ES delivered by insects following the CICES hierarchical structure.

3.1. Provisioning—nutrition and biomass

3.1.1. Wild and reared insects and their outputs for direct consumption by humans

Insects are part of human diet in many parts of the world, and it is estimated that at least 2 billion people worldwide regularly consume insects [22]. Most of this consumption occurs in central and southern Africa, Asia, Australia, and Latin America and can provide 5–10% of the annual animal protein consumed by various indigenous groups as well as fat and calories and various vitamins and minerals [23, 24]. Jongema [25] compiled a list of edible insects, where 1900 edible insect species were reported. In **Figure 1**, the percentage of insects eaten grouped by their orders is compiled.

Recently, several "entomophagic" movements have appeared in order to introduce insects in daily diets, with entomologists and chefs joining together to introduce insects as gourmet food items, e.g., [26]. These movements have the blessing and active participation of Food and Agriculture Organization (FAO), which supported several actions on edible insects worldwide in recent years. According to FAO, scarcities of agricultural land, water, forest, fishery and biodiversity resources, as well as nutrients and nonrenewable energy, are foreseen. In this context, insects appear as an alternate form of having a source of protein and other nutrients without the depletion of resources like in livestock production systems.

Besides direct consumption of insects, honey is the most widely insect output consumed by humans. Worldwide, the EU accounts for approximately 20–25% of global consumption (amounted to 310 thousand tonnes in 2007). The other two major consumers of honey in the world are China (15%) and the USA (10%) [27].

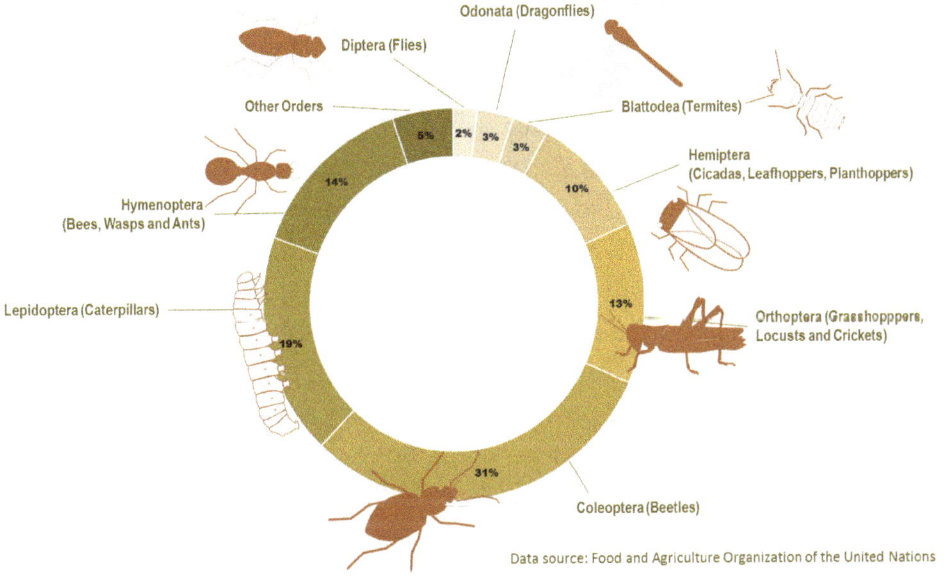

Figure 1. Percentage of insects eaten according to their order.

3.1.2. Wild insects for indirect consumption by humans

Several animals feed primarily or exclusively on insects; and some of these animals are important food sources for humans [28]. Numerous insect orders have been used as complementary food sources for poultry [29]. Terrestrial invertebrates can comprise more than half of energy ingested by stream fishes [30, 31] and are often the preferred prey of juvenile salmonids [32] with significant consequences for fish growth and abundance [33, 34]. Insects falling into streams are estimate to comprise about 30–80% of the diets of young salmon [31], 56.7% of masu salmon, 50.5% of white-spotted char, 46.3% of rainbow trout, 22.6% of Dolly Varden, and 12.0% of sculpin diets [35]. Even in systems where invertebrates made up only 10–15% of the terrestrial input, they comprised in total more than a third of the diet [36], which seems to be an indication that stream fish often prey selectively on terrestrial invertebrates falling into the water, e.g., [37].

Chicks of most game birds (e.g., quail, rouse, and pheasant) feed mainly on insects and other invertebrates, because of the high protein content, essential for the birds' rapid development [38]. Apparently, many chicks are not able even to digest plant material [39].

3.1.3. Insect products for direct use or processing

Bees produce six hive products—honey, pollen, royal jelly, beeswax, propolis, and venom—all of which are collected and used by people for various nutritional and therapeutic purposes. Honey is the most well-known and economically important hive product. The first written reference to honey appeared in a Sumerian tablet writing, dating back to 2100–2000 BC, that mentions honey's use as a drug and an ointment [40]. Bees deliver about 1.2 million tons of commercial honey per year [5]. Honey and beeswax, whose trade dates back to ancient Greece and Rome, were used in Medieval Europe as trade goods and as payment for taxes [41]. Beeswax is popular for making candles, adhesives, seals, and molds for casting [42]. It is also used as wood polish and is one of the most commonly used waxes in cosmetics.

Wax can also be produced by scale insects (Hemiptera, superfamily Coccoidea). The Chinese insect wax, or just insect wax, is produced by *Ceroplastes ceriferus* [43], which is common in China and India, or *Ericerus pela*, which is distributed in most parts of China, Japan, and the Korean peninsula, where it has been bred for over 1000 years [44]. Chinese insect wax has been used as a substitute for beeswax and is also used for candles, wax paint, polishing leather, and sizing paper [45]. Scale insects are also the main source of commercial shellac, also known as lac. The lac scale *Kerria lacca* (Kerr, 1782) can be found in the forests of India and Thailand and secretes shellac a natural mixture of resin and wax [46]. The propagation, care, and harvest of the raw lac product are largely dependent on tribal people scattered throughout forested areas. For these people, lac cultivation and harvest can be a substantial source of monetary income [47]. Lac has been an integral part of Indian history and culture for thousands of years, beautifying wood as a coating or used as a thermoplastic molding material [47]. It can also be used as a red colorant for dying natural fibers [48], e.g., wool in Persian carpets dated as early as 714 B.C [47]. Shellac wax is used in crayons and lipstick, as well as polishes and waxes for furniture [47]. Shellac can also be used as an electrical insulator, as well as a binder for man-made abrasive grinding wheels used in the crystal cutting and polishing industry [47]. Shellac was an ingredient in the production of gramophone records, until vinyl took its place [49].

As it is a natural, nontoxic resin, shellac can be used in the food industry as a colorant or coating [50]. In the perfume industry, it is used for its ability to microencapsulate, allowing a timed release of perfume in lotions and other beauty products [48].

Scale insects still provide carmine, also called cochineal, a red dye mainly produced by *Dactylopius coccus*. This dye is used since pre-Columbian civilizations, as early as 2000 years ago, and was once one of the most valuable exports from colonial Mexico [51]. Carmine is used to color foods, textiles, and pharmaceuticals [5]. For three centuries, the English, French, and Dutch resorted to espionage, piracy, bribery, and theft to learn the secret of this fabulous dye and break Spain's monopoly [52]. When the FDA banned Red Dye No. 2 in 1976, many food and cosmetics producers returned to cochineal, which is neither a toxin nor a carcinogen. Sometimes listed as carmine or E120, cochineal is the only natural red food coloring authorized by the FDA [52].

Before the discovery of *D. coccus*, red dyes were obtained from other scales species. *Kermes* (from *Kermes vermilio* Planchon) was used in Europe since Roman times [53]. In central and northern Europe, the Polish cochineal (from the *Porphyrophora polonica*) was used throughout middle age [54]. Besides these, Armenian cochineal (from *Porphyrophora hamelii* Brandt) was collected from the base of its hosts *Aeluropus littoralis* Gouan or *Phragmites communis* Trin., and occurred in Armenia, Turkey, and Iran [55].

One of the most important insect products is silk. Sericulture has been practiced at least since 2000–3000 B.C. in China and is among the most widely traded commercial products. The historic importance of sericulture is demonstrated by the Silk Roads that connected Europe, the Middle-East, and China [41]. Silkworms remain the only source of silk and produce more than 90,000 tons of this natural protein fiber per year [56]. However, silk production is not confined to China. Indeed, in Madagascar, wild silk commonly termed "Landibe" was exploited long before the introduction of "Chinese silk" from *Bombyx mori* during the 19th century [57, 58]. Wild silk in Madagascar has many origins, with many silk producing species being present on the island. One of such species is *Borocera cajani* [59]. Historically, the Madagascan population used *B. cajani* silk to create sumptuous shrouds. Although sericulture in Madagascar plays a minor role at international level [57], it remains a major source of income for more than 10,000 local families working in the silk industry [60].

Iron gall ink, also referred to as iron gallotannate ink, is one of the most important inks in the history of western civilization and was in widespread use from the middle age until the twentieth century, with a significant part of our cultural heritage recorded with it [61]. For instance, Bach used it to write down his music, Rembrandt used it for his drawings, while Thomas Jefferson wrote the Declaration of Independence with it [61]. However, this ink presents some corrosive properties, reason why several historical artifacts suffer from its corrosive action. Examples are the original complete works of Victor Hugo which are almost destroyed, while 60–70% of Leonardo da Vinci's work shows signs of deterioration [62].

The use of insects and derived products as medicines has a long tradition in many parts of the world and in many cultures, e.g., in Chinese medicine, Indian Ayurvedic medicine, Australian Aborigines, as well as in Africa and South America. Science has provided evidence of immunological, analgesic, antibacterial, diuretic, anesthetic, and antirheumatic properties [63].

Extensive reviews regarding the use of insects in medicine are provided by Cherniak [6] and Dossey [7], which include details of the use of medicinal insects throughout the World and aspects of their potential for drug discovery. Ratcliffe et al. [8] provided an overview of significant recent developments in bioengineering natural products from insects, describing use or potential use in modern medicine, as well as the use of insects as models for studying essential mammalian processes such as immune responses to pathogens. Examples of these "rediscovered" products come from bees. Honey has been used for a long time as wound dressing by several different cultures in different parts in the world (a good historical review is provided by Jones) [64]. Another review by Molan [65] demonstrates that honey, the oldest wound dressing material known to medicine, can give positive results where the most modern products are failing, mainly due to its reported antimicrobial and anti-inflammatory activities.

Several properties were reported for royal jelly as well, and these included anti-cancer metastasis [66], neurite growth stimulation [67], and anti-allergic responses [68],

Apart from these, a broad spectrum of biological activities have also been reported in studies using propolis, such as antioxidant [69], antibacterial, antiviral, antifungal [70], anti-inflammatory [71], and anticancer effects [72]. Currently, the pharmaceutical industry uses beeswax as a binding agent, time-release mechanism, and drug carrier [43].

Chinese wax has also been employed medically. Taken internally, it has been used as a remedy for hoarseness, worms, and nervousness and to aid the mending of broken bones. Externally, it has been used as an ointment for treatment of wounds [73].

If we think about cancer and the side effects of current treatments, we definitely have to account for knowledge of traditional medicine, such as in Chinese medicine where several kinds of insects have been traditionally used to treat tumors and cancer [10]. Insects produce a huge number of defensive secretions against predators, and these molecules are potentially medicinal drugs, e.g., [7]. For instance, cantharidin, a defensive alkaloid produced by blister beetles (Meloidae) that is used commercially to remove warts [74], has been shown to kill a variety of tumor cells in vitro as well as in animal models in vivo (reviewed by [8]).

Ant venom was the subject of U.S. patent 4,247,540. This came about through a medical study at the Miami School of Medicine, where ant venom had an effect on painful and inflamed joints [9]. In fact, besides ailment treatment, ants were also used for diagnosis, as in the case of an old observation, made by tribes in the Amazon forest that ants will swarm over the urine of diabetics but will leave ordinary urine alone [9]. In fact, ants are widely used worldwide: in Africa, they are used to treat whooping cough, in Morocco to overcome lethargy, Australian aborigines used them to treat headaches, in South America to cure colds and impotence in men, and in Russia, a tincture of ants known as "muraschkowka," was used to treat colds. In Tibetan medicine, ants were used for urinary retention and as a mild purgative, and in Central Asia, its main use was for curing arthritis [9]. Lockhart [9] refers to Old European literature that contains recipes for "eau de magnanamite" used to treat sore eyes and in Arabian medical literature, where ants were used to treat the victims of the rutaila spider bite, leprosy, and as an aphrodisiac is described.

Ant mandibles were used to stitch wounds of surgical incisions—a practice documented in India as early as 1000 B.C., using live ants to bite the pinched sides of a wound or incision, leaving the mandibles fastened until the wound heals [75], a practice also used in the Mediterranean area but discontinued due to some criticism of some European doctors [9].

One of the well-studied medical applications of arthropods is the use of maggots [6]. Blow fly (*Lucilia* spp.) maggots, used in traditional medicine to remove necrotic tissues without surgery, received renewed medical attention for wound debridement, as surgical treatments increase the risk of infection by antibiotic-resistant pathogens [76–78].

Besides its economic value as a product, silk has been prescribed in traditional medicine as well. Silkworm larvae infected with fungus are one of the most used sources of medicine in Korean medicine and, according to Pemberton [79], are used to treat stroke, tonsillitis, and rubella among other diseases, while adult male silkworm moths are used to treat impotence and premature ejaculation. Additionally, silkworm larvae frass were also used to treat diabetes, neuralgia, and skin rash, among other conditions [80]. In fact, powdered silkworm larvae were tested and shown in modern bioassays to inhibit absorption of glucose in human intestinal epithelium cells [81]. More recently, due to its unique properties, silk fibroin, a naturally occurring protein polymer, was shown to be a favorable matrix for the incorporation and delivery of a range of therapeutic agents [82]. Besides silk, other insect proteins are also used in medicine. For instance, resilin, a rubber-like protein that enables insects to jump (e.g., fleas), has been used in medicine to repair arteries taking advantage of its elastic properties [83].

Shellac has also been used for a long time in Ayurvedic medicine [47] and more recently in modern pharmaceuticals as a tablet coating [84] and in dental medicine [85].

The anticoagulants and other components in the salivary glands of blood-sucking insects, for example in horseflies, have been used for hundreds of years as anti-thrombosis treatments in Eastern Medicine [86].

Recently, the importance of insects as source of natural products has been recognized by several pharmaceutical companies which set up to exploit these substances using modern molecular and biochemical techniques [8]. Furthermore, since 2009, the Chinese Government began to invest more than 12 billion US dollars over 5 years in new drug development, and insects as potential sources were not neglected [8].

Not only humans use insects to treat their ailments, animals do it as well. The use of ants by bears suffering from abdominal complaints is referred by the Roman writer Plutarch [9]. In 1935, Erwin Stresemann invented the term "anting," which refers to bathing in ant nests [9]. Nowadays, the behavior involves the bird seizing one or more ants in the bill and then, while holding one or both wings outstretched and the tail bent forward between the legs, wiping the ants into the primary feathers of wings and tail [87]. It is believed that birds use ants to maintain their feathers free from parasites, such as feather lice *Mallophaga* species and feather mites *Acarina* species [9], although this is not a consensual belief [88]. Anting has been reported for over 200 species of birds, mostly passerines (reviewed by [87]). The ants are typically *Formicinae*, which spray formic acid as a defense, or *Dolichoderinae*, which exude a repugnant anal fluid. Anting was also reported in some mammals, e.g., for capuchin monkeys [89] and gray squirrels [90]. European farmers also feed chickens with large red ants, as well as ant eggs, since these are considered powerful stimulants of egg production. Farmers would feed these to broody hens and "lazy layers." Feeding ant eggs to chicken results in weight gain with fat loss and seems to prevent diseases like coccidiosis and enteritis (see [9]).

3.2. Regulation and maintenance

3.2.1. Mediation of waste, toxics, and other nuisances

When cattle was introduced to Australia in 1788, waste biodegradation became an immediate problem as endemic dung beetles, adapted to marsupial dung, were simply insufficient to decompose the increased amounts of manure [91]. To solve this problem, dung beetles were introduced to Australia from South Africa, Europe, and Hawaii. Of 46 introduced species, 23 successfully established [5]. These beetles, from the insect families *Scarabaeidae* and *Geotrupidae*, bury dung under the soil surface and reduce the associated nuisances with it. These nuisances include a decrease in palatability of plant material that is fouled with dung and increase of diseases, since many cattle parasites and pest flies require a moist environment such as dung to complete their development [92]. Losey and Vaughan [28] estimated the annual value of dung burial by dung beetles in the US in US$ 380 million. The use of dung is not restricted to beetles. For instance, the Hawaiian big-headed ant *Pheidole megacephala* accidentally introduced to Fiji also gets into the droppings of animals [9].

In aquatic ecosystems, insects falling into freshwater waterbodies influence the nutrient availability, since they add substantial amounts of carbon, nitrogen, and phosphorus to these nutrient-poor ecosystems and affect aquatic community dynamics [93–95]. In addition, the different functional groups (scrapers/grazers, shredders, and collector gatherers) living in aquatic systems contribute to the integration of nutrients from terrestrial sources.

3.2.2. Mediation of flows

3.2.2.1. Mass stabilization and control of erosion rates

It is well documented that the reduction in activity and biodiversity of the soil biota, which includes insects, leads to the degradation of the soil structure and increase of runoff and erosion. Soil biota influence soil properties through formation of stable aggregates and development of organomineral complexes and bonding through fungal hyphae and polysaccharides [96]. Soil properties can also be improved by mixing soil with organic residues and by turnover through burrowing and the formation of subterranean chambers and feeding galleries. These activities improve macroporosity and continuity of pores from surface to the subsoil and increase the infiltration rate, which contributes to the reduction of runoff rate and amount [96]. Some examples of insects that influence soil structure are, e.g., termites, ants, dung beetles, and other burring insects.

3.2.3. Maintenance of physical, chemical, and biological conditions

3.2.3.1. Pollination

The Convention on Biological Diversity (CBD) has recognized pollination as a key driver in the maintenance of biodiversity and ecosystem functions [97]. Insects play a central role in most plant species reproduction, since the majority of all known pollinator species (98%) are insects [98]. Over 90% of the 250,000 flowering plant species depend on pollinators. Not only natural ecosystems benefit from pollination, many cash crops are actually self-sterile and require cross-pollination to produce seeds and fruit [99]. But it is not only self-sterile varieties that benefit from cross-pollination, self-fertile varieties also produce more and better quality seeds and fruits, if they are cross-pollinated [99]. In fact, this is true for three-quarters of the 100 crop species that generate most of the world's food [98]. It is estimated that domesticated bees alone pollinate 15% of these species [5]. However, in recent years, there has been a worldwide decline in pollinator insects, mainly bees, which has made the world headlines, since this presents a serious threat not only to terrestrial ecosystems but also for agricultural production and ultimately for life on earth.

A good example of the lack of pollinators in agro-ecosystems can be seen in the Hindu Kush-Himalayan region (extends 3500 km over all or part of eight countries from Afghanistan in the west to Myanmar in the east). Typically, native wild bees, which, unlike the European honey bee *Apis mellifera*, are able to tolerate seasonal low temperatures and play a vital role in pollination of mountain crops and wild flora [100]. Among these are *A. dorsata*, *A. florea*, *A. laboriosa* (bees whose products can be collected but which cannot be kept in hives), and *A. cerana* (which can be kept in hives) [101]. However, these native bee species are imperiled nowadays due to changes in agricultural practices and due to competition with the introduced *Apis mellifera*. In the Hindu Kush-Himalayan region, agriculture is in a stage of transition from traditional cereal crop farming to high-value cash crops such as fruits and vegetables. This ongoing transformation poses a number of new challenges, including low production or crop failure due to inadequate pollination [101]. In fact, despite increasing agronomic inputs, the negative effects on pollinators have further contributed to the low yields [99]. To overcome this problem, in some areas (Maoxian County, China), farmers resorted to hand pollination of apples and pears [102]. However, in Pakistan, farmers started to cut down their trees [103].

More than 80% of the world's almonds are produced in California [104], which requires about half of the introduced *Apis mellifera* population in the US for pollination each year (NRDC, [105]). In addition, this provides beekeepers with lucrative pollination fees and a major source of income by renting their beehives [106], besides honey production.

Numerous other cash crops are 90% dependent on honey bee pollination, e.g., avocados, blueberries, cherries, cranberries, and sunflowers [107]. Pollinators are also required for seed production of many other crops such as soybeans, hay, and forage crops [108, 109], with a clear impact on beef and dairy products production.

A recent international study revealed that non-bee insects are valuable pollinators as well, with better performance in fruit set in crops and more resilient to landscape changes [110].

3.2.3.2. Seed dispersal by insects

One of the best examples of seed dispersal by insects is given by ants carrying seeds into their nests. Some plants even produce seeds with special coating with the purpose of being eaten by the ant or other phytophagous insect without damaging the seeds viability [111]. This behavior is not exclusive to ants, and it is also carried out by a phytophagous insects, mostly, coleopterans [78].

By burying seeds with dung, dung beetles are instrumental in sandy areas, in which they live [112]. Other insects can disperse seeds inadvertently, in which dispersal occurs with the help of gripping organs (hooks, teeth, viscous hairs, etc.). The seeds attach themselves to the insect's body and later fall to the ground, allowing new plants to germinate some distance away from the original plant [113]. By burying seeds, ants and dung beetles also protect them from being eaten [114].

3.2.3.3. Maintaining nursery populations and habitats

In natural ecosystems, besides fishes (Section 3.1.2), there are other riparian consumers that benefit from energy transfers gained from aquatic insects emerged from streams [115–117]. Aquatic insects provide significant seasonal subsidies to forest birds during the defoliation period, accounting for 50–90% of the monthly energy budget in half of the species. This contribution of aquatic prey to the annual, total resource budget differed among bird species, being estimated in 38.6% for great tit, 31.9% for nuthatch, 7.4% for marsh tit, and 9.5% for pigmy woodpecker [35]. Besides birds, emerged aquatic insects provide important subsidies for other predators such as spiders [118], lizards [119], and bats [119]. In fact, these groups do not rely only on aquatic insects, for instance, Lacki [120] found remains of 10 insect orders in fecal pellets of insectivorous bats.

This shows a huge potential for the incorporation of insects into the food production chain of food animals like poultry or in aquaculture [5].

3.2.3.4. Pest control

Biological control was practiced in China as early as the third century A.D., where ants were sold for the control of citrus insect pests [121]. However, the first widely successful biological control program was the introduction of the coccinellid, *Rodolia* (*Vedalia*) *cardinalis* which saved California's citrus industry, by controlling the cottony cushion scale, *Icerya purchasi* [122]. This example of classical biological control was the prelude for other introductions to fight pests all over the world, without resorting to pesticides. A good example of pest control is carried by dung beetles, which can colonize a dung heap within 24 hours, preventing flies from developing on them [5]. Similarly, by feeding on animal drops, the big-headed ant *Pheidole megacephala* killed eggs, larvae, and pupae of the common housefly *Musca domestica,* which resulted in reduced levels of flies and less fly-borne diseases [9].

Natural biological control has a world market value of 400,000 × 10^6 US$ billions [123, 124], much higher value then the 8.5 billion US$ annually spent on insecticides [125]. No less than 230 invertebrate natural enemies are used in augmentative biology, and the majority of species belongs to insects [123], representing a marked share of about 0.130 × 10^6 US$ billions [125] and thus with a high socioeconomic importance [126].

Another potential biocontrol undertaken by insects is control of weeds which compete with crops for nutrients and water. In fact, they are the target of much of the agrochemicals applied to crops worldwide. Herbicides comprise 47% of the world agrochemical sales compared with 29% of insecticides [127]. Weeding, usually by hand, accounts for up to 60% of total preharvest labor input in the developing world [128].

Introduced species can often become invasive, causing enormous environmental damage. For instance, in South Africa, the Indian fig opuntia, native to the southern United States, was introduced in 1659 as a drought fodder [129], by 1942 covered already 900,000 ha of rangelands, reducing the yield and accessibility of grass for cattle and small stock [130]. These negative effects were only reduced with the introduction of the scale insect *Dactylopius opuntiae* aiming its biocontrol, which markedly reduced *O. ficus-indica* density by 1948 [131]. Another good example was the use of snout beetles (*Neochetina* spp.) introduced from Australia that successfully controlled water hyacinths in Lake Victoria [132].

Seed predation, a particular case of weed biocontrol, can be an effective component on arable land, particularly at low weed densities [133]. By predating on seeds, ants may alter the abundance and local distribution of flowering plants in tropical and subtropical regions [134, 135]. In temperate regions, the most important seed eaters are most likely carabids (*Carabidae*: Coleoptera) [49]. Granivory by carabids has been confirmed by many authors (reviewed by [133]) and in arable fields can be as high as 1000 seeds.m^{-2}.day^{-1}, which can selectively influence the soil seed bank [133]. However, introduced species, even those used in biocontrol programs, can have important ecological effects on native species [136], and for this reason, deliberate introductions have generated great controversy [137, 138].

3.2.3.5. Weathering processes

In the soil, insects can have two major roles, they can be "litter transformers" or they can act as "ecosystem engineers" [139]. As litter transformers, insects fragment, or comminute, and humidify ingested plant debris, improving its quality as a substrate for later microbial decomposition. The feces of arthropods serve as nuclei for the accretion of soil aggregates, the basic unit of a soil's structure with a major role in maintaining its integrity, and are a significant factor in the formation of humus, which contributes to water and nutrient soil [140]. Termites and ants nests, with their incorporated fecal materials, waste dumps, or fungal gardens, provide rich substrates for the microbial degradation and mineralization of organic matter, resulting in the conversion of complex organic molecules into simpler, inorganic forms that can be used by plants [140]. As ecosystem engineers, they physically modify the habitat, directly or indirectly regulating the availability of resources to other species [141]. In the soil, this implies to alter the soil structure, as well as the mineral and organic matter composition and hydrology [142]. The tunneling and burrowing of arthropods provide channels for air passage and water infiltration and also serve to mix organic matter into the upper soil layers [140]. Some of the most important members of this guild are ants, termites, and dung beetles which, due to their dung burial activity, especially the digging tunnels functional types, are able to move large amounts of soil. Ants and termites are the pre-eminent earth movers in many regions of the world and may surpass earthworms in this capacity in some cases [140]. However, termites are probably the biggest contributors to plant litter breakdown among soil invertebrates and are the main agents of degradation, among the soil fauna, of the highly recalcitrant materials (cellulose and lignin) making up wood [140].

3.2.3.6. Decomposition processes

Insects play a vital role in waste biodegradation. Beetle larvae, flies, ants, and termites clean up dead plant matter, breaking down organic matter until it is fit to be consumed by fungi and bacteria. In this way, the minerals and nutrients of dead organisms become readily available in the soil for uptake by plants. Animal carcasses, for example, are consumed by fly maggots and beetle larvae [5]. Termites and leafcutter ants process large amounts of wood and leaves [143]. The decomposition of dead plant material can induce other services like a decline in the frequency and severity of forest fires [144].

Dung beetles are an important group of insects associated with the decomposition of animal manure. Their activity contributes to nutrient cycling. By burying dung under the soil surface, they prevent about 80% of nitrogen loss through ammonia (NH_3) volatilization [5, 145] and enhance soil fertility by increasing the amount of nitrogen available to plants through mineralization [146]. In their presence, carbon and minerals are recycled back to the soil, where they further decompose as humus for plants [5].

The role of insect herbivory in terrestrial ecosystems has only recently been considered an important and persistent control of ecosystem processes [41]. Severe insect outbreaks can provide mass pulses of nutrients, like for instance, cicada outbreaks provide mass pulses of nutrients that encourage growth of the forest trees they feed on. However, the less intense but more persistent insect herbivory present in many systems may also be an important influence on ecosystem processes [147, 148]. In addition, herbivores can influence ecosystem function by changing the nature of organic matter inputs to the soil [149, 150]. Such changes can alter litter quality parameters, such as carbon to nitrogen and lignin to nitrogen ratios, inducing changes in decomposition rates and subsequent nutrient transformations [151]. Furthermore, the outputs of their activity, such as frass, honeydew, or even their carcasses, have distinctive chemical properties that can alter decomposition rates and nutrient release [152–154].

In the long term, selective herbivory can alter plant community composition, thus altering litter compositions available for decomposition [155]. Herbivory reduces canopy cover and increases precipitation reaching the ground or flowing into streams. Soil and litter insects affect soil porosity, as well as decomposition rate, both of which influence the rate of water movement through the substrate [41].

3.3. Cultural

3.3.1. Physical and intellectual interactions with insects

3.3.1.1. Experiential use of insects and insect dependent wildlife

Several activities depend even indirectly from insects, e.g., bioblitz is an intense period of biological survey aiming to record all the living species, including insects, within a selected area. Other examples include wildlife observation not only of insects (e.g., butterflies) but mainly of birds and other insectivorous animals like amphibians, reptiles, and small mammals. According to Losey & Vaughan [28], Americans spend an estimated $49 billion annually on hunting, fishing, and observing insect dependent wildlife, and a substantial proportion of this spending goes directly to insect observation.

3.3.2. Intellectual and representative interactions

3.3.2.1. Scientific

Insects have inspired science and technology over time. When we think about insects and science, we immediately think about the fruit fly *Drosophila melanogaster*, which has been the source of many important scientific results over decades. Insects as biological model systems offer many advantages over mammalian models including ethical reasons, short generations, and exponential population growth. Another popular insect model is the wax moth *Galleria mellonella*, because it naturally hatches in bee hives, where it has to tolerate elevated temperatures. Reared at 37°C, it serves as an excellent model system especially for bacterial infections, which could not be followed well at room temperature. Moreover, as a good correlation exists between virulence of several bacterial species in insect and mammalian models, *G. mellonella* is increasingly used as an infection model for human pathogens [156]. The red flour beetle *Tribolium castaneum* is also a popular model, since its development is more representative of other insects than *D. melanogaster*, a fact reflected in gene content and function. For example, *T. castaneum* has retained more ancestral genes involved in cell-cell communication than Drosophila, some being expressed in the growth zone crucial for axial elongation in short-germ development [157].

A project at Harvard University has created robotic insects, the project Robobees (http://robobees.seas.harvard.edu/), inspired by bee's biology and hive behavior. There are many practical applications to these coordinated agile robotic insects, e.g., crop pollination, search, and rescue in the aftermath of a natural disaster, hazardous environment exploration, military surveillance, high resolution weather, and climate mapping or even traffic monitoring.

Similarly, insect products have received attention in order to develop new technology. For example, the unique structure of silk, its biocompatibility with living systems, its function as a tool for new materials engineering, and its thermal stability are only a few of the features that make it a promising material for many clinical functions [158]. Silk has been used in biomedical sutures for decades and has recently achieved US Food and Drug Administration approval for expanded biomaterials device utilities. The remarkable mechanical properties, versatile processing in an aqueous environment, biocompatibility, and controlled degradation suggest that silk (both native as well as recombinant) is an attractive biomaterial for controlled and sustained release, stabilization, and delivery of bioactive molecules [159]. Also the biopolymers chitin/chitosan, which have singular chemical and biological characteristics like biocompatibility, antibacterial properties, heavy metal ion chelation ability, gel-forming properties, hydrophilicity, and remarkable affinity to proteins, have been tested in for uses in medical and food industries [160].

Some useful models in building construction are termite mounds with their complicated network of tunnels and ventilation systems, in which several parameters (e.g., air quality, temperature and humidity) can be regulated in an efficient way [161].

Insects represent important tools in criminal investigations providing forensic information [162]. Forensic entomology is defined as the use of insects and other arthropods, such as mites, in medico criminal investigations [163].

The first recorded incident where insects were used in a criminal investigation was recorded in 13th century China [162]. The first application of forensic entomology in a modern court house occurred in 1850 in France, where entomological data were admitted as proof [164]. Later, Yovanovich [165] and Mégnin [166] were the first forensic examiners who attempted to evaluate insect succession on corpses, properly establishing the science of forensic entomology [167]. Insects allow the estimation of the time, at which a dead body was colonized [168]. Insects most often involved in the forensic investigations are true flies or Diptera. The predominant families in this order are *Calliphoridae* (blow flies), *Sarcophagidae* (flesh flies), and *Muscidae* (house flies). *Calliphoridae* and *Sarcophagidae* may arrive within minutes following death. *Muscidae* delay colonization until the body reaches bloat stages of decomposition [169].

Insects are highly sensitive to environmental changes, including those resulting from human activity, making them good indicators of ecological condition. Some species are capable of long distance dispersal, which enables them to find and colonize new patches. Other species are unable to fly and are therefore more vulnerable to disturbance. Due to some of their characteristics like small size, short life spans, and high reproductive rates and due to the fact that their abundances can change by several orders of magnitude on a seasonal or annual time scale (minimizing the time between disturbance and population adaptation to new conditions), insects are useful indicators of disturbance, more than are larger or longer-lived organisms with slower responses [41]. Studies have included the use of single species, higher taxa, assemblages and communities of, for example, dragonflies, ground beetles, tiger beetles, moths, butterflies, sawflies, and ants, and in habitats such as forests, grasslands, sand dunes, soils, urban areas, and mine sites [170]. However, there are some challenges, e.g., their taxonomy, which can pose an obstacle to their use in biomonitoring [171]. Aquatic insects are extremely important bioindicators of water quality and are frequently used following several well-established protocols. This fact is mainly due to the way how insects breathe in water, which makes them very sensitive to the amount of dissolved oxygen in water. Particular species are sensitive to dissolved oxygen, e.g., larvae of Plecoptera and Ephemeroptera which are pollution intolerant, whereas other taxa, such as particular species of chironomid midge larvae, can survive under deoxygenated conditions verging on anoxia [172].

3.3.2.2. Educational

Insects are frequently used as subjects of study in kindergartens and elementary schools. Several websites provide useful information regarding their biology and/or taxonomy. Among many others worldwide is the BBC website, several citizen science projects like the "UK ladybird survey" (http://www.ladybird-survey.org/ladybirds.aspx) or "the Lost Ladybug Project" in the U.S. (http://www.lostladybug.org/).

In the same way, insect zoos offer the opportunity to millions of people each year to observe live insects. Most insectariums and butterfly display offer some level of educational programming that promotes the appreciation and understanding of insect life, ranging from kindergarten to university level, and can also play a strong role in species conservation [173]. Although not currently updated, Saul-Gershenz [173] provides a list of selected insect exhibits built worldwide from 1797 to 2003. These are distributed as 26 in Europe, 9 in Asia, 29 in the US, 7 in Canada, 4 in Australia, 1 in New Zeeland, 6 in Latin America, 2 in Africa, and 1 in West India.

3.3.2.3. Heritage and cultural

Fossilized insects in amber provide priceless information, not only regarding extant species at prehistoric times but also about details of their daily lives; for instance, it was recently discovered a 100-million-year-old evidence of brood care among scale insects fossilized in amber [174]. Other extraordinary fossils are the *Meganeura* sp. from Carboniferous period (resembled and are related to present-day dragonflies though not belonging to the modern order Odonata). Paleoentomology has contributed to the understanding of planet Earth history. For example, paleontological data can provide a good record of how biodiversity has responded to previous climatic change events [175].

3.3.2.4. Entertainment

No doubt insects have their role in literature, movies, and visual art, as well as their place as collection items, ornaments, and more generally as inspiration for creative expression. Insects are mentioned frequently in the writings of poets and philosophers, and the folklore of nearly every country refers to them [176].

There are several children stories which have used insects as their main characters like the timeless La Fontaine tale and The cicada and ant and more recent children movies like Bugs life, Antz, Bee movie or the TV series Maya the bee, adapted from the German book The Adventures of Maya the Bee. But literature and movies using insects are not restricted to children tales, e.g., the book Metamorphosis by Franz Kafka or the movies Empire of the ants and The fly. The death's-head hawk moth (genus *Acherontia*) was popularized by the movie Silent of the lambs. The TV series C.S.I. has also contributed to the awareness of entomology, with its famous entomologist Grissom solving several mysteries using insects.

In Australia, the didjeridu (commonly spelt didgeridoo) or yiraki is a very ancient aborigine instrument, which consists of a more-or-less straight piece of tree trunk or branch, hollowed out by the successive action of fire and termites to produce a gently flaring tube [177]. Also, musical composition was influenced by insects as exemplified by Rimsky-Korsakov in his Flight of the Bumblebee, which reproduced the familiar hum of the bees, as well as Joseph Strauss' inspiration revealed in Dragonfly [176].

In Asian countries, cicadas and crickets often are caged and kept as pets for their songs [178]. There are also some references in ancient Greek poetry to the use of locusts and cicadas, which were prized for their singing. There are many references in Greek epigrams to the keeping of insects as pets [179]. In Portugal, it is common as child to have silk moths as pets and feed them with mulberry leaves, and in the north of the country, during a religious ceremony called "Senhor de Matosinhos," crickets are sold as pets.

Cricket fighting is a popular "sport" in China and in other countries with Chinese migrants [180]. Cocoons of several species of saturniid moths and other families have been used to make hand rattles, ankle rattles, necklaces, purses or other artifacts in several cultures around the world. Native Americans and indigenous peoples in Africa and Mexico still continue to use moth cocoons [181]. Although authentic, some Mexican and African ankle rattles are possibly being produced for sale as musical instruments and ethnic objects, as well as for the original intended use in ceremonial dances [181].

Insects are also the target of many internet groups on Facebook, LinkedIn, at national and international level, stimulating the discussion and the sharing of information among professional and amateur entomologists.

3.3.2.5. Aesthetic

Certain invertebrates, most notably butterfly and beetle species, have been used for various esthetic and decorative purposes. Designs based on invertebrates have been employed in art, jewelry, fashion, and other decorative motifs. Scarab beetles were represented widely in art and ornaments [178] and were depicted in hieroglyphs and pictograms. But not only in Egypt, insect forms were converted into pictograms, but also these were present in Mayan and Chinese writing [182].

In visual arts, there are also several artists who depicted insects. Salvador Dali with his famous ants was the artist responsible for the designing of the poster for the International Silk Congress in 1957. Also, the graphic artist M. C. Escher recurred insects as subjects in his work.

Another widespread depiction of insects is present in mail stamps. Almost every country in the world has produced stamps with insects. The website "Insects on Stamps" (http://www.asahi-net.or.jp/~CH2M-NITU/indexe.htm) provides an index of stamps and even coins portraying insects arranged according to country, classified by species.

3.3.3. Spiritual, symbolic, and other interactions with insects

3.3.3.1. Symbolic

In the U.S., most states have adopted at least one insect symbol as a state insect, state butterfly, state bug, or state agricultural insect. For instance, New York state insect is the ladybird (ladybug) *Coccinella novemnotata*, the Kentucky agricultural insect is the honeybee and the Oklahoma state butterfly is the black swallowtail butterfly (see more at https://statesymbolsusa.org/categories/insect).

3.3.3.2. Sacred and/or religious

Insects have made a clear impact on several cultural aspects of human societies. For instance, they were present in many aspects of ancient Egypt, where scarab beetles were considered to be symbols of eternal life [178] and part of a religion with the scarab cult. However, there are much older references of insects as religions symbols from the late Paleolithic epoch (10,000 to 20,000 years ago), in the form of shamanistic "pendants" with the shape of beetles. In fact, among shamanic societies, there are a series of myths relating the creation of the world to beetles [183]. Also in Asian countries, insects have their roles as religious symbols with cicadas, for instance, being used as symbols of rebirth [178].

In Madagascar, the native *B. cajani* silk is processed to produce ritual burial shrouds and clothing items throughout the highlands. The "Landibe" shroud is essential for any respected dead person and is a sign of supreme honor [184]. The "Landibe" shroud is preferred over the one made with *Bombyx mori* silk in Madagascan ritual burials and for exhumations (a Madagascan habit to renew the shroud of mortal remains every 5 to 10 years) [60].

In Yemen, the Khidab, a black waterproof gall ink obtained from oak-galls of *Quercus infectoria*, which result from wasp action, is used for body painting during traditional ceremonies [185].

3.3.3.3. Existence and bequest

It is difficult to account with all the entomological societies existing in the world since the information is scattered and some of them are inactive, no longer existing, or even changed their names. Some societies do not have a webpage, and others only have a website in their native language, which largely increases the difficulty of documenting them. We produce a list which can be fully downloaded (see [186]). Our search returned about 286 societies (including regional branches) distributed worldwide. These were distributed as 127 in Europe, 113 in America (103 in North America, 8 in South America and 2 considering North, Central and South America), 23 in Asia, 9 in Oceania, 5 in Africa, 4 in the Middle East, and 5 international societies. The oldest society in the world still active is the Entomological Club of London founded in 1826, while the most recent, the Society of Overseas Nepalese Entomologists, funded in 2009.

Collections of insects and related arthropods constitute an enormous resource for biological information and document Earth's natural history [187]. The best estimate of how many collections exist is given by Evenhuis [188] that maintains a continually updated list of the world's insect and spiders collections, based on a previous list by Arnett et al. [189], with substantial additions and corrections. Evenhuis [188] listed 1819 entries. In fact, in some parts of the world, the sustainable collection of insects for collections and insectariums can provide economic opportunities for local populations, in alternative to destructive resource extraction such as logging, mining, or conversion of forest land to agricultural enterprises, such as cocoa, coffee, or oil palm plantations [190].

4. Discussion

Undeniably, insects are responsible for many of the ES provided by the ecosystems, where they occur (see **Figure 2**), covering all services categories (Provisioning, Regulating and Maintenance, and Cultural). The use of indicators of the services provided by insects will support the use of ES concept in an integrated approach (e.g., related to habitat status, drivers of change, pressures, new potential services, trade-offs), acknowledging the paramount importance of the most abundant and diverse group of organisms in the planet. From our search, we extracted (see **Table 1**) a total of 73 indicators, divided as 17 Provisional indicators, 27 Regulation and Maintenance indicators, and 29 Cultural indicators. The use of indicators has been recognized as a useful communication tool that facilitates the simplification of human-environmental systems high complexity, e.g., [13], they can be selected for mapping and assessment of ES [14] and to support specific management purposes involving ES trade-offs. As we can only protect and manage what we know, the proposed indicators might in this way give a contribution to the global International Platform of Biodiversity and Ecosystem Services and to the UN Natural Capital Accounting. Looking at the indicators that emerged from this review, it is clear that there are fewer indicators for the provisional services provided by insects. This may result from a negative attitude toward insects, especially in

Figure 2. Different ecosystem services provided by insects. A—Lacewing adult: larvae are voracious predators of eggs and immature stages of many soft bodied insect pests; B—Syrphid fly: adults are pollinators, larvae are predators of agricultural pests; C—Cricket selling in the religious ceremony "Senhor de Matosinhos", Portugal; D—Bumblebee: pollinator; E—Damselfly: predator of other insects (e.g., mosquitoes); F—Ladybird: predator of agricultural pests; G— Butterfly: non-bee pollinator; H—Tiger beetle: ground predator.

Western countries. Public perception of insects needs to be improved and the profiles of beneficial species must be raised [200] to fully preserve essential ES. The opposite occurs in more nature orientated cultures, in which apart from being important food sources, insects are also providers of materials and medicines. However, this potential is at risk, threatened by current environmental changes which are contributing to the degradation of insect habitats. The lack of important ES providers has greatly affected human lives around the world, e.g., the decline of pollinator species. Its consequences can be devastating, especially in a context of current predictions of food shortage to feed an increasing world population, since abiotic pollination or pollination carried by other animals is not so effective, which makes insect pollination an irreplaceable service.

Citing the famous entomologist Edward Osborne Wilson: "If insects were to vanish, the terrestrial environment would soon collapse into chaos" and "More respect is due the little things that run the world."

Acknowledgements

This chapter was supported by Portuguese funds, through the national Foundation for Science and Technology – FCT to cE3c (UID/BIA/00329/2013). Thanks are also due, for the financial support to CESAM (UID/AMB/50017/2013), to FCT/MEC through national funds, and the co-funding by the FEDER, within the PT2020 Partnership Agreement and Compete 2020. A post-doctoral grant to OMCCA (SFRH/BPD/102965/2014) supported by FCT is also acknowledged.

Author details

Olga Maria Correia Chitas Ameixa[1,2*], António Onofre Soares[2], Amadeu M.V.M. Soares[1] and Ana I. Lillebø[1]

*Address all correspondence to: olga.ameixa@ua.pt

1 Biology Department and CESAM, University of Aveiro, Aveiro, Portugal

2 CE3C, Centre for Ecology, Evolution and Environmental Changes, Azorean Biodiversity Group, Department of Biology, University of the Azores, Portugal

References

[1] Thomas CD, Cameron A, Green RE, Bakkenes M, Beaumont LJ, Collingham YC, Erasmus BF, De Siqueira MF, Grainger A, Hannah L. Extinction risk from climate change. Nature. 2004;**427**(6970):145-148

[2] Hadfield MG. Introduction to the symposium: The crisis in invertebrate conservation. American Zoologist. 1993;**33**(6):497-498

[3] van Lenteren J. Ecosystem services to biological control of pests: Why are they ignored? Paper Presented at the Proc. Neth. Entomol. Soc. Meet; 2006

[4] Dunn RR. Global mapping of ecosystem disservices: The unspoken reality that nature sometimes kills us. Biotropica. 2010;**42**(5):555-557

[5] van Huis A, van Itterbeeck J, Klunder H, Mertens E, Halloran A, Muir G, Vantomme P. Edible insects: Future prospects for food and feed security. In: FAO Forestry Paper 171, editor. FAO Forestry Paper 171. Rome, Italy: Food and Agriculture Organization of the United Nations; 2003

[6] Cherniack EP. Bugs as drugs, part 1: Insects: The "new" alternative medicine for the 21st century. Alternative Medicine Review. 2010;**15**(2):124-135

[7] Dossey AT. Insects and their chemical weaponry: New potential for drug discovery. Natural Product Reports. 2010;**27**(12):1737-1757

[8] Ratcliffe NA, Mello CB, Garcia ES, Butt TM, Azambuja P. Insect natural products and processes: New treatments for human disease. Insect Biochemistry and Molecular Biology. 2011;**41**(10):747-769. DOI: 10.1016/j.ibmb.2011.05.007

[9] Lockhart GJ. Ants and other great medicines. Partially Published Online by Jacobson, AL; 2007

[10] Jiang SJ. Anti-cancer insect medicinal materials in China. The Journal of Chinese Medicine. 1990;**13**(11):11-14

[11] Kettunen M, Vihervaara P, Kinnunen S, D'Amato D, Badura T, Argimon M, Ten Brink P. Socio-Economic Importance of Ecosystem Services in the Nordic Countries. Synthesis in the Context of the Economics of Ecosystems and Biodiversity (TEEB). Copenhagen: Nordic Council of Ministers; 2012

[12] Norris K. Biodiversity in the context of ecosystem services: The applied need for systems approaches. Philosophical Transactions of the Royal Society, B: Biological Sciences. 2012;**367**(1586):191-199. DOI: 10.1098/rstb.2011.0176

[13] Müller F, Burkhard B. The indicator side of ecosystem services. Ecosystem Services. 2012;**1**(1):26-30. DOI: 10.1016/j.ecoser.2012.06.001

[14] Maes J, Liquete C, Teller A, Erhard M, Paracchini ML, Barredo JI, Grizzetti B, Cardoso A, Somma F, Petersen JE, Meiner A, Gelabert ER, Zal N, Kristensen P, Bastrup-Birk A, Biala K, Piroddi C, Egoh B, Degeorges P, Fiorina C, Santos-Matín F, Narusevicius V, Verboven J, Pereira H, Bengtsson J, Gocheva K, Marta-Pedroso C, Snäll T, Estreguil C, Miguel JS, Perez-Soba M, Grêt-Regamey A, Lillebø AI, Malak DA, Condé S, Moen J, Czúcz B, Drakou EG, Zulian G, Lavalle C. An indicator framework for assessing ecosystem services in support of the EU biodiversity strategy to 2020. Ecosystem Services. 2016;**17**:14-23

[15] Millennium Ecosystem Assessment. Ecosystems and Human Well-Being: Synthesis. Washington, DC: Island Press; 2005

[16] Haines-Young R, Potschin M. Common International Classification of Ecosystem Services (CICES): Consultation on Version 4, August–ecember 2012. EEA Framework Contract No EEA/IEA/09/003. 2013. [Accessed: 20 May 2014]

[17] Saarikoski H, Jax K, Harrison PA, Primmer E, Barton DN, Mononen L, Vihervaara P, Furman E. Exploring operational ecosystem service definitions: The case of boreal forests. Ecosystem Services. 2015;**14**:144-157. DOI: 10.1016/j.ecoser.2015.03.006

[18] Mononen L, Auvinen AP, Ahokumpu AL, Rönkä M, Aarras N, Tolvanen H, Kamppinen M, Viirret E, Kumpula T, Vihervaara P. National ecosystem service indicators: Measures of social–ecological sustainability. Ecological Indicators. 2015. DOI: 10.1016/j.ecolind.2015.03.041

[19] Benzies KM, Premji S, Hayden KA, Serrett K. State-of-the-evidence reviews: Advantages and challenges of including grey literature. Worldviews on Evidence-Based Nursing. 2006;**3**(2):55-61. DOI: 10.1111/j.1741-6787.2006.00051.x

[20] Feld CK, Da Silva PM, Sousa JP, De Bello F, Bugter R, Grandin U, Hering D, Lavorel S, Mountford O, Pardo I, Pärtel M, Römbke J, Sandin L, Bruce Jones K, Harrison P. Indicators of biodiversity and ecosystem services: A synthesis across ecosystems and spatial scales. Oikos. 2009;**118**(12):1862-1871. DOI: 10.1111/j.1600-0706.2009.17860.x

[21] Fagerholm N, Käyhkö N, Ndumbaro F, Khamis M. Community stakeholders' knowledge in landscape assessments - mapping indicators for landscape services. Ecological Indicators. 2012;**18**:421-433. DOI: 10.1016/j.ecolind.2011.12.004

[22] van Huis A. Potential of insects as food and feed in assuring food security. Annual Review of Entomology. 2013;**58**:563-583

[23] Gullan PJ, Cranston PS. The Insects: An Outline of Entomology. 3rd ed. Blackwell Science; 2004

[24] MacEvilly C. Bugs in the system. Nutrition Bulletin. 2000;**25**(4):267-268. DOI: 10.1046/j.1467-3010.2000.00068.x

[25] Jongema Y. List of Edible Insects of the World. Wageningen University. Laboratory of Entomology; 2014. http://www.wur.nl/en/Expertise-Services/Chair-groups/Plant-Sciences/Laboratory-of-Entomology/Edible-insects/Worldwide-species-list.htm [Accessed March 2015]

[26] van Huis A, van Gurp H, Dicke M. The Insect Cookbook: Food for a Sustainable Planet. Columbia University Press; 2014

[27] Centre for the Promotion of Imports from developing countries. CBI market survey: The honey and other bee products market in the EU. In C. M. I. Database Ed; 2009. pp. 32

[28] Losey JE, Vaughan M. The economic value of ecological services provided by insects. Bioscience. 2006;**56**(4):311-323. DOI: 10.1641/0006-3568(2006)56[311:tevoes]2.0.co;2

[29] Ravindran V, Blair R. Feed resources for poultry production in Asia and the Pacific. World's Poultry Science Journal. 1993;**49**:219-235

[30] Wipfli MS. Terrestrial invertebrates as salmonid prey and nitrogen sources in streams: Contrasting old-growth and young-growth riparian forests in southeastern Alaska, USA. Canadian Journal of Fisheries and Aquatic Sciences. 1997;**54**(6):1259-1269

[31] Allan JD, Wipfli MS, Caouette JP, Prussian A, Rodgers J. Influence of streamside vegetation on inputs of terrestrial invertebrates to salmonid food webs. Canadian Journal of Fisheries and Aquatic Sciences. 2003;**60**(3):309-320. DOI: 10.1139/f03-019

[32] Baxter CV, Fausch KD, Carl Saunders W. Tangled webs: Reciprocal flows of invertebrate prey link streams and riparian zones. Freshwater Biology. 2005;**50**(2):201-220. DOI: 10.1111/j.1365-2427.2004.01328.x

[33] Kawaguchi Y, Taniguchi Y, Nakano S. Terrestrial invertebrate inputs determine the local abundance of stream fishes in a forested stream. Ecology. 2003;**84**(3):701-708. DOI: 10.1890/0012-9658(2003)084[0701:TIIDTL]2.0.CO;2

[34] Baxter CV, Fausch KD, Murakami M, Chapman PL. Invading rainbow trout usurp a terrestrial prey subsidy from native charr and reduce their growth and abundance. Oecologia. 2007;**153**(2):461-470

[35] Nakano S, Murakami M. Reciprocal subsidies: Dynamic interdependence between terrestrial and aquatic food webs. Proceedings of the National Academy of Sciences of the United States of America. 2001;**98**(1):166-170. DOI: 10.1073/pnas.98.1.166

[36] Young MK, Rader RB, Belish TA. Influence of macroinvertebrate drift and light on the activity and movement of Colorado River cutthroat trout. Transactions of the American Fisheries Society. 1997;**126**(3):428-437

[37] Garman G. Use of terrestrial arthropod prey by a stream-dwelling cyprinid fish. Environmental Biology of Fishes. 1991;**30**(3):325-331. DOI: 10.1007/BF02028848

[38] Mobley JA. Birds of the World. Marshall Cavendish; 2008

[39] Liukkonen-Anttila T. Nutritional and genetic adaptation of galliform birds: Implications for hand-rearing and restocking. (PhD dissertation). Oulu, Finland: University of Oulu; 2001

[40] Crane E. History of honey. In: Crane E, editor. Honey, a Comprehensive Survey. Vol. 77. London: William Heinemann; 1975. pp. 439-448

[41] Schowalter TD. Insect Ecology: An Ecosystem Approach. 3rd ed. San Diego, CA: Elsevier/Academic Press; 2011

[42] Crane E. The World History of Beekeeping and Honey Hunting. New York, NY: Routledge; 1999

[43] Krell R. Value-added products from beekeeping. Food & Agriculture Organization; 1996

[44] Liu W-W, Yang P, Chen X-M, Xu D-L, Hu Y-H. Cloning and expression analysis of four heat shock protein genes in *Ericerus pela* (Homoptera: Coccidae). Journal of Insect Science. 2014;**14**:9. DOI: 10.1093/jisesa/ieu032

[45] Ward G, editor. The Grove Encyclopaedia of Materials and Techniques in Art. Oxford University Press.; 2008

[46] Derry J. Investigating shellac: Documenting the process, defining the product. A study on the processing methods of shellac, and the analysis of selected physical and chemical characteristics. (Master Thesis Project-Based Master Thesis), University of Oslo; 2012

[47] Baboo B, Goswani DN. Processing, Chemistry and Application of Lac. New Delhi, India: Chandu Press; 2010

[48] Rossman JM. Commercial manufacture of edible films. In: Huber KC, Embuscado ME, editors. Edible Films and Coatings for Food Applications. Springer New York; 2009. pp. 367-390

[49] Mills JS, White R. The Organic Chemistry of Museum Objects. Oxford: Butterworth - Heinemann; 1994

[50] Greig JB. WHO Food Additives Series 46: Cochineal Extract, Carmine, and Carminic acid. 1998. http://www.inchem.org/documents/jecfa/jecmono/v46je03.htm [Accessed: Nov 2014]

[51] Anelli CM, Prischmann-Voldseth DA. Silk batik using beeswax and cochineal dye: An interdisciplinary approach to teaching entomology. American Entomologist. 2009;**55**(2):95-105

[52] Theobald MM. Putting the red in redcoats. The Colonial Williamsburg Journal. 2012. http://www.history.org/foundation/journal/summer12/dye.cfm [Accessed: May 2015]

[53] Ferreira ES, Hulme AN, McNab H, Quye A. The natural constituents of historical textile dyes. Chemical Society Reviews. 2004;**33**(6):329-336

[54] Łagowska B, Golan K. Scale insects (Hemiptera, Coccoidea) as a source of natural dye and other useful substances. Aphids and Other Hemipterous Insects. 2009;**15**:151-167

[55] Cardon D. Le monde des teintures naturelles. Paris: Beeline; 2003

[56] Yong-woo L. Silk reeling and testing manual. FAO Agricultural Services Bulletin; 1999. p. 136

[57] Rafidiarimalala O. La sériciculture à Madagascar, technique d'élevage. Élevage (Madagascar): Mémoire de fin d'étude EESA: Université d'Antananarivo, Département; 1974

[58] Costa M. Manuel technique de sériciculture. Élevage des vers à soie. Vol. 2. Antananarivo: CITE, SAHA, Programme soie ONUDI Madagascar; 2004

[59] ONU. Développement de la sériciculture sur les Hauts plateaux, phase II, Madagascar. Rapport terminal. AG:D8:MAG :85 :006. Rome: Programme des Nations Unies pour le Développement; 1991

[60] CITE/BOSS Corporation. Étude nationale de la filière soie Analamanga, Ambalavao, Amoron'i Mania,Vakinakaratra, Itasy. Rapport final. Antananarivo, Madagascar: CITE; 2009

[61] Kolar J, Štolfa A, Strlič M, Pompe M, Pihlar B, Budnar M, Simčič J, Reissland B. Historical iron gall ink containing documents — Properties affecting their condition. Analytica Chimica Acta. 2006;**555**(1):167-174. DOI: 10.1016/j.aca.2005.08.073

[62] Schölten JS. Paper presented at the proc. European workshop on iron-gall ink corrosion. Amsterdam: Instituut Collectie Nederland; 1997

[63] Costa-Neto EM. Entomotherapy, or the medicinal use of insects. Journal of Ethnobiology. 2005;**25**(1):93-114. DOI: 10.2993/0278-0771(2005)25[93:EOTMUO]2.0.CO;2

[64] Jones R. Honey and healing through the ages. JAAS. 2009;**1**(1):2-5

[65] Molan PC. The evidence supporting the use of honey as a wound dressing. The International Journal of Lower Extremity Wounds. 2006;**5**(1):40-54. DOI:10.1177/1534734605286014

[66] Izuta H, Chikaraishi Y, Shimazawa M, Mishima S, Hara H. 10-Hydroxy-2-decenoic acid, a major fatty acid from royal jelly, inhibits VEGF-induced angiogenesis in human umbilical vein endothelial cells. Evidence-based Complementary and Alternative Medicine. 2009;**6**(4):489-494

[67] Hattori N, Nomoto H, Fukumitsu H, Mishima S, Furukawa S. AMP N-1-oxide, a unique compound of royal jelly, induces neurite outgrowth from PC12 cells via signaling by protein kinase A independent of that by mitogen-activated protein kinase. Evidence based Complementary and Alternative Medicine. 2010;**7**(1):63-68

[68] Okamoto I, Taniguchi Y, Kunikata T, Kohno K, Iwaki K, Ikeda M, Kurimoto M. Major royal jelly protein 3 modulates immune responses in vitro and in vivo. Life Sciences. 2003;**73**(16):2029-2045

[69] Kumazawa S, Ueda R, Hamasaka T, Fukumoto S, Fujimoto T, Nakayama T. Antioxidant prenylated flavonoids from propolis collected in Okinawa, Japan. Journal of Agricultural and Food Chemistry. 2007;**55**(19):7722-7725. DOI: 10.1021/jf071187h

[70] Kujumgiev A, Tsvetkova I, Serkedjieva Y, Bankova V, Christov R, Popov S. Antibacterial, antifungal and antiviral activity of propolis of different geographic origin. Journal of Ethnopharmacology. 1999;**64**(3):235-240

[71] Paulino N, Teixeira C, Martins R, Scremin A, Dirsch VM, Vollmar AM, Abreu SR, de Castro SL, Marcucci MC. Evaluation of the analgesic and anti-inflammatory effects of a Brazilian green propolis. Planta Medica. 2006;**72**(10):899-906. DOI: 10.1055/s-2006-947185

[72] Li F, Awale S, Zhang H, Tezuka Y, Esumi H, Kadota S. Chemical constituents of propolis from Myanmar and their preferential cytotoxicity against a human pancreatic cancer cell line. Journal of Natural Products. 2009;**72**(7):1283-1287. DOI: 10.1021/np9002433

[73] Encyclopaedia Britannica. 2015. http://www.britannica.com/EBchecked/topic/112805/Chinese-wax [Accessed: March 2015]

[74] Epstein WL, Kligman AM. Treatment of warts with cantharidin. Archives of Dermatology. 1958;**77**:508-511

[75] Gudger E. Stitching wounds with the mandibles of ants and beetles: A minor contribution to the history of surgery. American Medical Association; 1925

[76] Sherman RA, Hall MJ, Thomas S. Medicinal maggots: An ancient remedy for some contemporary afflictions. Annual Review of Entomology. 2000;**45**:55-81. DOI: 10.1146/annurev.ento.45.1.55

[77] Sherman RA, Stevens H, Ng D, Iversen E. Treating wounds in small animals with maggot debridement therapy: A survey of practitioners. Veterinary Journal. 2007;**173**(1):138-143

[78] Kerridge A, Lappin-Scott H, Stevens J. Antibacterial properties of larval secretions of the blowfly, *Lucilia sericata*. Medical and Veterinary Entomology. 2005;**19**(3):333-337

[79] Pemberton RW. Insects and other arthropods used as drugs in Korean traditional medicine. Journal of Ethnopharmacology. 1999;**65**(3):207-216

[80] Zimian D, Yonghua Z, Xiwu G. Medicinal insects in China. Ecology of Food and Nutrition. 1997;**36**(2-4):209-220. DOI: 10.1080/03670244.1997.9991516

[81] Han J, Inoue S, Isoda H. Effects of silkworm powder on glucose absorption by human intestinal epithelial cell line Caco-2. Journal of Natural Medicines. 2007;**61**(4):387-390

[82] Wenk E, Merkle HP, Meinel L. Silk fibroin as a vehicle for drug delivery applications. Journal of Controlled Release. 2011;**150**(2):128-141. DOI: 10.1016/j.jconrel.2010.11.007

[83] Elvin CM, Carr AG, Huson MG, Maxwell JM, Pearson RD, Vuocolo T, Liyou NE, Wong DCC, Merritt DJ, Dixon NE. Synthesis and properties of crosslinked recombinant proresilin. Nature. 2005;**437**(7051):999-1002

[84] Pearnchob N, Siepmann J, Bodmeier R. Pharmaceutical applications of shellac: Moisture-protective and taste-masking coatings and extended-release matrix tablets. Drug Development and Industrial Pharmacy. 2003;**29**(8):925-938. DOI: 10.1081/DDC-120024188

[85] Azouka A, Huggett R, Harrison A. The production of shellac and its general and dental uses: A review. Journal of Oral Rehabilitation. 1993;**20**(4):393-400

[86] Ma D, Wang Y, Yang H, Wu J, An S, Gao L, Xu X, Lai R. Anti-thrombosis repertoire of blood-feeding horsefly salivary glands. Molecular & Cellular Proteomics. 2009;**8**(9):2071-2079. DOI: 10.1074/mcp.M900186-MCP200

[87] Eisner T, Aneshansley D. "Anting" in blue jays: Evidence in support of a food-preparatory function. Chemoecology. 2008;**18**(4):197-203. DOI: 10.1007/s00049-008-0406-3

[88] Clayton DH, Koop JA, Harbison CW, Moyer BR, Bush SE. How birds combat ectoparasites. The Open Ornithology Journal. 2010;**3**:41-71

[89] Longino JT. True anting by the capuchin, *Cebus capucinus*. Primates. 1984;**25**(2):243-245

[90] Hauser DC. Anting by gray squirrels. Journal of Mammalogy. 1964;**136-138**

[91] Bornemissza GF. The Australian dung beetle project 1965-1975. A.M.R.C. Review. 1976;**30**:1-30

[92] Fincher GT. The potential value of dung beetles in pasture ecosystems. Journal of Entomological Science. 1981;**16**

[93] Carlton RG, Goldman CR. Effects of a massive swarm of ants on ammonium concentrations in a subalpine lake. Hydrobiologia. 1984;**111**(2):113-117

[94] Mehner T, Ihlau J, Dörner H, Hölker F. Can feeding of fish on terrestrial insects subsidize the nutrient pool of lakes? Limnology and Oceanography. 2005;**50**(6):2022-2031

[95] Pray CL, Nowlin WH, Vanni MJ. Deposition and decomposition of periodical cicadas (Homoptera: Cicadidae: Magicicada) in woodland aquatic ecosystems. Journal of the North American Benthological Society. 2009;**28**(1):181-195

[96] Lal R, Hawksworth DL. Soil conservation and biodiversity. Paper presented at the Biodiversity of microorganisms and invertebrates: Its role in sustainable agriculture. Proc. First workshop on the ecological foundations of sustainable agriculture (WEFSA 1), London, UK, 26-27 July 1990; 1991

[97] UNEP/CBD/COP/3/38. Report of the third ordinary meeting of the conference of the parties to the Convention on Biological Diversity. Argentina: Buenos Aires; 1996

[98] Ingram M, Nabhan GP, Buchmann SL. Our forgotten pollinators: Protecting the birds and bees. Global Pesticide Campaigner. 1996;**6**(4):1-12. http://www.apiservices.com/articles/us/pollinators.htm [Accessed: Jan 2015]

[99] Free JB. Insect pollination of crops. London, UK: Academic Press; 1993

[100] Corlett RT. Honey bees in natural ecosystems. In: Hepburn HR, Radloff SE, editors. Honey Bees of Asia. Berlin Heidelberg: Springer-Verlag; 2011. pp. 215-225

[101] Partap U, Partap T. Managed crop pollination: The Missing dimension of mountain agricultural productivity mountain farming systems' Discussion Paper. Kathmandu: ICIMOD; 1997

[102] Sheffield CS, Ngo H, Azzu N. Report of the workshop on natural pollination services for agricultural production in apple orchards in the Hindu-Kush Himalayan region. In F. Report (Ed.); 2014

[103] Ahmad F, Partap U, Joshi SR, Gurung MB. Indigenous honeybees: Allies for mountain farmers. LEISA Magazine. 2004;**20**(4):12-13. http://www.agriculturesnetwork.org/magazines/global/farming-with-nature/indigenous-honeybees [Accessed Dec 2014]

[104] USDA/NASS. California Almond Acreage Report, April 30; 2009, 2010

[105] Natural Resources Defense Council. Why we Need Bees Nature's Tiny Workers Put Food on our Tables. Bee Facts; 2011 https://www.nrdc.org/wildlife/animals/files/bees.pdf

[106] Ward R, Whyte A, James RR. A tale of two bees: Looking at pollination fees for almonds and sweet cherries. American Entomologist. 2010;**56**:172-179

[107] National Honey Board. National Honey Board Press Kit: Honey Industry Facts. 2. 2014. http://www.honey.com/images/uploads/general/NHB-Press-Kit-Honey-Industry-Facts_April_2014-1.pdf [Accessed Jan 2015]

[108] McGregor SE. Insect Pollination of Cultivated Crop-Plants. Vol. 496. U.S.D.A Agriculture Handbook; 1976. p. 411

[109] Crane E, Walker P. Pollination Directory for World Crops. International Bee Research Association; 1984

[110] Rader R, Bartomeus I, Garibaldi LA, Garratt MPD, Howlett BG, Winfree R, Cunningham SA, Mayfield MM, Arthur AD, Andersson GKS, Bommarco R, Brittain C, Carvalheiro LG, Chacoff NP, Entling MH, Foully B, Freitas BM, Gemmill-Herren B, Ghazoul J, Griffin SR, Gross CL, Herbertsson L, Herzog F, Hipólito J, Jaggar S, Jauker F, Klein A-M, Kleijn D, Krishnan S, Lemos CQ, Lindström SAM, Mandelik Y, Monteiro VM, Nelson W, Nilsson L, Pattemore DE, Pereira NO, Pisanty G, Potts SG, Reemer M, Rundlöf M, Sheffield CS, Scheper J, Schüepp C, Smith HG, Stanley DA, Stout JC, Szentgyörgyii H,

Taki H, Vergara CH, Viana BF, Woyciechowski M. Non-bee insects are important contributors to global crop pollination. Proceedings of the National Academy of Sciences. 2015. DOI: 10.1073/pnas.1517092112

[111] Li Vigni I, Melati MR. Policoria ed interazioni ecologiche in *Ricinus communis* L. Giornale Botanico Italiano. 1995;**129**(2):147

[112] van der Pijl L. Principles of Dispersal in Higher Plants. Berlin, Heidelberg, New York: Springer-Verlag; 1982

[113] Li Vigni I, Melati MR. Examples of seed dispersal by entomochory. Acta Botanica Gallica. 1999;**146**(2):145-156. DOI: 10.1080/12538078.1999.10515813

[114] MacMahon JA, Mull JF, Crist TO. Harvester ants (*Pogonomyrmex* spp.): Their community and ecosystem influences. Annual Review of Ecology, Evolution, and Systematics. 2000:265-291

[115] Jones C, Lawton JH. Linking Species & Ecosystems. Springer Science & Business Media; 1995

[116] Jackson JK, Fisher SG. Secondary production, emergence, and export of aquatic insects of a Sonoran Desert stream. Ecology. 1986;**67**(3):629-638

[117] Gray LJ. Response of insectivorous birds to emerging aquatic insects in riparian habitats of a tallgrass prairie stream. American Midland Naturalist. 1993:288-300

[118] Burdon FJ, Harding JS. The linkage between riparian predators and aquatic insects across a stream-resource spectrum. Freshwater Biology. 2008;**53**(2):330-346. DOI:10.1111/j.1365-2427.2007.01897.x

[119] Sabo J, Power M. River-watershed exchange: Effects of riverine subsidies on riparian lizards and their terrestrial prey. Ecology. 2002;**83**(7):1860-1869

[120] Lacki MJ, Johnson JS, Dodd LE, Baker MD. Prey consumption of insectivorous bats in coniferous forests of north-Central Idaho. Northwest Science. 2007;**81**(3):199-205

[121] Konishi M, Itô Y. Early entomology in East Asia. In: Smith RF, Mittler TE, Smith CN, editors. History of Entomology. Palo Alto, CA: Annual Reviews Inc.; 1973. pp. 1-20

[122] Hagen KS, Franz JM. A history of biological control. In: Smith RF, Mittler TE, Smith CN, editors. In History of Entomology. Palo Alto, CA: Annual Reviews Inc.; 1973. pp. 433-476

[123] van Lenteren JC. The state of commercial augmentative biological control: Plenty of natural enemies, but a frustrating lack of uptake. BioControl. 2012;**57**(1):1-20

[124] Costanza R, d'Arge R, de Groot R, Farberk S, Grasso M, Hannon B, Limburg K, Naeem S, O'Neill RV, Paruelo J. The value of the world's ecosystem services and natural capital. Nature. 1997;**387**:253-260

[125] van Lenteren JC, editor. IOBC Internet Book of Biological Control, Version 6. Wageningen; 2012

[126] Warner KD, Getz C. A socio-economic analysis of the north American commercial natural enemy industry and implications for augmentative biological control. Biological Control. 2008;**45**(1):1-10

[127] Woodburn AT. The market for agrochemicals: Present and Future. Paper Presented at the Proc. Brighton Crop Prot. Conf. - Weeds, Brighton, England; 1995

[128] Webb M, Conroy C. The socioeconomics of weed control on smallholder farms in Uganda. Paper presented at the Proc. Brighton Crop Prot. Conf. – Weeds; 1995

[129] Wells MJ, Balsinhas AA, Joffe H, Engelbrecht MW, Harding G, Stirton CH. A catalogue of problem plants in Southern Africa. Vol. 53. Memoirs of the botanical survey of South Africa; 1986. p. 658

[130] Du Toit R. The spread of prickly pear in the union. Farming in South Africa. 1942;**17**: 300-304

[131] Zimmerman HG, Moran VC, Hoffmann JH. Insect herbivores as determinants of the present distribution and abundance of invasive cacti in South Africa. In: Macdonald IAW, Kruger FJ, Ferrar AA, editors. The Ecology and Management of Biological Invasions in Southern Africa. Cape town, South Africa: Oxford University Press; 1986. pp. 269-274

[132] Wilson JR, Ajuonu O, Center TD, Hill MP, Julien MH, Katagira FF, Neuenschwander P, Njoka SW, Ogwang J, Reeder RH. The decline of water hyacinth on Lake Victoria was due to biological control by *Neochetina* spp. Aquatic Botany. 2007;**87**(1):90-93

[133] Honek A, Martinkova Z, Jarosik V. Ground beetles (Carabidae) as seed predators. European Journal of Entomology. 2003;**100**(4):531-544

[134] Brown JH, Reichman O, Davidson DW. Granivory in desert ecosystems. Annual Review of Ecology, Evolution, and Systematics. 1979:201-227

[135] Andersen A. Seed removal by ants in the mallee of northwestern Victoria ant-plant interactions in Australia. Springer; 1982. pp. 31-43

[136] Kindlmann P, Ameixa OMCC, Dixon AG. Ecological effects of invasive alien species on native communities, with particular emphasis on the interactions between aphids and ladybirds. BioControl. 2011;**56**(4):469-476. DOI: 10.1007/s10526-011-9384-4

[137] Simberloff D, Stiling P. Risks of species introduced for biological control. Biological Conservation. 1996;**78**(1):185-192

[138] Louda SM, Pemberton RW, Johnson MT, Follett P. Nontarget effects-the Achilles' heel of biological control? Retrospective analyses to reduce risk associated with biocontrol introductions. Annual Review of Entomology. 2003;**48**(1):365-396

[139] Lavelle P, Lattaud C, Trigo D, Barois I. Mutualism and biodiversity in soils. Plant and Soil. 1995;**170**(1):23-33. DOI: 10.1007/BF02183052

[140] Culliney T. Role of arthropods in maintaining soil fertility. Agriculture. 2013;**3**(4):629-659

[141] Jones CG, Lawton JH, Shachak M. Organisms as Ecosystem Engineers Ecosystem Management. Springer; 1996. pp. 130-147

[142] Lobry de Bruyn L, Conacher A. The role of termites and ants in soil modification-a review. Australian Journal of Soil Research. 1990;**28**(1):55-93

[143] Kremen C, Chaplin-Kramer R. Insects as providers of ecosystem services: Crop pollination and pest control. In: Stewart AJA, New T, Lewis OT, editors. Insect Conservation Biology. Wallingford, UK: CABI; 2005. pp. 349-382

[144] Losey J, Vaughan M. Conserving the ecological services provided by insects. American Entomologist. 2008;**54**(2):113-114

[145] Gillard P. Coprophagous beetles in pasture ecosystems. The Journal of the Australian Institute of Agricultural Science. 1967;**33**:30-34

[146] Kazuhira Y, Hdeaki K, Takuro K, Toshiharu A. Nitrogen mineralization and microbial populations in cow dung, dung balls and underlying soil affected by paracoprid dung beetles. Soil Biology and Biochemistry. 1991;**23**(7):649-653

[147] de Mazancourt C, Loreau M, Dieckmann U. Can the evolution of plant defense lead to plant-herbivore mutualism? The American Naturalist. 2001;**158**(2):109-123

[148] Yang LH. Periodical cicadas as resource pulses in north American forests. Science. 2004;**306**(5701):1565-1567

[149] Schultz JC, Baldwin IT. Oak leaf quality declines in response to defoliation by gypsy moth larvae. Science. 1982;**217**(4555):149-151. DOI: 10.1126/science.217.4555.149

[150] Agrawal AA, Tuzun S, Bent E. Induced Plant Defenses against Pathogens and Herbivores: Biochemistry, Ecology, and Agriculture. APS Press; 1999

[151] Scott NA, Binkley D. Foliage litter quality and annual net N mineralization: Comparison across North American forest sites. Oecologia. 1997;**111**(2):151-159

[152] Hollinger DY. Herbivory and the cycling of nitrogen and phosphorus in isolated California oak trees. Oecologia. 1986;**70**(2):291-297

[153] Lovett GM, Ruesink AE. Carbon and nitrogen mineralization from decomposing gypsy moth frass. Oecologia. 1995;**104**(2):133-138

[154] Stadler B, Solinger S, Michalzik B. Insect herbivores and the nutrient flow from the canopy to the soil in coniferous and deciduous forests. Oecologia. 2001;**126**(1):104-113. DOI: 10.1007/s004420000514

[155] Ritchie ME, Tilman D, Knops JM. Herbivore effects on plant and nitrogen dynamics in oak savanna. Ecology. 1998;**79**(1):165-177

[156] Harding CR, Schroeder GN, Collins JW, Frankel G. Use of *galleria mellonella* as a model organism to study *Legionella pneumophila* infection. Journal of Visualized Experiments. 2013;(81):e50964. DOI: 10.3791/50964

[157] The Tribolium Genome Sequencing Consortium. The genome of the model beetle and pest *Tribolium castaneum*. Nature. 2008;**452**(7190):949-955. DOI: http://www.nature.com/nature/journal/v452/n7190/suppinfo/nature06784_S1.html

[158] Vepari C, Kaplan DL. Silk as a biomaterial. Progress in Polymer Science. 2007;**32**(8):991-1007

[159] Arai T, Freddi G, Colonna GM, Scotti E, Boschi A, Murakami R, Tsukada M. Absorption of metal cations by modified *B. mori* silk and preparation of fabrics with antimicrobial activity. Journal of Applied Polymer Science. 2001;**80**(2):297-303. DOI:10.1002/1097-4628(20010411)80:2<297::AID-APP1099>3.0.CO;2-Z

[160] Krajewska B. Membrane-based processes performed with use of chitin/chitosan materials. Separation and Purification Technology. 2005;**41**(3):305-312. DOI: 10.1016/j.seppur.2004.03.019

[161] Turner JS, Soar RC. Beyond biomimicry: What termites can tell us about realizing the living building. Paper presented at the Proc. 1st Int. Conf. Industrialized, Intelligent Construction; 2008

[162] Catts E, Goff ML. Forensic entomology in criminal investigations. Annual Review of Entomology. 1992;**37**(1):253-272

[163] Hall RD. Perceptions and status of forensic entomology. In: Byrd JH, Castner JL, editors. Forensic Entomology: The Utility of Arthropods in Legal Investigations. Boca Raton: CRC Press; 2001. pp. 1-15

[164] Bergeret M. Infanticide. Momification naturelle du cadavre. Découverte du cadavre d'un enfant nouveau-né dans une cheminée où il s'était momifié. Détermination de l'epoque de la naissance par la présence de nymphes et de larves d'insectes dans le cadavre, et par l'étude de leurs métamorphoses. Annales d'hygiène publique et de médecine légale. 1855;**4**:442-452

[165] Yovanovitch GP. Entomologie appliquée à la médicine légale. Paris: Ollier-Henry; 1888

[166] Mégnin P. La faune de cadavres. Application de l'entomologie a la médicine légale. Paris: Encyclopdie scientifique des Aides-Mémoire, Masson, Paris, Gauthier-Villars; 1894

[167] Amendt J, Krettek R, Zehner R. Forensic entomology. Naturwissenschaften. 2004;**91**(2):51-65. DOI: 10.1007/s00114-003-0493-5

[168] Greenberg B. Flies as forensic indicators. Journal of Medical Entomology. 1991;**28**(5):565-577

[169] Joseph I, Mathew DG, Sathyan P, Vargheese G. The use of insects in forensic investigations: An overview on the scope of forensic entomology. Journal of Forensic Dental Sciences. 2011;**3**(2):89

[170] McGeoch MA. The selection, testing and application of terrestrial insects as bioindicators. Biological Reviews. 1998;**73**(02):181-201

[171] New T. Taxonomic focus and quality control in insect surveys for biodiversity conservation. Australian Journal of Entomology. 1996;**35**(2):97-106

[172] Hodkinson I, Jackson J. Terrestrial and aquatic invertebrates as bioindicators for environmental monitoring, with particular reference to mountain ecosystems. Environmental Management. 2005;**35**(5):649-666. DOI: 10.1007/s00267-004-0211-x

[173] Saul-Gershenz L. Insect zoos. In: Resh VH, Cardé RT, editors. Encyclopedia of Insects. 2nd ed. San Diego: Academic Press; 2009. pp. 516-523

[174] Wang B, Xia F, Wappler T, Simon E, Zhang H, Jarzembowski EA, Szwedo J. Brood care in a 100-million-year-old scale insect. eLife. 2015;**4**:e05447. http://www.ncbi.nlm.nih.gov/pmc/articles/PMC4378507/ [Accessed Feb 2015]. DOI: 10.7554/eLife.05447

[175] Ameixa OMCC Aphids in a changing world. In: Kindlmann P, Dixon AFG, Michaud JP, editors. Aphid biodiversity under environmental change. Dordrecht: Springer; 2010. pp. 21-40

[176] Frost SW. Insect Life and Insect Natural History. New York: Dover Publications, Inc.; 1959

[177] Fletcher N. The didjeridu (didgeridoo). Acoustics Australia. 1996;**24**(1):11-15

[178] Clausen LW. Insect fact and folklore. New York, NY: MacMillan Co.; 1954

[179] Fain GL. Ancient Greek epigrams: Major poets in verse translation. University of California Press; 2010

[180] Xing-Bao J, Kai-Ling X. An index-catalogue of Chinese Tettigoniodea (Orthopteroidea: Grylloptera). Journal of Orthopaedic Research. 1994:15-41

[181] Peigler RS. Non-sericultural uses of moth cocoons in diverse cultures. Paper presented at the Proc. Denver Museum of Natural History; 1994

[182] Hogue CL. Cultural entomology. Annual Review of Entomology. 1987;**32**(1):181-199. DOI: 10.1146/annurev.en.32.010187.001145

[183] Cambefort Y. Beetles as religious symbols. Cultural Entomology Digest. Issue 1; 1993. http://www.insects.org/ced1/beetles_rel_sym.html (Accessed May 2015)

[184] Rakotoniaina NS. Vers une démarche de gestion durable des ressources de la forêt sclérophylle de moyenne altitude d'Arivonimamo II- Madagascar. (Thèse de doctorat). Madagascar: Université d'Antananarivo; 2009

[185] Schönig H. Traditional cosmetics of women in Yemen. The black dye ẖiḍāb: Traditional and modern ways of fabrication. Paper presented at the Proc. Seminar for Arabian Studies; 1996

[186] Ameixa OMCC, Soares AOM, Soares AMVM, Lillebø AI. Worldwide list of entomological societies. 2017. Available at: www.cesam.ua.pt/index.php?tabela=pessoaldetail&menu=95&user=981

[187] Nishida GM. Museums and display collections. In: Resh VH, Cardé RT, editors. Encyclopedia of Insects. 2nd ed. San Diego: Academic Press; 2009. pp. 680-684

[188] Evenhuis NL. Abbreviations for insect and spider collections of the world. 2015. Available at: http://hbs.bishopmuseum.org/codens/codens-inst.html [Accessed: November 2015]

[189] Arnett RA Jr, Samuelson GA, Nishida GM. The insect and spider collections of the world. 2nd ed. Gainesville FL: Sandhill Crane Press; 1993

[190] Small RD. Becoming unsustainable? Recent trends in the formal sector of insect trading in Papua New Guinea. Oryx. 2007;**41**(03):386-389

[191] Leather SR. Taxonomic chauvinism threatens the future of entomology. Biologist. 2009;**56**(1):10-13

5

Commercial Harvesting of Marula (*Sclerocarya Birrea*) in Swaziland: A Quest for Sustainability

Alfred Francis Murye and André J. Pelser

Abstract

All across sub-Saharan Africa rural communities, and especially women, are harvesting natural products for income generation purposes. In recent times, the degradation of ecosystems in terms of loss of biodiversity has become a major concern, especially when the harvesting of the species in question has become commercialized as is the case of marula harvesting in Swaziland. This chapter reports on some of the findings of a study conducted to explore the impact of commercial harvesting on the future availability of marula. If current levels of harvesting are unchecked, overconsumption or even depletion of marula in Swaziland will deepen the existing poverty levels among the rural poor, particularly women, who are proportionately more dependent on marula for their livelihoods.

Keywords: marula, poverty, sustainability, commercial harvesting

1. Introduction

Many countries in sub-Saharan Africa encourage rural communities, especially women, to harvest natural products for income generation purposes [29, 43, 71]. This means that biodiversity has become the basis for ecosystem services that people primarily depend on for commercial harvesting to sustain their livelihoods and well-being. Hence, the increasing demands on these resources, the decrease in their availability, and the probable consequence of environmental degradation emphasize the need for a coordinated approach to their sustainable utilization. [62, 69] initiatives on biodiversity and natural resources conservation in the form of terrestrial and marine parks and protected areas, as well as commercial ventures into the harvesting of such resources, thus, often come into conflict with the livelihood strategies of local populations [5, 44]. Therefore, Hugo [24] rightly emphasizes that the sustainable management and use of such resources require an interdisciplinary approach and sound knowledge of each resource, while the ecological and socio-economic factors related to their use must be sorted out as well. Furthermore, since the long-term success of any initiative hinging on the utilization of natural resources depends on local benefit and support, it is imperative that local livelihoods and poverty reduction are considered during the formation and management of such initiatives [3, 31, 45, 74].

Currently, there are growing concerns about the effects of people on ecosystems and the reciprocal impact of ecosystem degradation on the quality of human life [30, 38, 72]. The degradation of ecosystems in terms of loss of biodiversity has particularly become a major concern, especially when the harvesting of the species in question and/or its products has become commercialized as is the case of marula harvesting in Swaziland. This chapter focuses on the recently introduced practice of commercial harvesting of marula products in Swaziland, and reports on some of the findings of a study conducted to explore the impact of commercial harvesting on the future availability of marula. The study aimed at identifying challenges faced by the harvesting of marula products and to explore opportunities for the sustainability of the marula species and harvesting of its products.

2. The ecology of marula trees in Swaziland

Marula is a wild indigenous fruit tree (see **Figure 1**) which is a member of the floral family *Anacardiaceae* [41]. *Sclerocarya birrea caffra* (the species found in Swaziland), is one of three species of *Sclerocarya*—the others being *S. gillettii* and *S. multifoliolata* [18, 60]. Marula is endemic to Africa and occurs in a diversity of vegetation types; mainly open woodland, deciduous savannah and semi-deciduous forest [14, 18, 22]. It grows well in altitudes ranging from sea level to 1800 m above sea level in areas that are frost-free, as the species is generally sensitive to frosty conditions [25, 47]. Marula trees also prefer areas where temperatures vary between 10°C

Figure 1. A marula tree (picture taken by first author, December 2013).

(in high altitudes) and 40°C (in low altitudes) [9]. Such wide-ranging temperatures are beneficial for marula seed germination, which happens usually at temperatures between 27 and 37°C [19]. The marula tree species is relatively drought-tolerant and grows well in areas that are humid to subhumid with wide-ranging amounts (200–1500 mm) of annual rainfall [18, 19].

In Swaziland, marula trees grow abundantly in the lowveld and in the lower parts of the middleveld. The UNDP [70] estimated that about 2 million marula trees are growing in Swaziland and that each tree can produce up to 500 kg of fruits per year. Marula is a dioecious tree that grows to about 15 m in height with spreading crowns and dense foliage [41, 60]. When ripe, the marula fruits (see **Figure 2**) and kernels are edible and are rich in minerals, proteins, oils and vitamins [17]. The kernels are normally eaten fresh, mixed and cooked with other foods or stored for later consumption. This, according to Shackleton et al. [60] and also Wynberg et al. [75], makes marula an integral part of the diet, tradition and culture of rural communities in many southern African communities. Apart from playing an important role in cultural practices in Swaziland, Marula is also sought after for many commercial initiatives across the African continent and beyond.

Marula plays a fundamental role in the ecosystems and provides a number of important services to human beings and other living organisms. Being a keystone species, marula plays a significant role in the ecology of other plants and animals [22]. The marula tree is large and, usually, it is considered a community-dominant species [27, 47]. It also provides valuable shade that produces large areas with cool subcanopy environments that are good habitats for other plant and animal species. Marula is also a favorite food of numerous birds, mammals and insects [15, 22, 25, 42]. Therefore, removal of this large dominant species can threaten biodiversity and result in loss of many important subcanopy species such as mistletoes which grow on marula branches and wood roses that are often used by rural curio traders [10].

The leaves of the marula trees are also browsed by domestic and game animals, such as cattle, goats, impala, zebras and kudu. These trees are also host to various insect species such as butterflies and moths. The larvae of the edible kinds of these insects such as the mopane

Figure 2. The marula fruit (picture taken by first author, December 2013).

worm (*Imbrasia belina*) and the African moon moth (*Argema mimosa)* are used as food by many tribes in southern Africa [16, 21, 42, 55, 76]. The Swazis and Zulus also traditionally use the tough silk pupa cases of the moon moth for anklet rattles used in tribal dances such as the annual reed dance and *Buganu* ceremonies practiced in Swaziland. The barks of the marula trees are used by elephants and rhinoceros. The marula fruits, which are rich in vitamin C, are also eaten by animals, such as elephants, warthogs, porcupine, monkeys, baboons, millipedes, domestic cattle and goats and, thus, creating an intricate ecological linkage between the marula trees and other animal species [21, 27, 55].

The water-filled holes in the trunks of marula trees are very important breeding grounds for mosquitoes and other animals such as amphibians. These interrelationships such as the over-browsing by animals can destabilize the population structure of marula trees as it affects the growth of marula seedlings and, thus, impacts on its successful regeneration and recruitment potential [19, 27, 55]. Marula seedlings are also susceptible to fire which also impact on its regeneration potential [23, 25].

3. Methodology

The study was conducted in the Lubombo region of the Kingdom of Swaziland—a small country in southern Africa which covers an area of 17,363 km^2 [77]. Both quantitative and qualitative research approaches to data collection and data analysis were employed. Two populations were targeted for the study, that is, the people of the Lubombo region and key informants from relevant stakeholder organizations. The quantitative research design was explanatory and descriptive in nature and was grounded in a cross-sectional socio-economic survey. The researchers selected respondents from all four chiefdoms of the Mpolonjeni constituency. A multistage probability sampling procedure was used to systematically select 411 households (based on a 95% confidence level and 5% confidence interval) from the population of 14,716 and then randomly select one adult individual from each of the selected households (or and oldest child in the case of child-headed households) [2, 53]. A structured filter questionnaire was used to collect data among the 411 household members. The questionnaire was pilot-tested at Ngculwini *Inkhundla,* which was not included in the study area. Both descriptive and inferential statistics were used to analyze the quantitative data. [39] the qualitative approach, on the other hand, entailed a purposive sampling method to select 20 key informants, while an in-depth interview schedule was used for data collection purposes. Data from the in-depth interviews were recorded verbatim after which the contents were transcribed and triangulated with the questionnaire findings [68].

In order to contextualize the findings of the study, we first need to reflect on the important role of marula in Swaziland, particularly from a social and economic perspective.

4. The social and economic importance of marula

As a low-middle-income country with a per-capita Gross Domestic Product of only US$ 5940 per annum (2014 estimates), the poverty rate in Swaziland is estimated at 69% and a large proportion of people live on an income below US$2 per day [40, 48, 57, 73].From time immemorial, people in Swaziland have harvested marula fruits and seeds for livelihoods in the form of food and for producing a traditional brew (*Buganu*). The brew is largely consumed during traditional festivals or sold to local customers to generate an income [34, 36, 49]. The people of Swaziland have used the marula kernels in relishes, eating the fresh fruits and seeds, using the bark for relieving stomach aches and also for spiritual reasons [64]. Marula products also feature in traditional ceremonies, such as the *Buganu* ceremony, celebrated annually where people gather with their King (The *Ingwenyama*) and Queen Mother (The *Indlovukazi*) to celebrate [34, 36, 49]. This ceremony normally marks the start of the *Buganu* season whereby, after the ceremony, every Swazi is free to drink the *buganu* brew [26, 63]. Recognizing its social, cultural and economic values, the Queen Mother in 2004 established two commercial processing plants (Swazi Secrets and Swaziland Marula) as initiatives to economically empower rural women through enterprising on marula harvesting and processing. In Swaziland, marula is traditionally harvested almost exclusively by women who pick up the fruits from the ground when ripe. The UNDP [70] has observed that the Swazi Secrets project alone is supporting 2400 rural women who earn a living through selling seed kernels from marula trees.

The establishment of commercial processing plants for marula is an initiative to augment the battle for economic growth and poverty alleviation in Swaziland. The challenges for economic growth in Swaziland have been compounded by the global economic meltdown that started in 2008, the reduction in Southern African Customs Union (SACU) revenue as of 2011, and the environmental challenge of global climate change which is impacting developing countries the most [4, 12, 58]. The consequences of climate change in turn present one of the most fundamental threats to biodiversity and the functioning of ecosystems [54]. Swaziland is therefore faced with two dilemmas: the grapple with the challenge of marrying economic growth and poverty alleviation; and at the same time, it is expected to exert efforts for environmental conservation and resource sustainability, among others of the marula species.

The past few decades have seen marula and its products gaining popularity in the international market due to demand for its many by-products, such as marula oil and the well-known Amarula cream liqueur [7, 8, 60, 64, 75]. The increased demand for marula products, coupled with the high incidence of poverty, especially among rural women in Swaziland, led to the establishment of the two marula processing plants in 2004. These initiatives have attracted a large number of Swazi women to scout the forests and fields around their homesteads in search of marula fruits and seeds to sell and earn an income. This commercially induced harvesting of marula fruits and seeds poses a potential threat to the marula tree species as more and more of the seeds that would support regeneration and recruitment of new marula trees are removed through increased harvesting.

According to Peters [56], when the bark, fruits, seeds, wood and other parts of a species are harvested for processing into various products at household and/or commercial levels, there may be significant impacts on the population structure and distribution of the species, depending on the nature and intensity of the harvest. This imposes a potential threat to the species in the sense that it might be driven to depletion and/or extinction if proper sustainability measures are not put in place. Therefore, it is clear that the potential depletion or extinction of marula in Swaziland could compromise the livelihood activities and well-being of large numbers of the rural poor in terms of income generation, as they derive a substantial part of their livelihoods from harvesting marula products.

5. Findings

The researchers wanted to first ascertain the sources of livelihoods for the people in the study area. The majority (73.2%; n = 255) of the respondents depended on subsistence crop production, while 36.6% (n = 127) relied, to a very low level, on rearing/breeding animals such as cattle, goats and chickens. Others sourced income from selling marula products (50.4%; n = 173), which emphasizes the importance of marula to the Swazi nation. Others relied on harvesting and selling other wild fruits and seeds except marula (54%; n = 54). Some (12.4%; n = 37) relied on other sources, such as collecting Mopani worms, wild fruits (wild berries and guava) and wild vegetables [like Amaranthus (*umbidvo*), Black Jack (*Imbuya*) and okra (*ligusha)*], and other respondents are housewives who depend mainly on their husbands' income.

6. Use of marula and its products by households in the study area

The majority of the respondents (76.4%; n = 216) used marula for brewing *buganu*. Others used it as a source of food (67.5%; n = 191) and others sold the kernels (51.9%; n = 147) for income. The findings concur with observations by Marula Natural Products [35] and Mabaya et al. [33] who highlight different uses of marula. The respondents also indicated that marula is used for medicinal purposes (12.5%; n = 35)—a finding that confirms those of several other authors [11, 16, 32, 37]. The findings also show that marula is used in carving artifacts, as animal fodder, fencing homesteads, for cultural practices, for spiritual purposes including chasing away Gremlins (*tokolosi)*, divination counters and for banishing evil spirits. These findings agree quite well with those of Nwongwu [51] and O'Brien [52] who reported similar uses of marula. The respondents also indicated that the marula kernel is used for pressing out the oil for cooking and for home-made body care lotion, which is in concert with other authors, such as Abdalbasit and Ibrahim [1], Glew et al. [13] and Wynberg et al. [75].

7. Changes in consumption patterns of marula

Respondents were asked to indicate the number of years they have been involved in marula harvesting (**Figure 1**). The data showed that, out of the 219 respondents who responded to the question, 38% (n = 83) have been harvesting marula for less than 10 years, while 31% (n = 68) have harvested marula for 10–15 years and 31% (n = 68) have been doing so for 16 years and above. The data suggest that the 38% who have been involved in marula trade for less than 10 years joined the marula enterprise after the introduction of the marula processing plants in 2004. These marula processing plants could have attracted more people to join the marula trade. This is a very significant number of harvesters added and, has increased the burden on marula resources. The increase in numbers of harvesters in this study agrees fairly well with the increase in the number of basket makers in the case of the introduction of basketry in Botswana in the 1970s [46] which contributed to the collapse of the basket industry due to resource depletion.

Also in Zimbabwe, Terry and Cunningham [66] reported that a weaving club that was started in 1986 with 20 members expanded to 500 people by 1988 due to the expansion in market demand and eventually contributed to the collapse of the weaving club. This means that the introduction of commercial market economy to natural products, such as marula, tends to attract more entrepreneurs to the enterprise which, in turn, leads to overharvesting and, subsequently, to depletion of the resource base. Our data indeed revealed that 53.8% (n = 171) of the respondents were of the opinion that marula is becoming less available than before, and attributed the decline to the increase in the number of harvesters. Once a communal resource is exposed to an improved market, it tends to attract many harvesters, which leads to overexploitation and depletion of the resource [20, 21]. These findings concur with those reported by Shackleton and Shackleton [61] where more than one-third of their respondents in the Bushbuckridge area in South Africa opined that there had been a decrease in marula following the introduction of improved market for marula products.

The Swaziland Indigenous Products (SIP) (2012) observed that the factors currently threatening the continued productivity and biodiversity in Swaziland are largely a result of socio-economic changes without appropriate adaption, exacerbated by repeated droughts and, possibly, climate change, which are interlinked and, in some cases, form negative synergies. Other factors threatening biodiversity in Swaziland (and thus the sustainability of marula harvesting) include population growth, which has resulted into changes in land use and more intense use of natural resources, overgrazing, deforestation, erosion, and an increase in invasive alien plants [65].

Infestation of marula trees and fruits by pests in the study area has also been reported. This was confirmed by the Swaziland Minister of Tourism and Environmental Affairs when he pointed out that marula is commonly being affected by worms [50]. According to the Minister, the worms make the marula trees dry out in a very strange way, leaving them with no leaves and yielding no fruits, which disturbs the production of marula fruits. The impact of the pests on the marula tree not only affects the popular *Buganu* ceremony but also the production of various products by the two local commercial companies—the Swazi Secrets and the Swaziland Marula—as well as the livelihoods of the poor rural communities. Some respondents indicated worries on the pest situation, alluding to the fact that some of the trees are no longer yielding fruits due to the impact of pests. This worry was also echoed by the spokesperson for the Swazi Secrets marula company. Besides, storms frequently occur annually in Swaziland and when they do, marula trees fall down and, thus, reduce fruit yield which in turn impact negatively on the commercial marula enterprises and rural livelihoods in the country.

With regard to expected future availability of marula, if levels of harvesting and selling were to continue at the current rate, 44.7% (n = 140) of the respondents opined that marula will only be available in the next 5 years, while 40.0% (n = 125) thought it might still be available for the next 10 years. These responses suggest that the respondents are aware that, with increased harvesting, marula will get exhausted in the near future. This seems to concur fairly well with Helm and Witkowski [21] who reported that marula is declining at an unprecedented rate in the Kruger National Park and other areas in South Africa. Aggravating the prospects for sustainable marula harvesting is the absence of dedicated leadership to start implementing strategies for sustaining marula, as there seems to be no government body or non-governmental organization helping in conservation activities in the area. In addition, it was observed that there is no overarching policy and no regulations were in place to govern access to and harvesting of marula products. If depletion of marula were to occur, it will adversely affect poor rural households that depend on it for income, food and other services. Observations around the households during the socio-economic interviews showed glaringly that the poorest households in the study area were food insecure and it was obvious that marula played a huge role as a source of food in those circumstances. Kirkland et al. [28] as well as Tibesigwa et al. [67] made a similar observation in their respective studies in South Africa.

8. Discussion

The study revealed that a large proportion of rural people in the study area relied mainly on marula harvesting for their livelihoods, in addition to subsistence crop production and subsistence animal rearing. The current production of value-added products from marula fruits and seeds in Swaziland has given impetus to rural people, especially women, to harvest marula products on commercial scale. The harvesting process itself, driven by market demand, is thus threatening the sustainability of marula harvesting and, inevitably, also the survival of the species. Every year during the marula season, the majority of rural people in Swaziland take up the opportunity of harvesting marula fruits and seeds for making an extra income to cater for multiple needs that the households face, ranging from payment of school fees to buying of electricity, medicines and food. As observed by Campbell et al. [6], as well as Shackleton and Shackleton [59], trade in marula products is one of the important local income-generating options available to the rural poor, and particularly women in southern Africa. The interface between poverty, rural livelihoods and marula harvesting in Swaziland reflects the dependency of impoverished rural populations on natural capital and, thus, demonstrates the intricate linkage between people and the ecosystems in which they live [69].

Apart from the commercialization of marula products in Swaziland, the study found that several socio-economic factors interlock to contribute to the intense harvesting of marula in Swaziland. Widespread conditions of poverty fuelled by low levels of education are well established in several places across southern Africa where a strong dependency on and wide range of uses of marula have led to the increased exploitation of the resource. All agricultural and grazing fields in the study area have been exploited for marula harvesting causing the level of regeneration of marula in those areas to be very low due to constant removal of the fruits and seeds. The multiple uses of marula products, combined with the introduction of the processing companies, have encouraged many people to harvest large amounts of marula in large quantities for income generation. This commercialized pattern of harvesting is considered a very important factor in accelerating the potential depletion of marula resources in Swaziland and is threatening the quest for its sustainability if appropriate measures are not put in place.

9. Conclusion

Marula plays a crucial role in the livelihoods of poor rural households in Swaziland. The current level of commercialized harvesting of marula products in Swaziland is unsustainable as it has negatively impacted on the regeneration potential of marula and, thus, requires deliberate monitoring and strict control of the harvesting process. Many socio-economic factors are interlocking to accelerate the harvesting of marula products beyond its sustainability thresholds to support the livelihoods of many poverty-stricken rural households in Swaziland. Therefore, overconsumption or even depletion of marula in Swaziland will deepen the existing poverty levels among the rural poor, particularly women, who are proportionately more dependent on marula, and therefore most vulnerable to the loss of a natural resource that will impact their livelihoods negatively.

10. Recommendations

To sustain the harvesting of marula in Swaziland, a combination of environmental and socio-economic policy interventions are imperative. The community members, for instance, should plant marula in the fields around the homesteads to replace the current old marula trees when they die out or stop producing fruits so as to ensure the reliable supply of marula products, and at the same time reduce the harvesting pressure in the agricultural fields and grazing areas. Therefore, the communities should dedicate the planting of marula in an agroforestry system to alleviate the pressure on marula and enhance its population size and structure, thus increasing the sustainable utilization of the marula species and the economic sustainability of the commercial marula enterprises in the country. The communities should furthermore restrict domesticated animals from accessing the agricultural fields to reduce the browsing of growing marula seedlings in those fields.

At a policy level, the local authorities should stop people from cutting down marula trees, engage the communities in education on the importance of marula conservation and the practice of not collecting all the marula fruits and seeds from the ground for purposes of regenerating new seedlings. Rural development policies in particular should be revisited and realigned towards encouraging job creation in the rural areas of Swaziland. The Ministry of Education should widen the base and access to quality educational opportunities in the rural areas through provision of adult education programmes which will increase the chances of rural dwellers, especially women, of finding employment opportunities. This would, in turn, empower rural communities and minimize overdependence on natural forest products, thus relieving the harvesting pressure on marula.

Author details

Alfred Francis Murye[1] and André J. Pelser[2*]

*Address all correspondence to: pelseraj@ufs.ac.za

1 Centre for Environmental Management, University of the Free State, Bloemfontein, South Africa

2 Department of Sociology, University of the Free State, Bloemfontein, South Africa

References

[1] Abdalbasit MA, Ibrahim AS. *Sclerocarya birrea* (Marula), an African tree of nutritional and medicinal uses: A review. Food Reviews International. 2012;**28**(4):375-388

[2] Al-Saleh MF, Al-Omari AI. Multistage ranked set sampling. Journal of Statistical Planning and Inference. 2002;**102**(2):273-286

[3] Baral N, Stern MJ, Heinen JT. Integrated conservation and development project life cycles in the Annapurna conservation area, Nepal: Is development overpowering conservation? Biodiversity and Conservation. 2007;**16**(10):2903-2917

[4] Basdevant O, Chikako B, Mircheva B. Macroeconomic vulnerabilities stemming from the global economic crisis: The case of Swaziland. Washington DC: International Monetary Fund. Africa Department (Series) 11/7; 2011

[5] Borrini-Feyerabend G, Kothari A, Oviedo G. Indigenous and Local Communities and Protected Areas: Towards Equity and Enhanced Conservation. Cambridge: IUCN; 2004

[6] Campbell BM, Jeffrey S, Kozanayi W, Luckert M, Mutamba M, Zindi C. Household Livelihoods in Semi-arid Regions: Options and Constraints. Jakarta: CIFOR; 2002

[7] Cant M, Machado R. Amarula Cream – The Spirit of Africa. Marketing Success Stories: South African Case Studies. 7th ed. Cape Town: Oxford University Press; 2010

[8] Castro AP, Nielsen E. Natural Resource Conflict Management Case Studies: An Analysis of Power Participation and Protected Areas. Rome: Food and Agriculture Organization of the United Nations; 2003

[9] Coetzee BJ, Englbrecht AH, Joubert SC, Retief PF. Elephant impact on *Sclerocarya caffra* trees in *Acacia nigrescens* tropical plains thornveld of Kruger National Park. Koedoe. 1979;**22**:39-60

[10] Dzerefos C. Distribution, Establishment, Growth and Utilization of Mistletoes (Loranthaceae) in the Mphumalanga Lowveld. Johannesburg: University of Witwatersrand; 1996

[11] Elijah R, Abdulkadir L, Adisa B. Investigation of extracted *Sclerocarya birrea* seed oil as a bioenergy resource for compression ignition engines. International Journal of Agriculture and Biological Engineering. 2012;**5**(3):59-69

[12] Funder M, Fjalland J, Ravnborg HM, Egelyng H. Low carbon development and poverty alleviation options for development cooperation in energy. In: Agriculture and Forestry. Copenhagen: DIIS; 2009

[13] Glew GS, Vanderjagt DJ, Huang YS, Chuang LT, Bosse R, Glew RH. Nutritional analysis of the edible pit of *Sclerocarya birrea* in the republic of Niger (daniya, Hausa). Jounal of Food Composition and Analysis. 2004;**17**(1):99-111

[14] Gouwakinnou GN, Kindomihou V, Assogbadjo AE, Sinsin B. Population structure and abundance of *Sclerocarya birrea* (a. Rich) Hochst subsp. *birrea* in two contrasting land-use systems in Benin. International Journal of Biodiversity and Conservation. 2009;**1**(6):194-201

[15] Hal HP. A review of the proximate composition and nutritional value of Marula (*Sclerocarya birrea* subsp. *caffra*). Phytochemistry Reviews. 2014;**13**:881-892

[16] Hal HV. Processing of Marula (Sclerocarya Birrea Subsp. Caffra) Fruits: A Case Study on Health-Promoting Compounds in Marula Pulp. Wageningen: Wageningen University; 2013

[17] Hall JB, O'Brien EM, Sinclair FL. Sclerocarya *b*irrea: A Monograph. Bangor: University of Wales; 2000

[18] Hall JB, O'Brien EM, Sinclaire FL. Sclerocarya *b*irrea: A Monograph. Bangor: University of Wales; 2002

[19] Hamidou A, Iro DG, Boube M, Malik TS, Ali M. Potential germination and initial growth of *Sclerocarya birrea* (a Rich.) Hoscht, in Niger. Journal of Applied Biosciences. 2014;**76**:6433-6443

[20] Harper CL. Environment and Society: Human Perspectives on Environmental Issues. 5th ed. Boston: Prentice Hall; 2012

[21] Helm CV, Witkowski ET. Continuing decline of a keystone tree species in the Kruger National Park, South Africa. African Journal of Ecology. 2013;**51**(2):270-279

[22] Helm CV, Witkowski TE. Continuing decline of a keystone tree species in the Kruger National Park. South African Journal of Ecology. 2012;**51**:270-279

[23] Helm CV, Wilson G, Midgley J, Kruger L, Witkowski ET. Investigating the vulnerability of an African savanna tree (*Sclerocarya birrea* ssp.*caffra*) to fire and herbivory. Austral Ecology. 2011;**36**:964-973

[24] Hugo ML. Environmental Management: An Ecological Guide to Sustainable Living in Southern Africa. Pretoria: Ecoplan; 2010

[25] Jacobs OS, Biggs R. The impact of the African elephant on Marula trees in the Kruger National Park. South African Journal of Wildlife Research. 2002;**2**(1):13-22

[26] Jele L. Marula trees place SD on world map – King. Times of Swaziland, March 04, 2013. p. 7

[27] Kgomoamagodi FP. Determination of fruit yield and fruit quality in marula *(Sclerocarya birrea subsp. caffra)* selections. M.Sc. thesis, University of Pretoria, Pretoria; 2008

[28] Kirkland TM, Kemp RJ, Hunter LM, Twine WM. Towards improved understanding of food security: A methodological examination based on rural South Africa. Food, Culture and Society – An International Journal of Multidisciplinary Research. 2013;**16**(1):65-84

[29] Ladefoged T, Hansen RB, Worsoe TA, Fredslund HM. Jatropha and sustainable livelihood of small-scale farmers. M.Sc. thesis, University of Roskilde, Teksam; 2009

[30] Levin P, Fogarty M, Murawski S, Fluharty D. Integrated ecosystem assessments: Developing the scientific basis for ecosystem-based management of ocean. PLoS Biology. 2009;**7**(1):23-28

[31] Lockwood M, Kothari A. Social context. In: Lockwood A, Worboys GL, Kothari A, editors. Managing Protected Areas. London: Earthscan; 2006. pp. 41-72

[32] Lombard C, Allanic B, Shilote B. Potential for the Development of Marula Products in the Bushbuckridge. Nelspruit: DANCED; 2000

[33] Mabaya E, Jackson J, Ruethling G, Carter CM, Castle J. Wild fruits of Africa: Commercializing natural products to improve rural livelihoods in Southern Africa. International Food Agribusiness Management Review. 2014;**17**(B):69-74

[34] Magagula M. Buganu makes them happy. Times of Swaziland, March 04, 2012

[35] Marula Natural Products [Online]. 2012 Available at: www.marula.org.za/prodfruit.htmlDBS/Marula.info.htm [Accessed November 06, 2014]

[36] Mathunjwa N. Over 5000 women attend the Buganu ceremony. Times of Swaziland; 2010. p. 1

[37] Mawoza T, Ojewole JA, Chiwororo WD, Owira PM. *Sclerocarya birrea* (A. Rich.) Hochst [Marula] (Anacardiaceae): A review of its phytochemistry, pharmacology and toxicology and its ethnomedicinal uses. Phytotherapy Research. 2010;**24**(5):633-639

[38] MEA. Ecosystems and Human Well-Being: A Framework for Assessment. Washington DC: World Resources Institute; 2005

[39] Microsoft, 2010. Microsoft SPSS Version 10, Boston: Microsoft Office

[40] Ministry of Economic Planning and Development. Poverty Reduction Strategy and Action Plan, Policy Brief. Mbabane: Kingdom of Swaziland; 2005

[41] Mojeremane W, Tshwenyane SO. The resource role of Marula (*Sclerocarya birrea*): A multipurpose indigenous fruit tree of Botswana. Journal of Biological Sciences. 2004;**4**(6):771-775

[42] Morris S, Humphreys D, Reynolds D. Myth, Marula, and elephant: An assessment of voluntary ethanol intoxication of the African elephant (*Loxodonta africana*) following feeding on the fruit of the Marula tree (*Sclerocarya birrea*). Physiological and Biochemical Zoology. 2006;**79**(2):363-369

[43] Myerson RB. 2015. home.uchicago.edu. [Online] Available at: http://home.uchicago.edu/rmyerson/research/decent.pdf [Accessed January 17, 2016]

[44] Naughton-Treves L, Holland MB, Brandon K. The role of protected areas in conserving biodiversity and sustaining local livelihoods. Annual Reviews of Environment and Resources. 2005;**30**:219-252

[45] Nepal RC. Integrated Conservation and Development Projects – A Case Study of Annapura Conservation Area Project. Akershus: Norwegian University of Life Sciences; 2005

[46] Neumann RP, Hirsch E. Commercialisation of Non-Timber Forest Products: Review and Analysis of Research. Bogor: Centre for International Forestry Research; 2000

[47] Ngorima GT. Towards Sustainable Use of Marula (*Sclerocarya birrea*) in the Savannah Woodlands of Zvishavane District. Johannesburg: University of the Witwatersrand; 2006

[48] Nindi AG, Odhiambo NM. Poverty and economic growth in Swaziland: An empirical investigation. Managing Global Transitions. 2015;**13**(1):59-74

[49] Nkambule S. Buganu Ceremony a Must Attend Event. Mbabane: Swazi Observer; 2015

[50] Ntiwane Q. Concerns Over Maganu as Trees Dry Up. Mbabane: Swazi Observer; 2015

[51] Nwongwu FO. The socio-cultural and economic relevance of the Marula tree and its sustainable use in South Africa. Africa Insight. 2006;**36**(3-4):249-269

[52] O'Brien E. Social and cultural values of trees and woodlands in Northwest and South East England. Forestry Snow Landscape Research. 2005;**79**(1-2):169-184

[53] Onwuegbuzie AJ, Leech NL. Sampling designs in qualitative research: Making the sampling process more public. The Qualitative Report. 2007;**12**(2):238-254

[54] Osman-Elasha B, Parrotta J, Adger N, Brockhaus M, Pierce CJ, Sohngen B, Dafalla T, Joyce LA, Nkem J, Robledo C. Future socio-economic impact and vulnerabilities. In: Seppälä R, Buck A, Katila P, editors. Adaptation of Forests and People to Climate Change. Vienna: International Union of Forest Research Organizations (IUFRO); 2009. pp. 101-122

[55] Pegg N. Antelope ingestion enhances germination of the Marula (*Sclerocarya birrea*), an important African savannah tree. African Journal of Ecology. 2014;**52**(4):499-505

[56] Peters CM. Observations on the sustainable exploitation of non-timber tropical forest products: An ecological perspective. In: Ruiz P, Arnold JM, editors. Current Issues in Non-Timbet Forest Products Research. Bogor: CIFOR; 1995. pp. 1-18

[57] Population Reference Bureau. World Population Data Sheet with Special Focus on Women's Employment. New York: USA Census Bureau, USAID; 2015

[58] SARUA. Climate change counts – strengthening university contributions to climate compatible development in Southern Africa. Swaziland Country Report, Johannesburg: Southern African Regional Universities Association (SARUA); 2014

[59] Shackleton CM, Shackleton S. The importance of non-timber forest products in rural livelihood security and as safety nets: A review of evidence from South Africa. South African Journal of Science. 2004;**100**(11-12):658-664

[60] Shackleton CM, Botha J, Emanuel PL, Ndlovu S. Inventory of Marula (Marula Subsp. *Caffra)* Stocks and Fruit yields in Communal and Protected Areas of the Bushbuckridge Lowvelds. Grahamstown: Rhodes University; 2002

[61] Shackleton S, Shackleton C. Use of Marula Products for Domestic and Commercial Purposes by Households in the Bushbuckridge District. Limpopo Province: Rhodes University; 2002

[62] Silvis HJ, van der Heide CM. Economic Viewpoints on Ecosystem Services. Wageningen: The Statutory Research Tasks Unit for Nature and the Environment; 2013

[63] Swazi Observer. 2015. observer.org.sz. [Online] Available at: www.observer.org.sz/ [Accessed July 15, 2015]

[64] Swazi Secrets. The Quarterly Newssheet from Swazi Indigenous Products. Mbabane: Swazi Sectrets; 2009

[65] Swaziland Indigenous Products. 2012. Justfare.org. [Online] Available at: www.justfare.org [Accessed July 10, 2015]

[66] Terry ME, Cunningham AB. The impact of commercial marketing on the basketry of Southern Africa. Journal of Museum Ethnography. 1993;**4**:25-48

[67] Tibesigwa B, Visser M, Twine W, Collinson M. Investigating the Sensitivis of Household Food Security to Agriculture – Related Shocks and the Implications of Informal Social Capital and Natural Resource Capital. Cape Town: SIDA, Environment for Development, Resources for the Future; 2014

[68] Tong A, Sainsbury P, Craig J. Consolidated criteria for reporting qualitative research (COREQ): A 32-item checklist for interviews and focus groups. International Journal for Quality in Health Care. 2007;**19**(6):349-357

[69] UNDP & UNEP. Mainstreaming Poverty-Environment Linkages into Development Planning: A Handbook of Practitioners. Nairobi: UNDP & UNEP; 2009

[70] UNDP. 2012. Rural Swazi woman turns childhood hobby into source of income. [Online] Available at: http://www.sz.undp.org/content/swaziland/en/home/ourwork/povertyreduction/successstories/rural-swazi-woman-turns-childhood-hobby-into-source-of-income.html# [Accessed June 05, 2015]

[71] UNEP & IISD. Human well-being. In: Poverty and Ecosystem Services – Exploring the Links. Nairobi: UNEP; 2004

[72] United Nations. 2002 Report of the World Summit on Sustainable Development, Johannesburg. New York: United Nations

[73] World Bank Group. 2011. worldbank.org. [Online] Available at: www.worldbank.org [Accessed August 07, 2012]

[74] World Resources Institute. Restoring Nature's Capital: An Action Agenda to Sustainable Ecosystem Services. Washington, DC: World Resources Institute; 2007

[75] Wynberg R, Laird S, Botha J, den Adel S, McHardy T. The Management, Use and Commercialization of Marula: Policy Issues. Wallingford: Centre for Ecology and Hydrology; 2002

[76] Xaba P, Moll E. Marula: Gardening with traditionally useful indigenous plants. SABINET. 2011;**97**(2):76-77

[77] Zwane PE, Masarirambi MT, Magagula TN, Dlamini AM, Bhebhe E. Exploitation of *Agave americana* L. plant for food security in Swaziland. American Journal of Food and Nutrition. 2011;**1**(2):82-88

6

Impacts of Climate Change and Climate Variability on Wildlife Resources in Southern Africa: Experience from Selected Protected Areas in Zimbabwe

Olga L. Kupika, Edson Gandiwa, Shakkie Kativu and Godwell Nhamo

Abstract

Climate change and variability pose a threat to wildlife resources in semi-arid savannahs. With examples from selected protected areas in Southern Africa, this chapter highlights studies on detected climate changes particularly rainfall and temperature, outlines the predicted and observed impacts of climate change and variability on wildlife resources in savannah ecosystems and highlights the adaptation and mitigation strategies and implications for conservation. Literature indicates that Southern Africa is characterised by highly variable, erratic and unpredictable rainfall and increasing temperature coupled with an increasing trend in climate-related extreme events such as frequent droughts, cyclones and heat waves. Drought, in particular, has led to death in several wildlife species. This has implications on long-term survival of the species. Changes in rainfall and temperature patterns influence habitat quality and consequently abundance of distribution of wildlife species. Large herbivores such as elephants and hippopotamus in particular are vulnerable to climate change due to their ecology, whereas other species are less vulnerable. Climate-related extreme events, coupled with other anthropogenic stressors, interact to influence changes in abundance and distribution of wildlife resources. Understanding the influence of these climatic factors on wildlife resources is vital for adaptive management and protection of biodiversity.

Keywords: adaptation, biodiversity, climate, management, mitigation, resilience

1. Introduction

Climate change poses major risks to global biodiversity in the twenty-first century [1] as it affects ecosystems processes, flora and fauna abundances and distribution [2]. Climate change refers to any change in the state of climate that is reflected in shifts in mean climatic variables over extended periods, typically decades or longer [3]. Climate change may result from natural internal processes within the climate system [4] or variations in natural or persistent anthropogenic external variability [5]. Climate results from fluctuations in the mean state or other climate statistics on temporal scales beyond those of individual weather events [6]. Climate change impacts, such as increased water shortages due to persistent droughts, present a threat to wildlife resources and consequently wildlife-dependent livelihoods in Africa. Thus, managing wildlife resources populations requires an understanding of the nature, magnitude and distribution of current and future climate impacts [7].

The term wildlife collectively refers to all forms of undomesticated flora and fauna found in terrestrial or aquatic environments [8]. In the present study, wildlife resources refer to faunal or floral species with potential benefit to human kind. Climate change has greater impacts on livelihoods of people in developing countries due to their low adaptive capacity [9]. However, this chapter focuses on terrestrial and semi-aquatic vertebrates, specifically mammals and their habitats. In Southern Africa, wildlife species promote the lives and livelihoods of local communities, particularly those living adjacent to protected areas [10]. Such communities generate revenue from activities such as eco-tourism, safari and consumptive hunting and bush meat trading. Therefore, there are huge economic losses associated with floods, droughts and wildfire. Against this backdrop, different mechanisms have been put in place to encourage climate change adaptation and mitigation in the biodiversity sector at global, regional and national levels [11].

The Intergovernmental Panel on Climate Change [12] projects a rise in surface temperature over the twenty-first century under all assessed emission scenarios. Global averaged temperatures are projected to increase by between 0.15 and 0.3°C per decade [13]. The frequency and intensity of heat waves are also likely to be more frequent and prolonged [14]. Many regions are likely to experience more frequent and extreme precipitation events. According to the Intergovernmental Panel on Climate Change (IPCC) [15], there is high likelihood that climate-related extremes, such as heat waves, droughts, floods, cyclones and wildfires, will expose some ecosystems, rendering them vulnerable to climate variability. Climate models project increased aridity and persistent droughts in the twenty-first century for most of Africa, Southern Europe, Middle East, Southeast Asia and Australia [16].

According to the Intergovernmental Panel on Climate Change [17], climatic changes are occurring at a faster rate than expected, particularly in Southern Africa. The IPCC predicts a 10 to 24% in mammalian species in sub-Saharan Africa national parks [13]. Climate change due to natural variability is therefore affecting terrestrial biological systems. Diffenbaugh and Field noted that there has been significant rise in temperatures of terrestrial ecosystems due to global warming [18]. Climate change directly affects ecosystems through seasonal changes in rainfall and temperature and indirectly through other disturbances such as fire and drought [11]. The IPCC [12] also notes that generally there is high confidence that several terrestrial, freshwater and marine species have shifted their geographic ranges, seasonal activities, migration patterns, abundances and species interactions in response to climate change. Based on available scientific literature, the IPCC Fourth Assessment Report (AR4) also reports that there is medium confidence that terrestrial ecosystems could have faced some impacts in Africa in recent decades due to climate change [12]. Additional threats from extensive land use and degradation, changes in frequency and severity of extreme events and interactions with other stresses [18] further threaten the resilience of terrestrial ecosystems. However, the

IPCC AR4 acknowledges that there is gap in knowledge on the climate change impacts across many regions.

Common effects of climate change on species and ecosystems include (1) changes in life-history events or phenology, (2) effects on demographic rates, such as survival and fecundity, (3) reductions in population size and (4) shifts in species distributions. Climate change poses direct and indirect effects on herbivore species [19] through changes in the fitness, survival and reproductive success [20]. In semiarid ecosystems, climatic changes in frequency and severity of droughts are likely to exacerbate the effects of drought on forage availability, which can feed back to regulate reproduction and offspring recruitment among ungulates [21]. Forrest et al. [22] highlighted that climate change is likely to affect the persistence of large, space-requiring species through habitat shifts, loss and fragmentation.

Understanding the effects of climate change and variability on wildlife species is vital in conservation biology and wildlife management [21], especially proactive management and formulation of conservation status decisions [23]. Large mammalian herbivores are key drivers of rangeland dynamics [24] hence assessing the effects of climate change and variability on these populations is essential for the stewardship of ecosystems and biodiversity [25]. The World Wide Foundation [26] noted that the impacts of climate change on global biodiversity and how biological species may (or may not) adapt are yet to be quantified. To date, few studies have assessed the impacts of climate change and variability on resources that support wildlife in Southern Africa [27].

Using case study examples, this chapter reviews the potential impacts of climate change and variability on wildlife resources in some protected areas in Southern Africa. The objectives of this chapter are to: (1) highlight studies on detected climate changes particularly rainfall and temperature, (2) outline the predicted and observed impacts of climate change and variability on wildlife resources in Southern Africa and (3) highlight climate change adaptation and mitigation strategies and implications for conservation.

2. Materials and methods

This study adopted both quantitative and qualitative methods to obtain data on the effects of climate change and variability on mammalian species on selected national parks in Zimbabwe. We conducted a review existing literature (1980–2015) from peer-reviewed journal articles, books, edited book chapters, electronic academic theses, technical reports from Google Scholar, Scopus and Web of Science covering issues on climate change, climate variability and wildlife resources in Southern Africa. Data were also obtained from unpublished internal scientific reports and management plans from the Zimbabwe Parks and Wildlife Management Authority (ZPWMA) and other and external reports from policy documents related to biodiversity and climate change in Southern Africa. Technical reports and public publications from key stakeholders, such as the International Union for Conservation of Nature (IUCN), World Wide Fund for Nature (WWF) and African Wildlife Foundation (AWF), were also reviewed. The following keywords or phrases were used: "climate change," "climate variability," "impacts/effects," "wildlife," "mammals," "biodiversity," "ecosystems," "drought," "rainfall," "temperature," "herbivores," "global environmental change," "plants," "flora" and "fauna" with a combination of "AND" between the keywords to retrieve the relevant literature.

Secondary data on rainfall, temperature and large mammal surveys for the period 1960–2015 were also collected from published and unpublished reports from the Zimbabwe Parks and Wildlife Management Authority. Secondary data were either adopted or used to plot graphs showing large mammal trends in relation to rainfall and temperature. Data were collected between August 2016 and February 2017. Literature obtained from all the documents was categorised under the following themes: (1) historical trends in selected mammals in relation to climate variables (1980–2015), (2) predicted and observed impacts of climate change and variability on wildlife resources in selected protected areas in Southern Africa and (3) implications for conservation. For further analysis, case studies from Zimbabwe (**Figure 1**) were used to highlight impacts of climate change on wildlife resources.

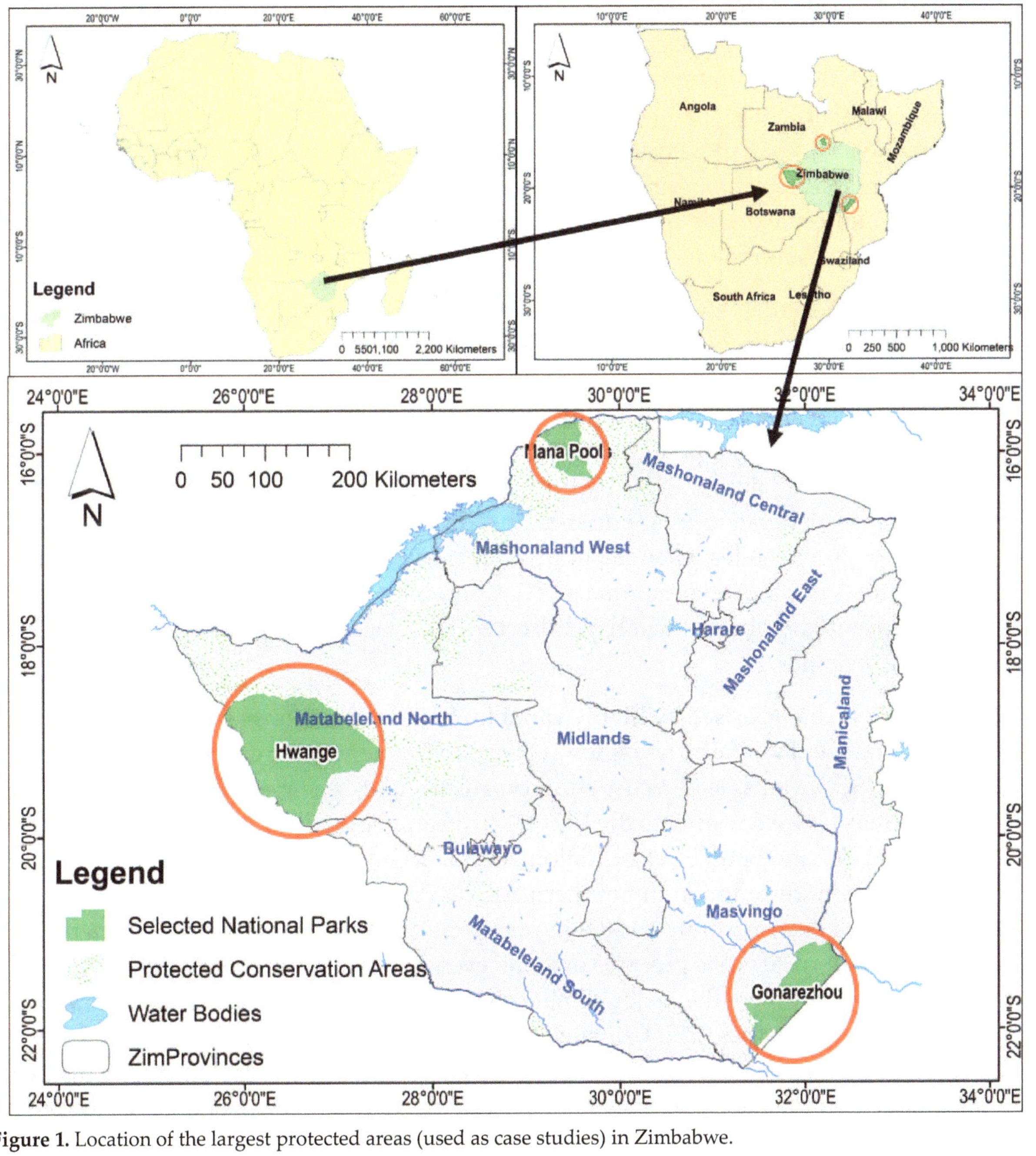

Figure 1. Location of the largest protected areas (used as case studies) in Zimbabwe.

3. Results and discussion

This section highlights predicted and observed trends in rainfall and temperatures across Southern Africa and their implications on wildlife resources. Impacts of climate change on selected wildlife species and some of the predicted and observed impacts of climate change and variability are subsequently highlighted for selected protected areas. Implications in terms of adaptation and mitigation for conservation are also discussed.

3.1. Rainfall and temperature trends: implications for wildlife resources

Global warming has caused a gradual reduction in annual rainfall across Africa over the past 50 years [24]. Projections of precipitation and runoff in Africa suggest a decrease of up to 10% in precipitation in most of Southern Africa (including Zimbabwe) by 2050 [28]. This reduction could reduce the distribution and availability of both food and surface water for animal species [24]. Rainfall is the main climatic factor governing herbivore population dynamics across Africa [29, 30]. African wildlife resources are therefore likely to be subjected to climate warming due to rising temperatures and extreme events, such as droughts and floods [27]. Thus, climate-induced extreme events threaten some of the large protected areas that have been designated to conserve much of Africa's magnificent biodiversity [31].

The IPCC [3] predicts that Africa is projected to have 'above-average' climate change in the twenty-first century. Thus, global warming will have the greatest effects on biodiversity in the continent [32]. Chamaille-Jammes et al. [33] noted that climate change is predicted to affect both the mean annual rainfall and its seasonal distribution over the African continent. Thus, climate warming has a potential to directly affect wildlife resources through shifts in onset and duration of rainy seasons and drought on wildlife species, reduction in species distribution ranges, alteration in abundance and diversity of mammals, changes in calving and population growth rates, changes in juvenile survival of most ungulates and changes in species richness of birds and mammals [27]. Consequently, these changes in wildlife species abundance and distribution will have direct serious negative impacts on ecotourism and game hunting activities.

Southern Africa is characterised by highly variable climatic conditions associated with fluctuating temperature and rainfall. The region is prone to frequent and intense El Nino-Southern Oscillation (ENSO) events, leading to widespread drought in some areas and widespread flooding in others [34]. According to the IPCC [12], mean seasonal temperatures are predicted to increase, and El Nino-Southern Oscillation effects, fires and severe weather anomalies are more likely to be more common in Southern Africa. Projections of precipitation and runoff in Africa suggest a decrease of up to 10% in precipitation in most of Southern Africa by 2050 [28]. Most of Southern Africa is prone to extreme events such as droughts and floods [35] and climate warming, which are the major climate change factors that are likely to affect wildlife resources in Southern Africa [27]. Climate can affect mammalian populations indirectly by excessive temperatures or rainfall, through bottom-up effects on food plant productivity or top-down effects on predator efficiency [36]. Bottom-up control mechanisms are based on the view that herbivore populations are limited by forage quality and quantity [37].

Climate variables, particularly rainfall and temperature, generally influence habitat quantity and quality within savannah ecosystems thereby affecting the structure, composition and dynamics of wildlife species. Temperature and rainfall display complex temporal variation changing from place to place across the earth. These key climatic factors determine plant productivity and hence animal food availability [38]. For instance, several studies have also reported on the influence of rainfall, especially during the dry season, on the availability of forage of adequate quality and large herbivore population performance [29, 30, 39–41]. Rainfall controls primary production [33] and ungulate grazer populations across the African savannah [42–45]. Strong relationships between abundance and rainfall suggest that rainfall underpins the dynamics of African savannah ungulates and that changes in rainfall due to global warming may markedly alter the abundance and diversity of these mammals [41].

Generally in African savannah, wildlife populations grow in wet season and decline in dry season such that there tends to be fluctuations from year to year depending on seasonal conditions. Variability in rainfall influences animal populations in the savannahs [46]. For instance, a study carried out in Kruger National Park sought to establish the extent to which the population composition of the common impala (*Aepyceros melampus* (Lichtenstein, 1812), Burchell's zebra (*Equus burchelli*), blue wildebeest (*Connochaetes taurinus*), greater kudu (*Tragelaphus strepsiceros)*, giraffe (*Giraffa camelopardalis*), common waterbuck (*Kobus ellipsiprymnus)*, warthog (*Phacochoerus aethiopicus*), sable antelope (*Hippotragus niger*), tsessebe (*Damaliscus lunatus)* and roan antelope (*Hippotragus equinus)* were associated with changes in rainfall and prey availability [30]. Multiple regression models fitted to survival estimates indicated that juvenile survival was sensitive to annual variability in rainfall for most of these species, especially in the dry season [30]. Rainfall components affected adult survival in several of the declining species, while negative density dependence in adult survival was evident for three of the four species that maintained high abundance [30]. Dixon et al. [47] also estimated that approximately 66% on nyala and zebra in Kruger National Park, South Africa are likely to be lost as a result of climate change.

Ungulates respond to rainfall fluctuations through movements, survival [46] and reproductive phenology [36]. Rainfall influences the composition of the herbaceous layer and its quality; forage vegetation growth and food production during the wet season and retention of green foliage during the dry season [30]. Ogutu et al. [48] investigated the influence of rainfall and temperature fluctuations on the dynamics and abundance of an insularised and compressed impala population. The study observed that whilst births negatively correlated with rainfall, high rainfall depressed reproductive success in impalas [43].

Episodic and local droughts help limit elephant (*Loxodonta Africana*) population numbers [24]. Shrader et al. [24] revealed that annual variations in rainfall and the confounding effects of water provision and fences influence elephant survival rates across different sites in Africa. In Southern Africa, wildlife resources also depend on water resources, which are derived from major basins such as Limpopo and Zambezi. For instance, the Zambezi basin contains a number of important national parks in Southern Africa that are likely to be negatively impacted by such climate-induced changes in the hydrological cycle. Analysis of the Zambezi system in Southern Africa has revealed that the basin has low runoff efficiency and a high dryness index, which indicates its high sensitivity to climate change. Climate warming is predicted to result in runoff decreases even when precipitation increases due to the large hydrological role played by evaporation, especially from wetlands. Thus, increased runoff from loss of forest cover will only worsen water losses from wetland areas due to evaporation that is predicted under a warmer climate.

3.2. Projected impacts of climate change on wildlife resources

General circulation models (GCMs) are numerical models that simulate the physical processes of climate [7]. The GCMs have been used in the Intergovernmental Panel on Climate Change (IPCC) Fourth Assessment Report (IPCC AR4) as a basis for forecasting (**Table 1**) future climate as a primary determinant of species distributions and ecosystem processes impacts [7].

Global models predict that climate in Southern Africa is warming at a faster rate than has been predicted although no clear trend in rainfall patterns has been observed. Southern Africa is characterised by extremely variable rainfall patterns coupled with drought episodes. On the other hand, extreme events, such as droughts and floods, appear to have increased in frequency in the recent past. The IPCC [12] predicts that specific areas likely to be affected by climate change in Africa include African Rift Mountains, the Zambian and Angolan Highlands and the Cape Province of South Africa. Under the high-end A2 scenario, 12–39% and 10–48% of the Earth's terrestrial surface may respectively experience novel and disappearing climates by 2100 AD [6]. Corresponding projections for the low-end B1 scenario are 4–20% and 4–20%. Terrestrial fauna biodiversity in Africa is concentrated in the moist forests, woodlands and savannahs. Loss or alterations of terrestrial habitats by climate change will likely impact these species as they struggle to adapt to changing conditions [6]. The following section outlines observed impacts of climate change on selected species in selected protected areas in Southern Africa.

Projected impacts	Model	Source
About 50,000 African plant species impacted: substantial reductions in areas of suitable climate for 81–97% of the 5,197 African plants examined, 25–42% lose all area by 2085	Africa Hadley Centre Third Generation Coupled GCM (HadCM3) for years 2025, 2055, 2085, plus other models—shifts in climate suitability examined	[50]
Future distribution in 2050 of 975 endemic plant species in Southern Africa distributed among seven life forms endemic flora of Southern Africa (Namibia and South Africa) on average decreases with 41% in species richness among habitats and with 39% on species distribution range for the most optimistic scenario	SDMs were fitted to climatic data using the BIOMOD package in SPLUS	[51]
Fynbos and succulent Karoo biomes: losses of between 51 and 61%	Africa for years 2025, 2055, 2085, plus other models—shifts in climate suitability examined	[52]
Projected losses by 2050, critically endangered taxa (e.g. Proteaceae) in South Africa: losses increase, and up to 2% of the 227 taxa become extinct increase, and up to 2% of the 227 taxa become extinct	(HadCM2 IS92aGGa)	[49]

Source: Adopted from IPCC [50].

Table 1. Projected impacts of climate change on wildlife resources in Africa.

3.2.1. Case study 1: Gonarezhou National Park, South Eastern Lowveld, Zimbabwe

Gonarezhou National Park (GNP), the second largest national park in Zimbabwe, is located in southeastern Lowveld between latitudes 21° 00′–22° 15′ S and longitudes 30°15′–32°30′ E [37]. The park was established in 1975 [50]. The park covers total area of approximately 5053 km^2 and has a mean altitude of 400 m above sea level [50]. GNP is part of the Great Limpopo Transfrontier Conservation Area (GLTFCA), Southern Africa [50]. The park has a hot and semi-arid climate, which is characterised by highly seasonal and unreliable rainfall quantity and duration [51]. The park receives an average annual precipitation of approximately 466 mm with high inter-annual variability [52]. GNP has three distinct climatic seasons namely hot and wet (November–April), cool and dry (May–August) and hot and dry (September–October). Average monthly maximum temperatures are 25.9°C in July and 36°C in January, whereas average monthly minimum temperatures range between 9°C in June and 24°C in January [53, 54]. Prevailing winds are southeasterly, with wind speeds of between 11 and 17 km per hour [50].

Vegetation communities in GNP can be broadly categorised into three macro groups namely the mopane, miombo and alluvial woodlands. The majority of major mopane dominated plant communities in GNP have a high capacity to resist drought [55]. However, *Androstachys johnsonni* thicket, *Spirostachys* woodland, *Terminalia prunoides* woodland and streams and pans with *Spirostachys africana* communities have low capacity to resist drought and hence are vulnerable to climate change and variability [55]. Climate-related extreme events have also affected the abundance and phenology of tree species. For instance, tree loss and mortality of many tree species occurred during the 1991–1992 drought [52]. Tafangenyasha [52] suggested that elephant density and drought are probably the major factors that influence tree mortality in Gonarezhou National Park.

The GNP has diverse vertebrate fauna that includes approximately 84 mammal species; 400 bird species; 76 reptile species; 28 amphibian species and 50 fish species. The mammal fauna include 23 large herbivores and 8 carnivore species [50]. The population of browsers/grazers, such as common eland (*Taurotragus oryx*, Pallas, 1766), giraffe, nyala (*Tragelaphus angasii*), waterbuck and wildebeest, are generally low in GNP, whereas elephants, hippopotamus (*Hippopotamus amphibius*), zebra, kudu (*Tragelaphus strepsiceros*) and buffalo (*Syncerus caffer*) are generally high [56]. Large herbivores, such as elephant, buffalo, eland, giraffe, kudu, nyala, waterbuck, zebra, wildebeest and hippopotamus, have a low capacity to resist drought [55]. Extreme events such as drought also directly affect primary productivity and consequently the predator-prey populations. For instance, in GNP, carnivore populations can increase during drought periods, but soon after that, the lack of prey species can lead to aggravated human-wildlife conflict due to overlap between human and wildlife habitats [55].

Drought in particular has caused large herbivore death in GNP during the 1982/1983 and 1991/1992 season [51, 52, 55, 59]. Large mammals, such as elephant, buffalo, hippo and large antelopes, have a low capacity to resist drought [55] and likely to be threatened by climate change [57]. Buffalos in particular are vulnerable to drought [58]. Approximately 1500 elephants died during the 1991–1992 drought [52] and numerous species also suffered significant mortality. Gandiwa [59] investigated the annual rainfall patterns and associated fluctuations of wild large herbivore species data collected from multi-species aerial surveys across five sites in the Great Limpopo Transfrontier Conservation Area (GLTFCA), Southern Africa. Findings from this study suggest that rainfall does have a strong influence on large herbivore population dynamics especially in really dry years in African savannah ecosystems [59]. Using data from aerial surveys, Dunham [39] also assessed the abundance of large herbivores and elephants in GNP for the period between 1980 and 2009. The study revealed that elephant, buffalo, eland, kudu, nyala, waterbuck, wildebeest and zebra generally increased in number, after population declines during the 1992 drought [39].

Drought conditions reduce the available aquatic habitat for water-dependent species such as the hippopotamuses. This could potentially lead to adverse effects on the species, including population declines, increased crowding and disease and more intraspecific violence due to increased contact between animals [83]. For instance, Zisadza et al. [53] observed an overall decline in common hippopotamus along the Mwenezi and Save River in the GNP (**Figure 2**), primarily to past droughts over the past four decades (e.g. 1982–84 and 1991–92) and other factors such as siltation and persecution in adjacent communal areas. Thus, climate change and variability interact with other stressors to influence the abundance and distribution of wildlife resources.

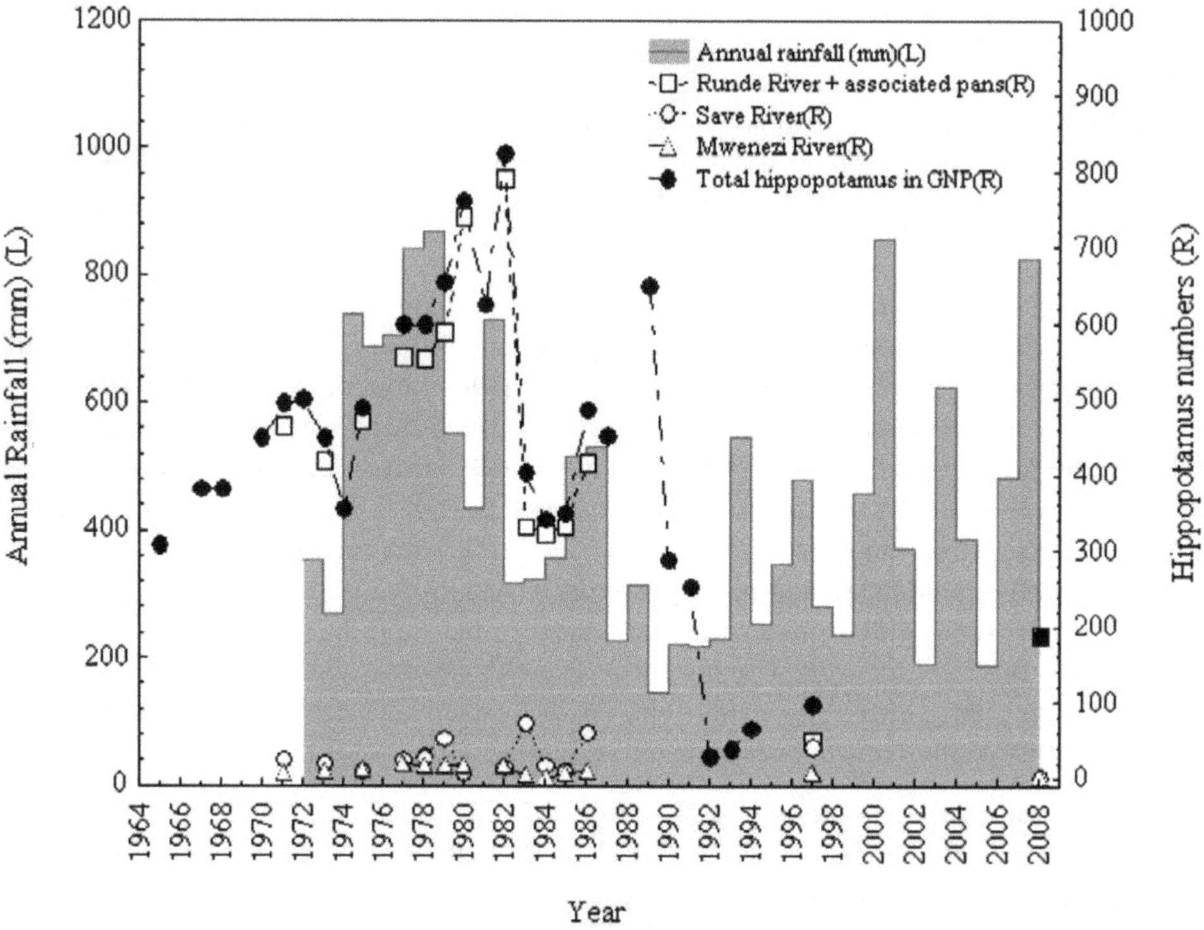

Figure 2. Trends in annual rainfall (1972–2008) and hippo population in the Gonarezhou National Park, Zimbabwe, and its three major rivers for the period 1965–2008. Source: Reproduced with permission from Zisadza et al [53].

Apart from mammals, Gandiwa and Zisadza [55] also noted that water birds, for example pelicans, plovers, storks and fish owls, associated with aquatic habitats such as larger river systems are also under threat from climate change due to reductions in precipitation and resultant changes in the flow regime changes and water holding capacity of water sources.

3.2.2. Case study 2: Mana Pools National Park, Middle Zambezi Valley, Zimbabwe

Mana Pools National Park (MPNP) is located in the Middle Zambezi Valley between (15° 98′ South and 29° 44′ East) and (16° 00′ South and 29° 90′ East) [60]. The park which covers an area of 2196 km^2 was designated a UNESCO World Heritage in 1984 and is also a core area of the Middle Zambezi Biosphere Reserve (MZBR) designated by UNESCO in 2010 [61]. MPNP lies within the proposed Lower Zambezi Mana TFCA. The park is located within the lower Zambezi basin at an altitude of about 350 m within the Zambezi floodplain to about 1100 m over the Zambezi escarpment [60]. Rainfall in the mid-Zambezi valley follows the regional mono-modal pattern with the bulk of the annual average of 706 mm falling between November and March [62]. December and January receive an average of 180 mm each and these two months account for almost half of the annual average in a normal year [63]. Rainfall in MPNP is characterised by monthly and intra-annual variability with a mean annual rainfall of 724 mm [64]. Rainfall variability has a distinct effect on the primary productivity of the area and hence the wildlife that it can support. Average maximum temperatures are 40°C towards the rainy season, whereas mean minimum temperatures are above 10°C [63].

Changes in large herbivore densities have also been reported in the MZBR particularly in the Mana Pools National Park and surrounding communal areas [62]. Climate change induced drought [63] coupled with the closure of floodgates at Kariba Dam and poaching as well as increasing human encroachment into the park [62]. In addition, during the dry season, water availability is restricted to riverine woodlands (**Figure 3**) and a few springs near the escarpment [62].

Apart from seasonal drying up, the quantity and quality of water at some sources have changed (**Figure 3b**, personal observations, 2015) probably due to incessant prolonged drought periods. For example, Chitake spring (**Figure 3a**) near the escarpment is reported to be drying up due to periodic droughts. Aggregation of animals around the springs exerts pressure on forage resources due to over browsing and trampling (**Figure 3b**). For example, Regassa [65] noted that animals tend to aggregate in areas with strategic and scarce resources such as water and pasture in the dry season. In MPNP, inland seasonal water pans (**Figure 4**) and springs also provide water for buffalo herds and other mammals inland.

Studies by Bosongo et al. [66] have shown that the Zambezi Valley is experiencing climate variability and an increase in these events is expected. Rainfall in the middle Zambezi valley is very variable (**Figure 5**) and this could have an effect on the abundance and distribution of large mammals in MPNP. An analysis of the annual rainfall data for Nyamepi shows a high of over 1600 mm in 1978 and a low of slightly above 200 mm in 1974. Large herbivores, in particular, elephant, buffalo, roan, zebra (*Equus quagga*), eland and sable could be under threat

Figure 3. (a) Image of the permanent but dying up Chitake spring near the Zambezi escarpment in MPNP. (b) Buffalo herd aggregate around a permanent but drying natural water source during the drought periods. (c) Natural water pans (Long Pool) close to Nyamepi Camp within the riverine woodlands drying up in October 2015. (d) Image of hippo water bathing along the Long Pool. Photo credit: Olga Laiza Kupika.

Figure 4. Image showing dried up section of Green Pool in MPNP. Photo credit: Olga Laiza Kupika.

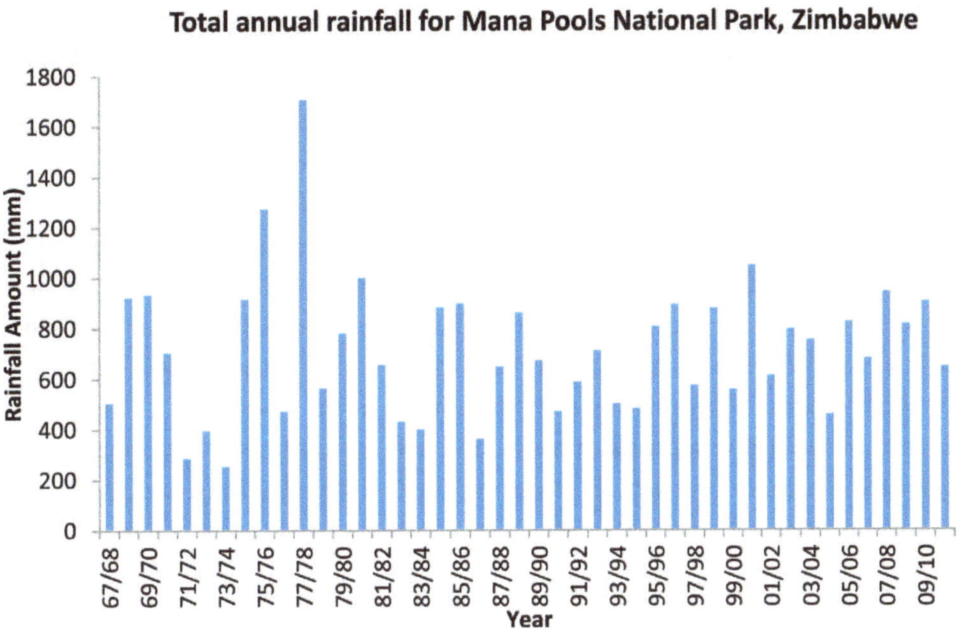

Figure 5. Seasonal rainfall recorded at Nyamepi in Mana Pools National Park, Zimbabwe, located along the Zambezi River for the period 1967–2010. Data are from unpublished records kept at Nyamepi.

from the periodic droughts. Noticeable changes in large mammal populations in the riverine woodlands and adjacent areas due to drought have been observed (**Figure 6**).

An analysis of the annual rainfall data for Nyamepi shows a high of over 1600 mm in 1978 and a low of slightly above 200 mm in 1974. Large herbivores, in particular, elephant, buffalo,

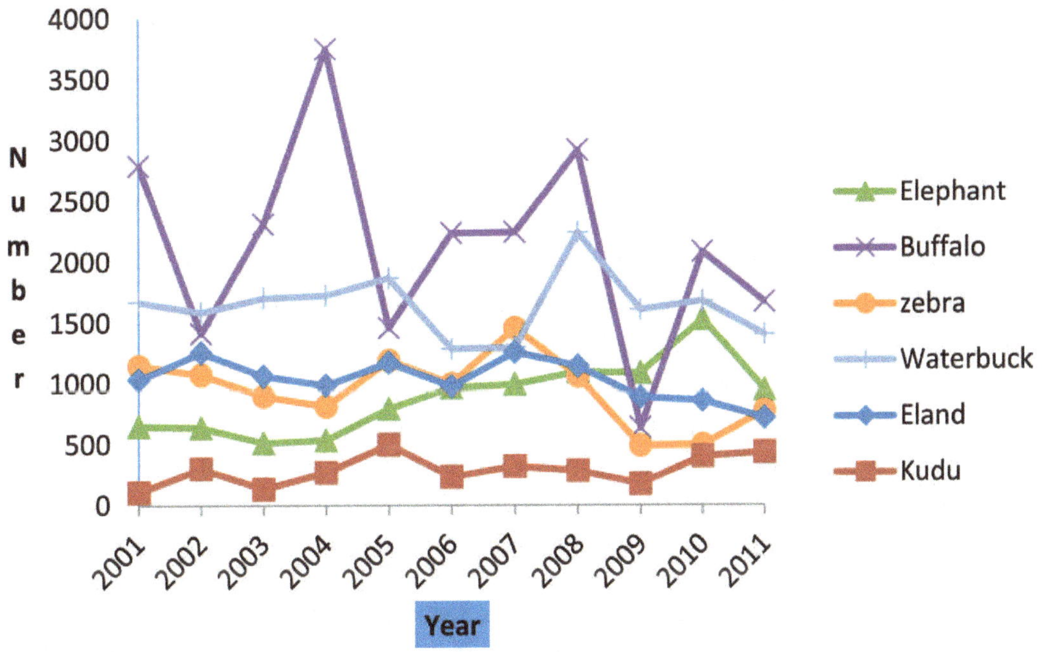

Figure 6. Large herbivore trends in MPNP floodplain (2001–2011). Source: Compiled using secondary data extracted from unpublished reports by research officers of the Department of National Parks and Wildlife Management.

roan, zebra (*Equus quagga*), eland and sable could be under threat from the periodic droughts. Noticeable changes in large mammal populations in the riverine woodlands and adjacent areas due to drought have been observed (**Figure 6**).

Although the abundance of selected herbivores in MPNP shows a fluctuating trend (buffalo), there has been a general decline for certain species such as eland and zebra. For example, Dunham [63] observed that between 1982 and 1984, buffalo population within the Zambezi riverine woodlands declined during the drought years due to high mortality and changes in habitat utilization. Macandza et al. [67, 68], Ryan and Wamsley [68] and Megaze et al. [69, 70] also noted that savannah buffaloes moved into riverine habitats during the dry season such that in areas such as Mana Pools National Park, provision of feed supplementation in the floodplain during the dry season could also be attributed to the seasonal increase in buffalo numbers. Additionally, such changes could be attributed to other non-climatic environmental changes within the Zambezi floodplain.

In MPNP, vegetation and other environmental changes have also been observed within the floodplain section partly due to changes in the hydrology of the Zambezi River and periodic droughts [70, 71]. Future studies should focus on monitoring the ecology of the floodplain to promote sustainable utilization or wildlife resources under a changing climate.

3.2.3. *Case study 3: Hwange National Park, Zimbabwe*

Hwange National Park (HNP), the largest in Zimbabwe, is located on the north-west border of Zimbabwe (19[1]00°S, 26[1]30°E) within a semi-arid dystrophic savannah [71, 72, 73]. The elevation of the park ranges between 900 and 1000 m [72, 73]. HNP is located within the Kavango-Zambezi Transfrontier Conservation Area and covers an area of approximately 15,000 km^2 wooded savannah with patches of grassland [33]. The park receives mean annual rainfall of approximately 600 mm with the most rainfall between October and May. Rainfall is generally erratic and unevenly distributed spatially and temporally [72, 73].

Studies by Chamaillé-Jammes et al. [2] revealed that between 1928 and 2005, dry years became even drier and droughts worsening in HNP thereby affecting large herbivore population dynamics. Surface-water availability was strongly influenced by annual rainfall with consequences on large herbivore population dynamics [2]. Drought-related mortality of elephants has also been observed in HNP between 1980 and 1984; 1987 and 1993–1995 prolonged drought period [72]. However, despite the mortalities, rainfall amounts during the 1993–1995 period was higher than the other years [72]. Loveridge et al. [73] carried out a study on the influence of drought on predation of elephant calves by lions (*Panthera leo*) in HNP for the period 1998–2004. The study revealed that high-density aggregations of elephants around limited water sources during the dry season may result in the depletion of local food resources. Additionally, elephant herds were forced to travel long distances between water and forage during which the elephant calves appeared more vulnerable to prey by lions [73, 74].

Crosmary et al. [74, 75] noticed that sable antelope densities declined across the greater HNP region between 1990 and 2001 most probably due to unfavourable rainfall conditions. Crosmary et al. [75, 76] also observed that large herbivores trends particularly selective grazers, generally declined between 1930 and 2000 in HNP and adjacent areas. This decline was attributed to probably greater sensitivity of selective grazers to variation in rainfall compared with other herbivores. During this period, rainfall certainly declined coupled with frequent droughts during the 1990s [75, 76]. Apart from climatic factors, generally large herbivore dynamics in HNP and its surrounding habitats are influenced by surface water provision coupled with other anthropogenic activities such as hunting [2, 75, 76, 77]). Artificial water provision has ecological implications on the structure of vegetation communities and other large herbivores such as elephants [71].

3.3. Climate change adaptation, mitigation, vulnerability and resilience in the wildlife sector

In Southern Africa, adaptation is a major concern in terms of climate change agenda. Climate change adaptation can be defined as the process of taking actions to help communities and ecosystems cope with changing climate conditions [77]. Adaptation can either be reactive in that it can be triggered by historical or current observed climatic events while anticipatory is motivated by the need to respond to future predicted impacts [78, 79]. This implies that wildlife resources have developed short- and long-term mechanisms of coping with the impacts of climate change and variability. Although wildlife has the potential to naturally adapt to the effects of climate change, some scholars contend that the observed and projected effects due to projected climate change might deserve mitigation measures [79, 80]. On the other hand, wildlife managers have also intervened as part of adaptive management either through the manipulation of wildlife habits or wildlife resources in response to the changing climate. Wildlife managers are faced with the daunting task of determining the nature and extent of the wildlife impacts of past climate change and variability [80, 81]. There is therefore need to identify practical strategies that could help to reduce the anticipated effects of climate change on wildlife species [81, 82].

Mitigation is also another area of concern in the context of climate change. Mitigation is defined as interventions to reduce the sources or enhance the sinks of greenhouse gases with the aim of reducing global temperatures. Mitigation measures in the wildlife sector include the adoption of more efficient uses of fossil fuels, the conversion to renewable energies such as solar and wind power and the expansion of forest areas and other sinks to remove greater amounts of carbon dioxide from the atmosphere. If nations worldwide fail to implement such measures, mean global temperatures will continue to increase and the adverse impacts of climate change on societies and economies will worsen.

The Heinz Center [82, 83] carried out extensive literature review that examined information about the ecology and physiology of several megafauna and how each of these species may be vulnerable in the face of climate change and suggested adaptation strategies. The review indicates that there are several key areas of vulnerability, which include the need for surface water; water-dependent; lack of habitat connectivity and heat stress as some of the common vulnerabilities shared across many of the African wildlife species [82, 83]. Mega herbivores, such as the African elephant, black and white rhino and hippopotamus, in particular, are prone to most habitat and species-specific elements of vulnerability [82, 83]. In addition, anthropogenic factors such as illegal harvesting of wildlife resources and poaching also continue to threaten endangered species such as the two rhinoceroses and the African elephant [82]. Climatic and non-climatic factors interact within a social ecological system to influence the vulnerability and resilience of savannah species to climate change. Thus, the capacity of wildlife species to adapt to climate change also depends on their ability to cope with several non-climatic stressors.

The findings from literature reveal that the large carnivore community (lion, leopard, cheetah and wild dogs) is mainly prone vulnerable to disease particularly anthrax and rabies and distemper. For instance, it is observed that climate-related factors in conjunction with other factors,

Strategy	Description
Land and water protection and management	• Increase amount and area covered by Protected Areas • Improve Representation and Replication within Protected-Area Networks • Improve Management and Restoration of Existing Protected Areas to Facilitate Resilience • Design New Natural Areas and Restoration Sites to Maximize Resilience • Protect Movement Corridors, Stepping Stones, and Refugia • Manage and Restore Ecosystem Function Rather than Focusing on Specific Components (Species or Assemblages) • Improve the Matrix by Increasing Landscape Permeability to Species Movement • Reduce non-climate stressors on natural areas and ecosystems
Direct Species Conservation/Management	• Focus Conservation Resources on Species that Might Become Extinct • Translocation or assisted dispersal of species at Risk of Extinction • Establish Captive Populations of Species that Would Otherwise Go Extinct • Reduce Pressures on Species from Sources Other than Climate Change
Monitoring and Planning	• Evaluation of existing monitoring programs for wildlife and key ecosystem components to determine: a) how these programs will need to be modified to provide management-relevant information on the effects of climate change b) what new monitoring systems will need to be established in order to address gaps in our knowledge of climate effects • Incorporate Predicted Climate-Change Impacts into Species and Land-Management Plans, Programs, and Activities • Ensure that wildlife and biodiversity needs are considered as part of the broader societal adaptation process • Develop Dynamic Landscape Conservation Plans
Law and policy review	• Review and Modify Existing Laws, Regulations, and Policies Regarding Wildlife and Natural Resource Management • Review existing laws, regulations and policies regarding wildlife and natural resource management to insure that these instruments provide managers with maximum flexibility in addressing the effects of climate change • Propose new legislation and regulations as needed to provide managers with the flexibility, tools and approaches needed to effectively address climate change impacts

Source: Adopted from Mawdsley et al [81, 82].

Table 2. Strategies for addressing the impacts of climate change on wildlife resources.

such as hunting and disease control impacts by boundary fencing of the park, culling operations and water provisioning, have influenced the dynamics of large herbivore population trends in the Kruger National Park (KNP) over the past century [83, 84]. Thus, the interactive effects of disease and climate-induced factors have the potential to affect many iconic African wildlife species under a changing climate. Changing climatic conditions are therefore likely to affect safari hunting and the wildlife tourism sector due to changes in the abundance of such iconic species.

Mawdsley et al. [81, 82] suggest that effective climate change adaptation incorporates protectionist and interventionist approaches (**Table 2**).

Generally, in Southern Africa, augmenting natural water supplies by providing artificial water points is an intervention commonly adopted by managers of national parks and other large protected areas. In addition, some of the most commonly used adaptation strategies to deal with impacts of climate change in the wildlife sector include:

i. Expansion of protected areas to include migration corridors and seasonal feeding areas [27].

ii. Improving connectivity of habitats to facilitate dispersal to appropriate habitats [27].

In view of the magnitude of impacts, in future it will be important to incorporate predicted and observed climate change impacts into overall wildlife management plans and to review and modify existing laws, regulations and policies regarding wildlife management. Kupika and Nhamo [84] noted that most countries in Southern Africa do not mainstream climate change in their biodiversity-related policies. Thus, lack on enabling policy framework retards progress in as far as responding to climate-related disasters is concerned. Other countries, e.g. Zimbabwe, rely on a centralized system towards disaster risk reduction. Although the implementation on the ground is multi-sectoral, disaster response is hampered by bureaucracy within the Parks and Wildlife Management Authority. Thus, protected area managers are not flexible to directly implement any short-term or long-term adaptation strategies to ensure improved species conservation.

Apart from mainstreaming enabling laws and regulations, consideration should also be given to managing protected areas for climate change mitigation through carbon storage and sequestration, as well as for other ecosystem services. Despite all efforts to adapt to the changing climate, efforts are hampered by several factors related to technical and financial constraints to implement the strategies. Africa as a continent is highly vulnerable to the impacts of climate change because of other multiple stressors such as poverty; illiteracy and lack of skills; weak institutions; lack of political will to confront climate change challenges; limited infrastructure; lack of technology and information; poor health care; armed conflicts; rampant corruption and host of other human anthropogenic activities [13]. Thus, managing threats to the wildlife resources consequently include managing other confounding stressors to biodiversity.

4. Conclusion

Generally, climate-induced extreme events such as prolonged and frequent drought periods influenced the abundance and distribution of wildlife resources across Southern

Africa. In addition, temperature related extreme events such as heat waves also threaten the ecology of heat sensitive species. Thus wildlife resources are under threat from the confounding effects of rainfall and temperature changing climate. The abundance of both flora and fauna is influenced by the interactive effects of climate change and other non-climatic multiple stressors. Wildlife species have different levels of vulnerability and resilience to climate change induced impacts. Stakeholders in the wildlife sector are therefore faced with a task to continuously monitor wildlife resources under a changing climate. Since responses of wildlife resources to climate variability and change are both location- and species-specific [85] protected area managers should come up with different strategies to mitigate the impacts of climate change on vulnerable habitats and specific species depending on their ecology.

Acknowledgements

This work was conducted within the framework of the Research Platform "Production and Conservation in Partnership (HYPERLINK "http://www.rp-pcp.org" www.rp-pcp.org). This research was supported by funding from the European Union under the Delivering Innovation and technology through the Reinforcement of Agricultural and Multidisciplinary research capacity (DREAM) project & the Chinhoyi University of Technology (CUT), Zimbabwe.

Author details

Olga L. Kupika[1*], Edson Gandiwa[1], Shakkie Kativu[2] and Godwell Nhamo[3]

*Address all correspondence to: olgal.kupika@gmail.com

1 Department of Wildlife Ecology and Conservation, Chinhoyi University of Technology, Chinhoyi, Zimbabwe

2 Department of Biological Sciences, Faculty of Science, University of Zimbabwe, Harare, Zimbabwe

3 Institute for Corporate Citizenship, University of South Africa, Pretoria, South Africa

References

[1] Dawson TP, Jackson ST, House JI, Prentice IC, Mace GM. Beyond predictions: Biodiversity conservation in a changing climate. Science (80-). 2011;**332**:53-58

[2] Chamaillé-Jammes S, Fritz H, Murindagomo F. Climate-driven fluctuations in surface-water availability and the buffering role of artificial pumping in an African savanna: Potential implication for herbivore dynamics. Austral Ecology. 2007;**32**:740-748

[3] IPCC. Climate change 2007: Impacts, adaptation and vulnerability. Contribution of working group II to the fourth assessment report of the Intergovernmental Panel on Climate Change. IPCC, Geneva; 2007. pp. 1-976

[4] AMCEN Secretariat. Addressing Climate Change Challenges in Africa; A Practical Guide Towards Sustainable Development. Addis Ababa, Ethiopia: African Union; 2011. p. 272

[5] IPCC. Climate Change and Biodiversity. IPCC Technical Paper V Group. 2002;**24**:77

[6] IPCC. Climate Change: Synthesis Report Summary Chapter for Policymakers. IPCC; Geneva; 2014. p. 31

[7] Wilsey CB, Lawler JJ, Freund JA, Hagmann K, Hutten KM, Sciences F, Wildland P, Sciences F, Street N. Tools for Assessing Climate Impacts on Fish and Wildlife. Journal of Fish and Wildlife Management. 2013;**4**:220-241

[8] Yarrow G. Wildlife and Wildlife Management. Clemson Extension; 2009. p. 2

[9] Collier P, Conway G, Venables T. Climate change and Africa. Oxford Review of Economic Policy. 2008;**24**:337-353

[10] Ntuli H, Muchapondwa E. A bio-economic analysis of community wildlife conservation in Zimbabwe. Journal for Nature Conservation. 2015;**37**:106-121

[11] Barker T. IPCC (Intergovernmental Panel on Climate Change) Climate Change 2007: An Assessment of the Intergovernmental Panel on Climate Change; 2007. p. 12-17

[12] IPCC. Climate Change 2014: Synthesis Report. Contribution of Working Groups I, II and III to the Fifth Assessment Report of the Intergovernmental Panel on Climate Change. Geneva; 2014

[13] IPCC. Climate Change 2007: Impacts, adaptation and vulnerability: Working Group II contribution to the Fourth Assessment Report of the IPCC Intergovernmental Panel on Climate Change; Geneva; 2007. p. 976

[14] IPCC. Summary for Policymakers. Climate Change 2014: Impacts, Adaptation, and Vulnerability—Contribution of Working Group II to the Fifth Assessment Report; Geneva; 2014. p. 1-32

[15] IPCC. Summary for Policymakers. Climate Change 2014: Impacts, Adaptation, and Vulnerability. Part A: Global and Sectoral Aspects. Contribution of Working Group II to the Fifth Assessment Report of the Intergovernmental Panel on Climate Change; Geneva; 2014. p. 1-32

[16] Alig JR. Effects of Climate Change on Natural Resources and Communities: A Compendium of Briefing Papers; 2011. p. 169

[17] Solomon S. Climate Change 2007—The Physical Science Basis: Working Group I Contribution to the Fourth Assessment Report of the IPCC. Vol. 4. Cambridge University Press; 2007

[18] Diffenbaugh NS, Field CB. Changes in ecologically critical terrestrial climate conditions. Science (80). 2013;**341**:486-492

[19] Texeira M, Baldi G, Paruelo J. An exploration of direct and indirect drivers of herbivore reproductive performance in arid and semi arid rangelands by means of structural equation models. Journal of Arid Environment. 2012;**81**:26-34

[20] Milligan SR, Holt WV, Lloyd R. Impacts of climate change and environmental factors on reproduction and development in wildlife. Philosophical Transactions of the Royal Society of London. Series B, Biological Sciences. 2009;**364**:3313-3319

[21] Koons DN, Terletzky P, Adler PB, Wolfe ML, Ranglack D, Howe FP, Hersey K, Paskett W, du Toit JT. Climate and density-dependent drivers of recruitment in plains bison. Journal of Mammalogy. 2012;**93**:475-481

[22] Forrest JL, Wikramanayake E, Shrestha R, Areendran G, Gyeltshen K, Maheshwari A, Mazumdar S, Naidoo R, Thapa GJ, Thapa K. Conservation and climate change: Assessing the vulnerability of snow leopard habitat to treeline shift in the Himalaya. Biological Conservation. 2012;**150**:129-135

[23] Molnar PK, Derocher AE, Thiemann GW, Lewis MA. Predicting survival, reproduction and abundance of polar bears under climate change. Biological Conservation. 2010;**143**:1612-1622

[24] Shrader AM, Pimm SL, van Aarde RJ. Elephant survival, rainfall and the confounding effects of water provision and fences. Biodiversity and Conservation. 2010;**19**:2235-2245

[25] Saba VS, Stock CA, Spotila JR, Paladino FV, Tomillo PS. Projected response of an endangered marine turtle population to climate change. Nature Climate Change. 2012;**2**:814-820

[26] Advani N. Mountain gorilla WWF Wildlife and Climate Change Series. 2014. 4 pp

[27] Chidumayo E, Okali D, Kowero G, Larwanou M, editors. Climate Change and African Forest and Wildlife Resources. Nairobi, Kenya: African Forest Forum; 2011. p. 252

[28] Levina E, Tirpak D. Adaptation to Climate Change: Key Terms. Organisation for Economic Co-operation and Development; 2006. p. 1-25

[29] Ogutu JA, Owen-Smith RN. ENSO, rainfall and temperature influences on extreme population declines among African savannah ungulates. Ecology Letters. 2003;**6**:412-419

[30] Owen-Smith N, Mason DR, Ogutu JO. Correlates of survival rates for 10 African ungulate populations: Density, rainfall and predation. The Journal of Animal Ecology. 2005;**74**:774-788

[31] Reid, H. and Swiderska, K. Biodiversity, Climate Change and Poverty: Exploring the Links. IIED Briefing. 2008. 6 pp

[32] Simmons RE, Barnard P, Dean W, Midgley GF, Thuiller W, Hughes G. Climate change and birds: perspectives and prospects from Southern Africa. Ostrich. 2004;**75**:295-308

[33] Chamaille-Jammes S, Fritz H, Murindagomo F. Spatial patterns of the NDVI–rainfall relationship at the seasonal and interannual time scales in an African savanna. International Journal of Remote Sensing. 2006;**27**:185-200

[34] Washington R, Preston A. Extreme wet years over Southern Africa: Role of Indian Ocean sea surface temperatures. Journal of Geophysical Research – Atmospheres. 2006;**111**:1-15

[35] Davis CL. Climate Risk and Vulnerability: A Handbook for Southern Africa. Pretoria, South Africa: Council for Scientific and Industrial Research; 2011. p. 92

[36] Berteaux, D, de Blois S, Angers J F, Bonin J, Casajus N, Darveau M, Fournier F, Humphries M M, McGill B, Larivée J, Logan T, Nantel P, Périé C, Poisson F, Rodrigue D, Rouleau S, Siron R, Thuiller W and Vescovi L 2010 The CC-Bio project: Studying the effects of climate change on Quebec biodiversity. Diversity. **2**:1181-1204

[37] Gandiwa E. Vegetation factors influencing density and distribution of wild large herbivores in a southern African savannah. African Journal of Ecology. 2014;**52**:274-283

[38] Tokolyi J, Schmidt J, Barta Z. Climate and mammalian life histories. Biological Journal of the Linnean Society. 2014;**111**:719-736

[39] Dunham KM. Trends in populations of elephant and other large herbivores in Gonarezhou National Park, Zimbabwe, as revealed by sample aerial surveys. African Journal of Ecology. 2012;**50**:476-488

[40] Du Toit JT, Biggs H, Rogers KH. The Kruger Experience: Ecology and Management of Savanna Heterogeneity; 2003. p. 519

[41] Ogutu JO, Owen-Smith N, Piepho HP, Said MY. Continuing wildlife population declines and range contraction in the Mara region of Kenya during 1977-2009. Journal of Zoology. 2011;**285**:99-109

[42] Fritz H, Duncan P. On the carrying capacity for large ungulates of African savanna ecosystems. Proceedings of the Biological Sciences. 1994;**256**:77-82

[43] Muthoni FK, Groen TA, Skidmore AK and van Oel P. 2014. Ungulate herbivory overrides rainfall impacts on herbaceous regrowth and residual biomass in a key resource area. Journal of Arid Environments. **100-101**:9-17

[44] Riginos C. Climate and the landscape of fear in an African savanna. The Journal of Animal Ecology. 2015;**84**:124-133

[45] Wigley BJ, Fritz H, Coetsee C, Bond WJ. Herbivores shape woody plant communities in the kruger national park: Lessons from three long-term exclosures. Koedoe. 2014;**56**:1-12

[46] Ogutu JO, Piepho HP, Dublin HT, Bhola N, Reid RS. Rainfall influences on ungulate population abundance in the Mara-Serengeti ecosystem. Journal of Animal Ecology. 2008;**77**(4):814-829

[47] Dixon RK, Smith J, Guill S. Life on the edge: Vulnerability and adaptation of African ecosystems to global climate change. Mitigation and Adaptation Strategies for Global Change. 2003;**8**:93-113

[48] Ogutu JO, Piepho H-P, Kanga E. Dynamics of an insularized and compressed impala population: rainfall, temperature and density influences. Open Ecology Journal. 2012;**5**:1-17

[49] IPCC. Cambio climático 2014. Impactos, adaptación y vulnerabilidad—Resumen oara responsables de políticas. Contrib. del Grup. Trab. II al Quinto Inf. Evaluación del Grup. Intergub. Expert. sobre el Cambio Climático; Geneva; 2014. p. 34

[50] Zimbabwe Parks and Wildlife Management Authority [ZPWMA]. Gonarezhou National Park General Management Plant 2011-2021, Harare, Zimbabwe. 2011

[51] Gandiwa E, Magwati T, Zisadza P, Chinuwo T, Tafangenyasha C. The impact of African elephants on Acacia tortilis woodland in northern Gonarezhou National Park, Zimbabwe. Journal of Arid Environments. 2011;**75**:809-814

[52] Tafangenyasha C. Tree Loss in Gonarezhou National Park (Zimbabwe) between 1970 and 1983. Journal of Environmental Management. 1997;**3**:355-366

[53] Zisadza P, Gandiwa E, Der WHV, Van E, Westhuizen D, Bodzo V, Zisadza P, Gandiwa E, Der WHV, Der WEV, Bodzo V. Abundance, distribution and population trends of hippopotamus in Gonarezhou National Park, Zimbabwe. South African Journal of Wildlife Research. 2010;**40**:149-157

[54] Gandiwa E, Kativu S. Influence of fire frequency on *Colophospermum mopane* and *Combretum apiculatum* woodland structure and composition in northern Gonarezhou National Park, Zimbabwe. Koedoe. 2009;**51**:1-13

[55] Gandiwa E, Zisadza P. Wildlife management in Gonarezhou National Park, Southeast Zimbabwe: Climate change and implications for management. Nature and Faune. 2010;**25**:96-106

[56] Dunham KM, Van Der Westhuizen E, Van Der Westhuizen H F, Gandiwa E. Aerial Survey of Elephants and other Large Herbivores in Gonarezhou National Park (Zimbabwe), Zinave National Park (Mozambique) and surrounds: 2009. 2010

[57] Magadza CHD. Magadza 2000.pdf. Environmental Monitoring and Assessment. 2000;**61**: 193-205

[58] Scholte P. Towards understanding large mammal population declines in Africa's protected areas: A West-Central African perspective. Tropical Conservation Science. 2011;**4**:1-11

[59] Gandiwa E. Rainfall variability and its impact on large mammal populations in a complex of semi-arid African savanna protected areas. Tropical Ecology. 2016;**57**:163-180

[60] Muposhi VK, Muvengwi J, Gandiwa E, Chinake N, Muboko N, Maponga T, Kuvaoga P, Maponga T. Vegetation characteristics of white-browed sparrow weaver nesting sites in Mana Pools. Journal of Biodiversity Management & Forestry. 2015;**4**:2-7

[61] UNEP. Mana Pools National Park, Sapi and Chewore Safari Areas, Harare, Zimbabwe. 1984

[62] Zimbabwe Parks and Wildlife Management Authority [ZPWMA]. Mana Pools National Park General Management Plan Part 2 : Background. Harare, Zimbabwe. 2009. pp. 104

[63] Dunham KM. The effect drought on the large mammal populations of Zambezi riverine woodlands. Zoological Society of London. 1994;**234**:489-526

[64] Dunham KM, du Toit A J. Using citizen-based survey data to determine densities of large mammals: A case study from Mana Pools National Park, Zimbabwe. African Journal of Ecology. 2013;**51**:431-440

[65] Regassa R. Distribution, abundance and population status of Burchells zebra (*Equus quagga*) in Yabello Wildlife Sanctuary, Southern Ethiopia. Journal of Ecology and the Natural Environment. 2013;**5**:40-49

[66] Bosongo GB, Longo JN, Goldin J, Muamba VL. Socioeconomic impacts of floods and droughts in the middle Zambezi river basin. International Journal of Climate Change Strategies and Management. 2014;**6**:131-144

[67] Macandza VA, Owen-Smith N, Cross PC. Forage selection by African buffalo in the late dry season in two landscapes. South African Journal of Wildlife Research-24-month delayed open access. 2004;**34**(2):113-121

[68] Ryan RL, Wamsley MB. Perceptions of wildfire threat and mitigation measures by residents of fire-prone communities in the Northeast: survey results and wildland fire management implications. 2006

[69] Megaze A, Balakrishnan M, Belay G. Diurnal activity budget of African buffalo (Syncerus caffer Sparrman, 1779) in Chebera Churchura National Park, Ethiopia. African Journal of Ecology. 2016

[70] Ncube S, Beevers L, Hes EMA. The Interactions of the Flow Regime and the Terrestrial Ecology of the Mana Floodplains in the Middle Zambezi River Basin. Ecohydrology; 2012

[71] Chamaillé-Jammes S, Valeix M, Fritz H. Managing heterogeneity in elephant distribution: interactions between elephant population density and surface-water availability. Journal of Applied Ecology. 2007;**44**(3):625-633

[72] Dudley JP, Criag GC, Gibson D, Haynes G, Klimowicz J. Drought mortality of bush elephants in Hwange National Park, Zimbabwe. African Journal of Ecology. 2001;**39**(2):187-194

[73] Loveridge AJ, Hunt JE, Murindagomo F, Macdonald DW. Influence of drought on predation of elephant (*Loxodonta africana*) calves by lions (*Panthera leo*) in an African wooded savannah. Journal of Zoology. 2006;**270**:523-530

[74] Crosmary WG, Chamaillé-Jammes S, Mtare G, Fritz H, Côté SD. Decline of sable antelope in one of its key conservation areas: The greater Hwange ecosystem, Zimbabwe. African Journal of Ecology. 2015;**53**(2):194-205

[75] Crosmary WG, Côté SD, Fritz H. Does trophy hunting matter to long-term population trends in African herbivores of different dietary guilds? Animal Conservation. 2015;**18**:117-130

[76] Chamaillé-Jammes S, Charbonnel A, Dray S, Madzikanda H, Fritz H. Spatial distribution of a large herbivore community at waterholes: An assessment of its stability over years in Hwange National Park, Zimbabwe. PLoS One. 2016;**11**:1-13

[77] Chong J. Ecosystem-based approaches to climate change adaptation: progress and challenges. International Environmental Agreements: Politics, Law and Economics. 2014;**14**:391-405

[78] Osbahr H, Twyman C, Adger W, Thomas D. Evaluating successful livelihood adaptation to climate variability and change in Southern Africa. Ecology and Society. 2010;**15**(2)

[79] Sheikh PA, Corn ML, Leggett JA, Folger P. Global Climate Change and Wildlife. Annual Review of Ecology, Evolution, and Systematics. 2006;**37**:637-669

[80] Prato T. Evaluating and managing wildlife impacts of climate change under uncertainty. Ecological Modelling. 2009;**220**:923-930

[81] Mawdsley JR, O'Malley R, Ojima DS. A review of climate-change adaptation strategies for wildlife management and biodiversity conservation. Conservation Biology. 2009;**23**:1080-1089

[82] The Heinz Center. Climate-Change Vulnerability and Adaptation Strategies for Africa's Charismatic Megafauna. Washington, DC, The H. John Heinz III Center for Science, Economics and the Environment; 2012

[83] Owen-Smith N. Spatial ecology of large herbivore populations. Ecography (Cop.). 2014;**37**:416-430

[84] Kupika OL, Nhamo G. Mainstreaming biodiversity and wildlife management into climate change policy frameworks in selected east and southern African countries. Jàmbá: Journal of Disaster Risk Studies. 2016;**8**(3):1-9

[85] Thompson ID. An overview of the science–policy interface among climate change, biodiversity, and terrestrial land use for production landscapes. Journal of Forest Research. 2015;**20**(5):423-429

7

Synopsis of Mangle Species in Mexico

Basáñez Muñoz Agustín de Jesús,
Serrano Solis Arturo, Martínez Cortés Esmeralda,
Cuervo López Liliana,
Capistrán Barradas Ascención and
Naval Ávila Celina

Abstract

The objective of this documentary research work is to contribute to better knowledge of the mangrove species that are located in our country, as well as to provide readers with written and illustrated information on these species. The species described are *Avicennia bicolor* Standl., *Avicennia germinans* (L.) L., *Conocarpus erectus* L., *Laguncularia racemosa* (L.) Gaertner, *Rhizophora harrisonii* Leechm., and *Rhizophora mangle* L. The mangroves present in Mexico comprise three families, four genera, and six species. They have a distribution in the 17 coastal states of the country, the largest number of species in the state of Chiapas. The data obtained that are included in the information are identity, distribution, taxonomy, diagnosis, distribution, uses, and protection. In the case of *Avicennia bicolor* Standl. and *Rhizophora Harrisonii* Leechm., the information is more scarce, since their distribution is limited to the state of Chiapas.

Keywords: knowledge, Mexican, mangroves

1. Introduction

Regarding mangroves, taxonomic uncertainties persist despite the fact that there are currently many works that refer to these tree and shrub species; this is due to the fact that the characteristics that separate mangrove species are diffuse as they belong to different families, some of them more related to terrestrial environments. In addition, identifications in herbaria are erroneous even by specialists. As a consequence of these situations, the systematic and taxonomic knowledge of the mangroves is recent, despite the fact that their study goes back 300 years [47]. It is important to have a synopsis that outlines the main characteristics that distinguish mangrove species, in this chapter, referring to those found on the coasts of Mexico.

The material examined consisted of printed books and information obtained on the Internet concerning the species *Rhizophora mangle* L., *Avicennia germinans* (L.) L., *Laguncularia racemosa* (L.) C.F. Gaertn., *Conocarpus erectus* L., *Rhizophora harrisonii* Leechm., and *Avicennia bicolor* Standl.

The mangroves present in Mexico comprise three families, four genera, and six species. They are distributed in the 17 coastal states of the country, the largest amount of species in the state of Chiapas. In 16 states of the country there are four species (*R. mangle, A. germinans, L. racemosa,* and *C. erectus*); in Chiapas, in addition to the previous ones, *R. harrisonii* and *A. bicolor* are presented.

The family Rhizophoraceae comprises one genus, *Rhizophora,* and two species, *R. mangle* and *R. harrisonii;* the family Acanthaceae is represented by one genus, *Avicennia,* and two species *A. germinans* and *A. bicolor*; and the family Combretaceae has two genera, *Laguncularia,* with the species *L. racemosa* and the genus *Conocarpus,* with the species *C. erectus.*

2. Development

2.1. *Avicennia bicolor* Standl.

Valid name

Avicennia bicolor Standl. Journal of the Washington Academy of Sciences 13 (15): 354. 1923. (J. Wash. Acad. Sci.) By Paul Carpenter Standley [51].

Synonym

Tomlinson [46] considered *Avicennia tonduzii* as a synonym for this species. *A. tonduzii* was described by Mondelke in 1938 in Phytology 1: 273–4. Tomlinson [47], states that *A. tonduzii* appears to be only a variant of *A. bicolor,* distinguishing itself by its narrow leaves and the particular shape of its panicles with pairs of individual flowers separated therefrom.

Taxonomy

Affinities

Suprageneric

Kingdom Plantae

Subkingdom Tracheobionta

Superdivision Spermatophyta

Division Magnoliophyta

Class Magnoliopsida

Subclass Asteridae

Order Lamiales

Family Acanthaceae (Verbenaceae; Avicenniaceae)

Subfamily Avicennioideae

Generic

Avicennia L. (1753) is currently derived from the Acanthaceae family, as per recent phylogenetic studies [5, 36, 45]. Although, it has been placed in the families Verbenaceae or Avicenniaceae in some classifications [33]. The genus can be treated as the subfamily Avicenniaceae [21].

The genus *Avicennia* was named in honor of the famous doctor and scientist, Abu-Ali al-Husayn ibn-Sina, known as Avicenna (980–1073 BC) [37].

Specifics

It is a bicolor species—bi meaning two. The corolla of the flower is white and may have a yellow hue at its base so that its flower is considered bicolor [46].

Type specimens

For *Avicennia bicolor* Standl. A, holotype has been reported in tropics.org [51] as follows: PANAMA: Province of Coclé: collected in mangrove swamp at Aguadulce, December 5, 1911. Collector and Number: H. Pittier 4968. Institution (s): HT: US-715142.

Type-protologue: distribution

Panama: Province of Coclé: collected in mangroves in Aguadulce, December 5, 1911 [51].

Diagnosis

It is a small tree with an irregular crown. Its bark is whitish or light gray-brown (in contrast to the dark grayish coffee of *Avicennia germinans*). The leaves are elliptical, with a bright beam and totally without hairs, characteristic with which it can be differentiated from *A. germinans*, since this one has crystals in the beam and the back. The underside is somewhat hairy and is usually covered by salt crystals. The inflorescence is a terminal or axillary panicle of white, hairless flowers. The fruit is a rounded and a smooth capsule. It has vertical, spongy roots that project over the mud, absorb the air, and ventilate the support root system [4].

Common and vernacular names

Palo de sal [46]

Other common names in Mexico: Mangle negro [34].

Common names in other countries: Curumo blanco—Honduras; Madresal—El Salvador; mangle gateador—Panama; mangle salado—Panama; and Palo de sal (Guanacaste-Costa Rica) [4, 26].

Distribution

Moldenke [40], Tovilla-Hernández et al. [48], and, later, Nettel et al. [41] report that the geographical distribution of *Avicennia bicolor* Standl., at its northernmost end, is in Tonalá, Chiapas, in Mexico (9° 51 'N 84° 41' W). Recent studies by the Southern Border College of Tapachula Unit have located a mature *A. bicolor* forest between the municipal boundaries of Tonalá and Pijijiapan, Chiapas, in addition to finding new distribution sites for *Rhizophora harrisonii* Leechm.(Tovilla, 2012 com. in [44]).

Flowers and propagules of *Avicennia bicolor*. Cistian Tovilla Hernández. Diplomado Internacional en ecología, manejo, restauración y legislación en sistemas de manglares (2017). ECOSUR.

2.2. *Avicennia germinans* (L.) L.

Valid name

Avicennia germinans (L.) L. Linnaeus, Carl Von. Published in: Species Plantarum, ed. 3, 2: 891. 1764. Sp. Pl (ed.3) [52].

Basionym

Bontia germinans was published in Systema Naturae, Editio Decima 2: 1122. 1759. Syst. Nat. (Ed.10). 19 **Synonym**; 1 **Homonym** *Avicennia germinans* (L.) Stearn; 6 **Ifraspecific categories** (form and variety) [52].

Taxonomy

Affinities

Suprageneric

Kingdom Plantae

Subkingdom Tracheobionta

Superdivision Spermatophyta

Division Magnoliophyta

Class Magnoliopsida

Subclass Asteridae

Order Lamiales

Family Acanthaceae (Verbenaceae; Avicenniaceae)

Subfamily Avicennioideae

Generic

Avicennia L. (1753) is derived from the Acanthaceae family, as per recent phylogenetic studies [5, 36, 45], although it has been placed in the families Verbenaceae or Avicenniaceae in some classifications [33]. The genus can be treated as subfamily Avicennioideae [21].

The genus *Avicennia* was named in the honor of the famous doctor and scientist, Abu-Ali al-Husayn ibn-Sina, known as Avicenna (980–1073 BC) [37].

Specifics

Germinans is a Latin word meaning "sprouting" in reference to its particular form of reproduction; this is due to the early germination of the seed within the fruit [4].

Type specimens

For *Avicennia germinans* (L.) L., a lectotype, related to its basionym *Bontia germinans*, has been reported in the Atlas of Florida Plants and Tropicos ® as follows: JAMAICA: Without data, Browne s.n. (lectotype: LINN 813.2). Lectotypified by Stearn, Kew Bull. 1958: 34, 1958 [64] and LT: Browne s.n.; Jamaica (LINN-813.2) LT designated by Stearn, Kew Bull. 13: 35 (1958) [53].

Type-protologue: distribution

Jamaica: No data, Browne s.n. (lectotype: LINN 813.2). Lectured by Stearn, Kew Bull. 1958: 34, 1958 [64].

Diagnosis

It is a tree or shrub that reaches a size of 3–10 m in height [6]. Leaves excrete salt through specialized glands and can be covered by salt, thus contributing to salty leaf litter [35]. Panicle inflorescence in the form of spikes or panicles composed, 9 cm long and 2–5 cm wide. Fruit ovate-oblique, apiculate, 1.5–2 cm long and 1–1.5 cm wide, is sparsely sericeous [6].

It grows on sandy, muddy, or argillaceous soils. It is found in heavily oxidized clays or in soils with high concentrations of pyrite [3]. The species is sensitive to frost but is considered to be the most tolerant mangle species to low temperatures [11].

Common and vernacular names

Mangle negro, Mangle prieto [12].

Other common names in Mexico: mangle blanco—Veracruz, Oaxaca, Tabasco, Yucatan; mangle negro, madre de sal—Acapetagua, Chis; Mangle prieto—Yucatán; Puyeque—Sinaloa [14]. Maya: tab che ', taab che' tat xiül [12].

Black mangrove (Belize); Culumate (Costa Rica); Curumo negro (Honduras); istatén (Costa Rica and El Salvador); mangle negro (El Salvador, Guatemala, Nicaragua, and Panama); Mangle prieto, Mangle salado (Panama); Mangle salsa (Costa Rica); Palo de sal (Costa Rica, Nicaragua) [4]; mangle iguanero (Colombia and Ecuador); mangle rosado (Venezuela) [8].

Distribution

It is found on both coastlines of the country, from the state of Tamaulipas to the Yucatan Peninsula in the Gulf of Mexico and the Caribbean Sea and from the states of Baja California and Sonora to Chiapas in the coast of the Pacific Ocean [42]

Flowers and propagules of *Avicennia germinans*. Agustín de Jesús Basáñez Muñoz (2006). Universidad Veracruzana

2.3. *Conocarpus erectus* L.

Valid name

Conocarpus erectus L. Linnaeus, Carl Von. Published in: Species Plantarum 1: 176. 1753. (1 May 1753) (Sp. Pl.) As erecta [54].

Basionym

The species *Conocarpus erectus* L. has as basionym *Terminalia erecta* (L.) Baill *, which means that this species was initially described by Linnaeus in the genus *Conocarpus* and later reported by Baillon in the genus *Terminalia*. 10 **Synonyms**; 8 **Ifraspecific name** (form and variety) [54].

Taxonomy

Affinities

Suprageneric

Kingdom Plantae

Subkingdom Tracheobionta

Superdivision Spermatophyta

Division Magnoliophyta

Class Magnoliopsida

Subclass Rosidae

Order Myrtales

Family Combretaceae

Generic

The genus *Conocarpus L. comes* from the Greek word "konos", cone and "carpos" fruit because the fruits resemble a conical shape [31].

Specifics

erectus.- masculine erectus, feminine *erecta,* neuter *erectum*; means erect, right; by the erect habit of the plant [63].

Type specimens

For *Conocarpus erectus* L., the Atlas of Florida Plants and Tropicos ® have reported as follows: JAMAICA: Without data (lectotype: Sloane, Voy Jamaica t 161 (2) 1703.). Lectotipified by Wijnands, Bot. Commelins 66, 1983 [64] and LT: Sloane, Voy. Jamaica t. 161, f. 2 (1725); LT designated by Wijnands, Bot. Commelins 66 (1983); TOP: Sloane Herb., 5: fol. 63 TT designated by C.E. Jarvis [54].

Type protologue: distribution

Jamaica: No data (lectotype: Sloane, Voy Jamaica, 161 (2) 1703.). Lectotipified by Wijnands, Bot. Commelins 66, 1983 [64].

Diagnosis

Conocarpus erectus L.: differs from mangroves in their reproductive strategy; it is reproduced through seeds [30].

It has an erect trunk or several trunks but may assume a prostrate body; the crust is gray or brown, wrinkled, fibrous, and moderately thin; the inner crust is dark cream color. The leaves are alternate, simple, and oblong, 2–7 cm in length (rarely 10 cm long) and 1–3 cm in width, with a decrease in the tip; they are dark green and bright in the bundle and pale in tone and with fine silky hairs on the underside. The inflorescences are terminal or axillary panicles, of small flowers of greenish-white color grouped in spheroidal heads of 3–5 mm in diameter. Fruits, 4-mm winged nuts, are added in globose brown heads, 1–1.3 cm in diameter. *Conocarpus erectus* L. is intolerant to shade [28].

Common and vernacular names

Mangle botoncillo [15].

Other common names in Mexico: mangle negro, mangle prieto—Veracruz, Tabasco, Campeche, Oaxaca, Guerrero; mangle botoncillo—Yucatan and Veracruz [42]. Maaya: k'an che '[27].

Botoncillo (El Salvador, Honduras, Nicaragua); buttonwood (Belize); mangle boton (Costa Rica, Panama); mangle botoncillo (Guatemala); mangle gris (Honduras); mangle negro (Costa Rica); palo boton (Honduras) [4]. Mangle zaragoza (Costa Rica, Panama) [29].

Distribution

It is found on both coastlines of the country, from the state of Tamaulipas to the Yucatan Peninsula in the Gulf of Mexico and the Caribbean Sea and from the states of Baja California and Sonora to Chiapas in the Pacific Ocean [42].

Flowers and fruits of *Conocarpus erectus*. Agustín de Jesús Basáñez Muñoz (2006). Universidad Veracruzana.

2.4. *Laguncularia racemosa* (L.) C.F. Gaertn.

Valid name

Laguncularia racemosa (L.) C.F. Gaertn. Gaertner, Carl (Karl) Friedrich Von. Published in: Supplementum Carpologiae 209. 1807. (Suppl Carp) [55].

Basionym

The basionym of this species is *Conocarpus racemosus* L. *, initially described in the genus *Conocarpus* by Linnaeus and translated by C.F. Gaertner in 1807 to the genus *Laguncularia*. The first publication of the species *Conocarpus racemosus* was in Systema Naturae, Editio Decima 2: 930. 1759. Syst. Nat. (Ed.10). 5 **Synonym**; 2 **Ifraspecific categories** (form and variety) [55].

Taxonomy

Affinities

Suprageneric

Kingdom Plantae

Subkingdom Tracheobionta

Superdivision Spermatophyta

Division Magnoliophyta

Class Magnoliopsida

Subclass Rosidae

Order Myrtales

Family Combretaceae

Generic

Laguncularia Gaertn (1807) is a monotypic genus (a single species) from tropical America and Africa [22].

Its name (*Laguncularia*) comes from the Latin term "laguncula," diminutive of "lagena," which means bottle; because the limbus of the chalice, when closing, constitutes the shape of the fruit (propagule), it is shaped like a small bottle [31] (and Personal Commentary, 2018).

Specifics

racemosa comes from the Latin racemosa, which means cluster, which alludes to the type of inflorescence that the plant presents (cluster type) [39].

Type specimens

The Atlas of Florida Plants and Tropicos ® [2] report a lectotype for *Laguncularia racemosa* (L.) C.F. Gaertn. rather related to his basionym *Conocarpus racemosus*: Without data (lectotype: LINN 237.2). Lectotypified by Bornstein, in R. A. Howard, Fl. Antill., Dicot. 2: 459, 1989 [64] and LT: Anon.; (LINN-237.2) LT designated by Bornstein, Fl. Lesser Antilles 5: 459 (1989) [56].

Type protologue: distribution

Jamaica: Without data (lectotype: Sloane, Voy, Jamaica, 161 (2), 1703). Lectotypified by Wijnands, Bot. Commelins 66, 1983 [64].

Diagnosis

This species presents as shrubs or trees, which reach a size of up to 10 m high, frequently with pneumatophores. Its trunk is straight with ascending branches, rounded, and a dense cup. The leaves of the white mangrove are opposite, elliptical, and rounded at both the base and the apex; they measure from 4–10 cm in length and from 2–4 cm in width; the top of the leaf (beam) is bright dark green and the bottom (underside) is yellowish green. The flowers appear in axillary and terminal panicles, are fragrant, and measure 1.5 mm in diameter [4].

Its silky and fleshy fruits have a flattened bottle shape, measure between 1 and 2.5 cm in length, and have several longitudinal grooves. They contain a seed; the seed often begins to germinate inside the fruit when it is still attached to the tree [42].

Common and vernacular names

Mangle blanco [16].

Other common names in Mexico: Mangle blanco—Tamaulipas, Oaxaca; mangle bobo—Yucatán; Mangle chino—Sinaloa [14] Sak-okom (Mayan language)—Yucatan; tzakol-kon—Yucatan Peninsula (Martínez, 1979 in [14]).

Patabán (Cuba); White mangrove (United States); Cincahuite, Palo de Sal (Costa Rica); akira (Suriname); jeli de mangle (Peru). [16].

Distribution

It is found on both coastlines of the country, from the state of Tamaulipas to the Yucatan Peninsula in the Gulf of Mexico and the Caribbean Sea and from the states of Baja California and Sonora to Chiapas in the Pacific Ocean [42].

Flowers and propagules of *Laguncularia racemosa*. Agustín de Jesús Basáñez Muñoz (2006). Universidad Veracruzana.

2.5. *Rhizophora harrisonii* Leechm.

Valid name

Rhizophora harrisonii Leechm: Bulletin of Miscellaneous Information Kew 1918 (1): 8, f. A. 1918. (Bull Misc.Kew Inform) [57].

Synonym

Rhizophora brevistyla Salvoza [57].

Taxonomy

Affinities

Suprageneric

Kingdom Plantae

Subkingdom Tracheobionta

Superdivision Spermatophyta

Division Magnoliophyta

Class Magnoliopsida

Subclass Rosidae

Order Malpighiales

Family Rhizophoraceae

Generic

Rhizophora: The word "rhiza" (root) and "phoros" (bearer or bearer) means in Greek "bearing the roots" and refers to its aerial (willow) roots' characteristics of the genus [25].

Specific

Harrisonii: In the honor of Professor J.B. Harrison, C.M.G., M.A. Director of Science and Agriculture in British (English) Guiana [32].

An important aspect to emphasize is the use of taxonomic status:

Rhizophora x harrisonii was proposed by Tomlinson in 1986 [18] in his book "The Botany of Mangroves." The author mentions that "in relation to *Rhizophora x harrisonii,* there is circumstantial evidence that the species is a hybrid between *Rhizophora mangle* and *Rhizophora racemosa.*"

In this regard, a study of hybridization and introgression between the species of the genus *Rhizophora* of the New World (*R. mangle, R. racemosa* and *R. harrisonii*) carried out by Cerón et al. [10], concludes that "our data support an ancient and persistent hybridization of the *Rhizophora* genus and propose a complete review of the group's systematic relationships based on finer morphological, ecological and genetic analyzes." "However, we found no genetic evidence that *R. harrisonii* is a hybrid species". Rather, *R. harrisonii* appears to represent a morphotype produced by a process of hybridization and backcrossing between *R. mangle* and *R. racemosa* [10].

Type specimens

Rhizophora harrisonii Leechm. has a lectotype as follows: LT: Leechman s.n.; Guyana (K) LT designated by Barrie, Fl. Mesoamer. 4 (1): ined. [57].

Type-protologue: distribution

Guyana: British Guiana: Two-mile stretch of coast, in the vicinity of Georgetown [57].

Diagnosis

Rhizophora harrisonii Leechm. are trees up to 25 m with leaves from 9–13 cm in length and from 3–8 cm in width, elliptic, acute apex, and the petiole measures are from 1–3 cm. It presents dichotomically branched inflorescences of 5–6 times or the first tricotómica branch; the branches are thin, laxas, and arranged in acute angles. The flower buds are 3–9 mm, ellipsoid, and the apex is slightly attenuated; sepals are of 8–10 mm compared to their petals of 5–6 mm; they are deciduous, abaxially glabrous, villous margins; they present 8 stamens of 4.5–5 mm and sessile and are apiculated. The fruit is 28–33 mm long with its hypocotyl from 11–40 cm, straight or curved [58].

Common and vernacular names

Mangle caballero, Mangle zapatero [7].

Common name in other countries: Mangle rojo—Venezuela [20].

Distribution

Rico-Gray [43] said that *Rhizophora harrisonii* Leechm. "is a new record for the coasts of Mexico", reporting to the species for the State of Chiapas. Recent studies by the South Border College of Tapachula Unit have located a mature forest of *Avicennia bicolor* between the municipal boundaries of Tonalá and Pijijiapan, Chiapas, in addition to finding new distribution sites for *Rhizophora harrisonii* (Tovilla, 2012 com. [44]).

Flowers and propagules of *Rhizophora harrisonii*. Cistian Tovilla Hernández. Diplomado Internacional en ecología, manejo, restauración y legislación en sistemas de manglares (2017). ECOSUR.

2.6. *Rhizophora mangle* L.

Valid name

Rhizophora mangle L. Linnaeus, Carl Von. Species Plantarum 1: 443. 1753. (1 May 1753) (Sp. Pl.) [59].

Synonym

Rhizophora americana Nutt.

Rhizophora mangle var. samoensis Hochr.

Rhizophora samoensis (Hochr.) Salvoza.

Name accepted

Nasir and Ali in 1972 [38], consider *Rhizophora mucronata* Lam. as an accepted name (it is the name which can be used to refer to species (or subspecies, varieties, or forms) of *Rhizophora mangle* L. 2 **Ifraspecific categories** (variety).

Taxonomy

Affinities

Suprageneric

Kingdom Plantae

SubKingdom Tracheobionta

Superdivision Spermatophyta

Division Magnoliophyta

Class Magnoliopsida

Subclass Rosidae

Order Malpihiales

Family Rhizophoraceae

Generic

Rhizophora: The word "rhiza" (root) and "phoros" (bearer or bearer) means in Greek "bearing the roots" and refers to its aerial (willow) roots characteristics of the genus [25].

Specific

Mangle: The word mangle is derived from Guarani and means "twisted tree" [1].

Type specimens

The tropicos.org ® website reports an epithet such as, ET: Jamaica (F; IET: DUKE, MICH) ET designated by Barrie, Fl. Mesoamer. 4 (1): ined. As well as a Lectotype: LT: Plumier, Nov. Pl. Amer. t. 15 (1703) LT designated by Barrie, Fl. Mesoamer. 4 (1): ined., And another related to: LT: Herb. Sloane 6: 62; (BM) LT designated by Keay, Kew Bull. 8 (1): 123 (1953) [59].

Type protologue: distribution

Jamaica (F; EET: DUKE, MICH) ET designated by Barrie, Fl. Mesoamer. 4 (1): ined. [59].

Diagnosis

The trees of *Rhizophora mangle* are from 4 to 10 m in height, their shape is like a tree or and evergreen shrub. The leaves are simple, opposite, petiolate, with rounded leaf, and they are elliptic to oblong; these are agglomerated at the tips of the branches, their color is dark green in the bundle, and yellowish on the underside. The flowers are small, 2.5 cm in diameter with four speared sepals, thick and leathery. The flower has four yellowish white petals. It has two to four flowers per stem or peduncle. The fruits are presented in the form of a brown, coriaceous, hard, piriform, farinous berry. A seed is developed, rarely two, per fruit [62].

Common and vernacular names

Common names in Mexico: Candelón—Veracruz, Colima, Sinaloa; Mangle; Mangle colorado; Mangle dulce—Baja California, Oaxaca; Mangle rojo; Mangle tinto—Veracruz [17, 62]. Maya: Tabché, Tapché, Xtabché [17, 62].

Mangle colorado, mangle (Honduras); mangle rojo (Costa Rica); mangle salado (Panama); Candelin, Mangle dulce (Mexico); mangle caballero, mangle gateador (Colombia); Mangle verdadero, mangle zapatero (Ecuador); purgua (Venezuela); apareiba, mangle zapateiro, mangle vermelho (Brazil); mang wouj (Haiti) [8].

Distribution

It is found on both coastlines of the country, from the state of Tamaulipas to the Yucatan Peninsula in the Gulf of Mexico and the Caribbean Sea and from the states of Baja California and Sonora to Chiapas in the Pacific Ocean [42].

Flowers and propagules of *Rhizophora mangle*. Agustín de Jesús Basáñez Muñoz (2006). Universidad Veracruzana

3. Importance

Mexico is among the five countries in the world with the largest extension of mangroves distributed; by 2015, 7,75,555 ha of mangroves have been registered in both coasts of the country covering at least 60% of the coastline [61].

In 2005, the National Commission for the Knowledge and Use of Biodiversity (CONABIO) initiated the bases for what is now called the Mangrove Monitoring System of Mexico. The aim of the SMMM is to generate information about changes in the mangrove ecosystem through the evaluation of its spatial distribution and condition over time. From this information, we also look for the identification of existing, latent threats and trends of changes (loss, deterioration, or recovery), in such a way as to support their conservation, understanding, and management. Based on the information generated, the threats and trends of change that through analysis have been incorporated into the conservation of this ecosystem have been identified. The results show a strong occurrence of both natural processes and human activities, which influence the loss of coverage or disturbance of the mangrove. Of these, those of anthropic origin stand out for their importance. In this category, two classes are presented: the agricultural-livestock and a pattern of occupation of land use derived from development. The first one is related to primary economic activities, that is to say agriculture (both irrigation and seasonal), livestock, and forestry. The second class encompasses land uses, such as rural areas, urban areas, industrial zones, aquaculture farms, ports, tourist infrastructure, and so on [61].

The tendency of loss of coverage occurred mainly in the period from 1970 to 2005 (up to 27,557 ha), there being a drastic change from 2005 to 2015, with the last period reported from 2010 to 2015 with a loss of 1090 ha but a profit of 1296 ha. Within the categories identified by the SMMM as threats in the loss of mangrove are the areas under construction, aquaculture farms and artificial ponds, hydraulic infrastructure (canals or dams), industrial zones (oil wells, salt pans, thermoelectric plants, complexes), and communication routes [61].

In Mexico, 6 of the world's 70 mangrove species (8.5%) have been reported [50] and contribute 5.4% of the total mangrove area, after Indonesia with 22.6%, Australia with 7.1%, and Brazil with 7% [60].

Its biological importance lies in being places of rest and nesting of birds (*Egretta caerulea, E. rufescens, E. tricolor, Ardea herodias, Aramides cajaneus, Sula leucogaster, Phalacrocorax auritus, Fregata magnificens, Ajaia ajaja,* among others). They represent an important habitat for species with some risk category (NOM-059-SEMARNAT-2010) such as Mexican *Tamandua mexicana, Buteogallus anthracinus, Rostrhamus sociabilis, Mycteria americana, Vireo pallens, Megascops cooperi, Crocodylus acutus, C. moreletii, Ctenosaura pectinata, C quinquecarinata, C. similis, C. acanthura, C. hemilopha,* and *Iguana iguana* [50].

The biological and ecological importance of mangroves has led CONABIO to establish 81 priority sites, determined by specialists in the subject; of these 29 are located on the Pacific coast, 27 in the Gulf of Mexico, and 25 in the Yucatan Peninsula. Each of these sites of interest has a characterization sheet with information on site location, physical characteristics, socioeconomic, uses, importance, impacts, and threats and transformation processes [13].

4. Uses

They are used in firewood and coal; poles for fences, piles, railway sleepers, piers, boats, telegraph poles, and electricity; and furniture, cabinets, door frames, musical instruments, handles for tools, and agricultural implements. Newborn seedlings are edible if cooked but raw seeds are poisonous. The bark contains tannin and is used to tan skins. The infusion of its cooking is drunk as a remedy for diarrhea, intestinal irritation and colic, washing or bathing to heal wounds and hemorrhoids, or in bleeding gums rinses. The flowers are rich in honey and in nectar. The honey obtained is white, clear, and of excellent quality [4].

5. Protection

In Mexico, the mangrove is considered a commonly used asset, although the General Law of National Assets prevents a private or public entity from making use of them; it will require the processing of a concession or assignment, which will never generate real rights, exclusively and without prejudice to third parties; it will only generate the right to use or exploit the assets with the limits set forth by the laws and concession. Revocation of a concession may exist, for example, if fixed constructions are made that damage the present ecosystems. From this regulation, in 1996, the General Law of Ecological Equilibrium and Environmental Protection (LGEEPA) determines, in its Article 28, that the environmental impact assessment procedure should start if works or activities that can be carried out are carried out. A delay can cause ecological imbalances in coastal ecosystems, coastal wetlands, mangroves, lagoons, rivers, lakes, and estuaries connected to the sea. In 2000, the General Wildlife Law was enacted, which excludes timber resources from sustainable use and species whose livelihood is water, if they are considered as species or populations at risk. To consider those species that were considered at risk, in 2001, Official Mexican Standard NOM-059-ECOL-2001 was published, listing the species referred to in the General Wildlife Law. In this Official Mexican Standard the four mangrove species with a national presence (*Rhizophora mangle, Avicennia germinans, Laguncularia racemosa,* and *Conocarpus erectus*) are integrated with the status of special protection. In the reform of the Official Mexican Standard (NOM-059-SEMARNAT-2010 [24]) mangrove species are considered as threatned.

Parallel to these instruments of environmental policy, since 2000, work began on an Official Mexican Standard (NOM) that established the specifications for the preservation, conservation, sustainable use, and restoration of coastal wetlands in mangrove areas; in 2003, when it is published, the NOM refers to a series of provisions that are mandatory for those responsible for carrying out works or activities that are intended to be located in coastal wetlands or whose characteristics may negatively influence them. At the same time, with the enactment of the General Law of Sustainable Forestry Development in 2003, legal uncertainty was created in the protection of mangroves when considering, in Article 28 of its regulation (published in 2005), that the areas with mangrove vegetation are areas of conservation and restricted use.

To give greater legal certainty to the protection of mangroves, in 2007, a reform to the General Wildlife Law was published, Article 60 TER, which specifies that any activity that affects the integrity of the species and its productivity is prohibited, as well as the hydrological flow and environmental services provided by the mangroves [9, 49]. Actions in favor of mangroves in the face of climate change are considered in the General Law on Climate Change published in 2012, which in its Article 26 mentions as the fundamental principle of the Law the conservation of ecosystems and their biodiversity, giving priority to wetlands, mangroves, reefs, dunes, coastal zones, and lagoons, that provide environmental services, fundamental to reduce vulnerability.

To support the protection of mangroves, the government of Mexico, through NOM-022-SEMARNAT-2003 [23], empowers the Federal Environmental Protection Agency (PROFEPA) to monitor the provisions stipulated. PROFEPA has established a policy of inspection and surveillance for the conservation of mangroves in which non-compliance with the provisions of prevention, conservation, sustainable use, and restoration is considered an environmental crime. In turn, the Secretariat of the Navy (SEMAR) through its institutional program for protection, conservation, restoration, and reforestation of the mangrove collaborates in the protection and conservation of the mangrove in places that are outside of some kind of special protection (e.g., that are not within protected natural areas) to diminish their deterioration.

In Mexico there are no regulations for the species *Avicennia bicolor* Standl and *Rhizophora harrisonii* Leechm. The International Union for the Conservation of Nature (IUCN) has them in the category of vulnerable species, given that there are only reports of few individuals (without specifying how many) present of these species within the mangrove (dominated by the other species of mangrove), without forming extensive pure masses.

6. Discussion

The current situation of the mangroves of Mexico is presented in a favorable context due to the fifth place that occupies worldwide by surface covered in mangrove, the existence of six species whit the representativeness of four of them in the 17 coastal states and they covered 60% of the coastal surface. Although there are threats that can cause loss of mangrove areas, 9.4% in 45 years (1970–2015) due mainly to the lack of urban, industrial, and tourist development planning, as well as the advancement of the agricultural frontier and the activities related to aquaculture, there is legislation and regulations for its long-term protection. The protection strategies hitherto employed have yielded good results, although there is still a need for more constant vigilance and not only through programmed operations. In relation to their conservation, the decrees of Natural Protected Areas and Ramsar Sites have under their protection more than 50% of the mangroves of the country; the actions aimed to generate greater decrees of mangrove zones are continued. The support to the rehabilitation and restoration of the mangroves has become a national strategy of government agencies that are related to these communities, including the Secretary of Environment and Natural Resources (SEMARNAT), the National Commission of Natural Protected Areas (CONANP), the National Forestry Commission (CONAFOR), the National Commission for the Knowledge and Use of Biodiversity (CONABIO), as well as decentralized public agencies such as Petroleos Mexicanos (PEMEX) and the Federal Electricity Commission (CFE).

The conservation of mangroves not only ensures the perpetuity of mangrove species but also the species that inhabit or rest in these environments—both at the level of those that are under some risk status and at the level of those that maintain the fishing production of the coastal zones.

The documented material consisted of printed books [9] and information obtained on the Internet concerning the species *Rhizophora mangle* L., *Avicennia germinans* (L.) L., *Laguncularia racemosa* (L.) C.F. Gaertn., *Conocarpus erectus* L., *Rhizophora harrisonii* Leechm., and *Avicennia bicolor* Standl. It was observed that the information obtained through written means is more truthful and reliable than the information obtained via the Internet. The information obtained from the Internet is more practical and easy to find, but when reviewing and analyzing this information we can see a repetition of pages, with the same information but with different authorship for different species of mangroves and in most cases the author of the informative text does not appear (reason why it was discarded).

It should be noted that the most reliable and quoted information was found on websites of other countries, mainly the United States and Australia. In addition, translations from English, French, and Portuguese into Spanish had to be done. On other occasions the original documents of the description of the species had to be reviewed, such as *R. harrisonii* and *A. bicolor*, using the search engine Biodiversity Heritage Library (biodiversitylibrary.org), which is a consortium of natural history and libraries of botanical topics that have come together to digitize the legacy of the literature on the biodiversity of their collections and make the available open access literature and for responsible use as part of the "global commons of biodiversity."

7. Conclusion

With the proposed synopsis, we have a broad overview of the systematic and taxonomic information of mangrove species in Mexico, which is not easily found in a single compendium, so that their contribution is of special interest to students of the upper levels, as well as the people who are interested in the topic of mangroves. A thorough investigation was undertaken on the meaning of their scientific names, an exercise that is not very common in the disclosure of species fills that gap of information. The topics of its importance, uses, and protection were approached with the most recent knowledge available, complemented by the authors' opinions. In relation to the importance of the mangroves, the species that are found listed in irrigation in the NOM-059-SEMARNAT-2010 [24] and that inhabit the mangroves stand out. The uses of mangrove species have not yet been addressed by citizen participation research in which use values are discussed; only references are made in relation to surveys among the inhabitants, and there is a lack of information. Legislation that protects mangroves is effective in the written word, but greater vigilance is needed in their compliance; their conservation strategies are carried out through decrees of Natural Protected Areas and Ramsar Sites but there are still strategic mangrove sites that they must be incorporated. The support granted to the rehabilitation and restoration of the mangroves is used by the inhabitants who adjoin these communities; the non-governmental organizations and the researchers of universities and institutes are to work in favor of this precious resource.

Author details

Basáñez Muñoz Agustín de Jesús[1]*, Serrano Solis Arturo[1], Martínez Cortés Esmeralda[2], Cuervo López Liliana[1], Capistrán Barradas Ascención[1] and Naval Ávila Celina[1]

*Address all correspondence to: abasanez@uv.mx

1 Cuerpo Académico Manejo de Ambientes Marinos y Costeros, Facultad de Ciencias Biológicas y Agropecuarias-Tuxpan, Universidad Veracruzana, Veracruz, México

2 Maestría en Manejo de Ecosistemas Marinos y Costeros, Facultad de Ciencias Biológicas y Agropecuarias-Tuxpan, Universidad Veracruzana, Veracruz, México

References

[1] Agraz-Hernández CM, Noriega-Trejo R, López-Portillo J, Flores-Verdugo FJ Jiménez-Zacarías JJ. Guía de campo: Identificación de los manglares en México. Universidad Autónoma de Campeche, EPOMEX, CFE, CONAFOR, UNAM, CEDESU, INECOL; 2006. p. 56. ISBN: 9685722455

[2] Atlas of Florida Plants. Institute for Systematics Botany; 2018. http://www.florida.plantatlas.usf.edu/plant.aspx?ID=3621

[3] Augustinus PGEF, Slager S. Soil formation in swamp soils of the coastal fringe of Surinam. Geoderma. 1971;**6**:203-211

[4] Barrance A, Beer J, Boshier DH, Chamberlain J, Cordero J, Detlefsen G, Finegan B, Galloway G, Gómez M, Gordon J, Hands M, Hellin J, Hughes C, Ibrahim M, Leakey R, Mesén F, Montero M, Rivas C, Somarriba E, Stewart J. Descripciones de especies de árboles nativos de América Central; (Arboles de Centroamérica un Manual para el Extensionista). San Salvador, El Salvador: CATIE; 2013. www.fundesyram.info

[5] Borg AJ, McDade LA, Schönenberger J. Molecular phylogenetics and morphological evolution of Thunbergioideae (Acanthaceae). Taxon. 2008;**57**:811-822

[6] Brako L, Zarucchi JL, editors. Catalog of the flowering plants and gymnosperms of peru. Monographs in Systematic Botany from the Missouri Botanical Garden. 1993; **45**(i–xl):1-1286

[7] Caballero R. La etnobotánica en las comunidades negras e indígenas del delta del Río Patia. Universidad Nacional de Colombia. Medellín, Colombia in: Enciclovida.Conabio. 2017. 1995. http://www.enciclovida.mx/especies/6058219

[8] CATIE. Manejo de semillas de 100 especies forestales de America Latina. Rodolfo Salazar, coord. Técn., Carolina Soihet, José Miguel Méndez. Comps. Téc. Turrialba, Costa Rica: CATIE, Proyecto de Semillas Forestales: Danida Forest Seed Centre. Serie Técnica. Manual técnico/CATIE; No. 41 2000

[9] CEMDA. Marco jurídico de los humedales costeros con presencia de manglar. Centro Mexicano de Derecho Ambiental y Fondo para la Conservación del Golfo de California. México. 32 p. 2011. http://www.cemda.org.mx/wp-content/uploads/2011/12/marco-jur%C3%ADdico-de-los-humedales-costeros-con-presencia-de-manglar1.pdf

[10] Cerón-Souza I, Rivera-Ocasio E, Medina E, Jiménez JA, Owen McMillan y W, Bermingham E. Hybridization and introgression in new World red mangrove, Rhizophora (Rhizophoraceae). American Journal of Botany. 2010;**97**(6):945-957

[11] Chapman VJ. 1976. Mangrove vegetation. Vaduz, Lichtenstein: J. Cramer. p. 447

[12] CONABIO. 2009. Mangle negro. Fichas de Especies Mexicanas. Comisión Nacional para el Conocimiento y Uso de la Biodiversidad. México, D.F., México. Compilado por Elizabeth Torres bahena; Revisado por Carlos Galindo leal. Marzo 2009. http://www.biodiversidad.gob.mx/v_ingles/species/especies_priori/fichas/pdf/mangleNegro.pdf

[13] CONABIO. 2009. Sitios de manglar con relevancia biológica y con necesidades especiales de rehabilitación ecológica. Comisión nacional para el Conocimiento y Uso de la Biodiversidad. México, D.F. http://www.biodiversidad.gob.mx/ecosistemas/manglares2013/sitiosPrioritarios.htmlhttp://www.conabio.gob.mx/institucion/proyectos/resultados/J084_Fichas%20de%20Especies.pdf

[14] CONAFOR. 2014. Fichas técnicas para reforestación. SEMARNAT. http://www.conafor.gob.mx:8080/documentos/docs/13/888Avicennia%20germinans.pdf

[15] CONAFOR. 2014. Fichas técnicas para reforestación. SEMARNAT. http://www.conafor.gob.mx:8080/documentos/docs/13/904Conocarpus%20erectus.pdf

[16] CONAFOR. 2014. Fichas técnicas para reforestación. SEMARNAT. http://www.conafor.gob.mx:8080/documentos/docs/13/937Laguncularia%20recemosa.pdf

[17] CONAFOR-CONABIO-SIRE. 2014. Paquetes tecnológicos. Rhizophora mangle. http://www.conafor.gob.mx:8080/documentos/docs/13/997Rhizophora%20mangle.pdf

[18] Cornejo X. Lectotypification and new status for Rhizophora X harrisonii (Rhizophoracea), a natural hybrid between *R. mangle* and *R. racemosa*. Harvard Papers in Botany. 2013; **18**(1):37

[19] Cortes Rojas G, García Santiago S. Investigación Documental. Guía de Autoaprendizaje. Secretaria de Educación Pública. Escuela Nacional de Biblioteconomía y Archivonomìa; México; 2003. 96p

[20] Cumana L, Sanabria M, Leopardi C, Guevara de Franco Y. Plantas vasculares de los manglares del Estado Sucre, Venezuela. Acta Botanica Venezuelia [online]. 2010;**33**(2):273-298

[21] Daniel T. Avicennia (Acanthaceae: Avicennioideae) in North America and Mesoamerica. In: Proceedings of the California Academy of Sciences. 2016 ser. 4, 63. pp. 163-189

[22] Davidse G, Sousa Sánchez M, Knapp S, Chiang Cabrera F. Cucurbitaceae a Polemoniaceae. Fl. Mesoamer. 2009;**4**(1):1-855. http://www.cicy.mx/Sitios/flora%20digital/indice_tax_especies.php?genero=Laguncularia

[23] Diario Oficial de la Federación. Norma Oficial Mexicana NOM-022-SEMARNAT-2003, Que establece las especificaciones para la preservación, conservación, aprovechamiento sustentable y restauración de los humedales costeros en zonas de manglar. (Primera Sección). 22p. 2003. http://www.profepa.gob.mx/innovaportal/file/3281/1/nom-022-semarnat-2003.pdf

[24] Diario Oficial de la Federación. NORMA Oficial Mexicana NOM-059-SEMARNAT-2010, Protección ambiental-Especies nativas de México de flora y fauna silvestres-Categorías de riesgo y especificaciones para su inclusión, exclusión o cambio-Lista de especies en riesgo. (Segunda Sección). 77p. 2010. http://biblioteca.semarnat.gob.mx/janium/Documentos/Ciga/agenda/DOFsr/DO2454.pdf

[25] Enciclovida.Conabio. 2017. http://www.enciclovida.mx/especies/6058219

[26] Espinoza R, Guadamuz A, Perez D, Chavarria y F, Masís A. Species Page de Avicennia bicolor (Verbenaceae), 5 amyo 2000. Species Home Pages, Area de Conservación Guanacaste, Costa Rica. 2000. http://www.acguanacaste.ac.cr

[27] Flora Digital: Península de Yucatán Herbario CICY. 2010. Unidad de Recursos Naturales. http://www.cicy.mx/sitios/flora%20digital/ficha_virtual.php?especie=1181

[28] Francis JK, editor. Wildland shrubs of the United States and its territories: Thamnic descriptions. vol. 1. General Technical Report IITF-GTR-26. San Juan, PR: U.S. Department of Agriculture, Forest Service, International Institute of Tropical Forestry, and Fort Collins, CO: U.S. Department of Agriculture, Forest Service, Rocky Mountain Research Station; 2004. p. 830

[29] Grandtner MM, Chevrette J. Dictionary of Trees, Volume 2: South America: Nomenclature, Taxonomy and Ecology. Amsterdam: Academic Press; 2013. p. 1172

[30] IFAS. South Florida native plants fact sheets. Buttonwood (*Conocarpus erectus*) U.S. Department of Agriculture, Cooperative Extension Service, University of Florida, IFAS, Florida A. & M. 9/2013; 2013. http://lee.ifas.ufl.edu/Hort/GardenPubsAZ/Buttonwood.pdf

[31] Institut colonial de Marseille. Annales de l'Institut colonial de Marseille. Quatriéme aneé, Troisieme volume. Flore pharénogamique des Antilles françaises (Guadeloupe et martinique). Marseille: Institut colonial de Marseille; 1897. p. 656

[32] Leechman Alleyne. The genus Rhizophora in British Guiana. Bulletin of miscellaneous information. Royal Botanic Gardens, KEW (1) Art. II: 4-8. 1918. http://biodiversitylibrary.org/page/11494646

[33] Li X, Duke NC, Yang Y, Huang L, Zhu Y, Zhang Z, et al. Re-evaluation of phylogenetic relationships among species of the mangrove genus Avicennia from Indo-West Pacific based on multilocus analyses. PLoS ONE. 2016;**11**(10):e0164453. DOI: https://doi.org/10.1371/journal.pone.0164453

[34] López Portillo J, Ezcurra E. Los manglares de México: una revisión. Revista Madera y Bosques, Vol. 8. Num. Esp. 1: 25-51. 2002

[35] Lugo AE, Snedaker SC. Properties of a mangrove forest in southern Florida. In: Walsh G, Snedaker S, Teas H, editors. Proceedings of international symposium on biology and management of mangroves. Gainesville: Institute of Food and Agricultural Sciences, University of Florida; 1975. pp. 170-212

[36] McDade LA, Daniel TF, Kiel CA. Toward a comprehensive understanding of phylogenetic relationships among lineages of Acanthaceae s.l. (Lamiales). American Journal of Botany. 2008;**95**:1136-1152

[37] Mangrove. 2015. watch.org http://www.mangrovewatch.org.au/index.php?option=com_content&view=category&layout=blog&id=34&Itemid=300172

[38] Nasir E, Ali SI. Flora of Pakistan. Departament of Botany. University of Karachi, Karachi University. Karachi: Printing Press; 1972

[39] Mazza G. Lumnitzera racemosa. 2015. http://www.photomazza.com/?Lumnitzera-racemosa&lang=es

[40] Moldenke HN. Materials toward a monograph of the genus Avicennia. Phytologia. 1960;**7**:123-168

[41] Nettel A, Dodd R, Afzal-Rafii Z, Tovilla-Hernández C. Genetic diversity enhanced by ancient introgression and secondary contac in East Pacific black mangroves. Molecular Ecology. 2008;**17**:260-2690

[42] Pennington T, Sarukhán J. Árboles Tropicales de México. Manual para la identificación de las principales especies. Universidad Nacional Autónoma de México y Fondo de Cultura Económica; 2005. p. 523

[43] Rico-Gray V. Rhizophora harrisonii (Rhizophoraceae), un nuevo registro para las costas de México. Boletín de la Sociedad Botánica de México. 1981;**41**:163-165

[44] Rodríguez-Zúñiga MT, Troche-Souza C, Vázquez-Lule AD, Márquez-Mendoza JD, Vázquez- Balderas B, Valderrama-Landeros L, Velázquez-Salazar S, Cruz-López MI, Ressl R, Uribe-Martínez A, Cerdeira-Estrada S, AcostaVelázquez J, Díaz-Gallegos J, Jiménez-Rosenberg R, FueyoMac Donald L, Galindo-Leal C. Manglares de México/ Extensión, distribución y monitoreo. Comisión Nacional para el Conocimiento y Uso de la Biodiversidad. México D.F; 2013. p. 128

[45] Schwarzbach AE, y McDade LA. Phylogenetic relationships of the mangrove family Avicenniaceae based on chloroplast and nuclear ribosomal DNA sequences. Systematic Botany. 2002;**27**:84-98

[46] Tomlinson PB. The botany of mangroves. Cambridge: Cambridge University Press; 1986

[47] Tomlinson PB. The botany of mangroves. 2nd edn. New York: Cambridge University Press; 2016. p. 417

[48] Tovilla-Hernández C, De la Presa Sl, Romero JC, Ovalle EI, Gómez F, Hernández R, Cruz JE, Hernández A. Inventario Forestal de los bosques de manglar de la costa de Chiapas. El Colegio de la Frontera Sur; 2007. p. 92

[49] Travieso AC. La Legislación aplicable a los Sitios ramsar en México. Letras Jurídicas No. 2009;**20**:1-14. https://agua.org.mx/wp-content/uploads/2017/08/La-legislaci%C3%B3n-aplicable-a-los-sitios-Ramsar-en-M%C3%A9xico.pdf

[50] Troche-Souza C, Rodríguez-Zúñiga MT, Velázquez-Salazar S, Valderrama-Landeros L, Villeda-Chávez E, Alcántara-Maya A, Vázquez-Balderas B, Cruz-López MI, Ressl R. Manglares de México: extensión, distribución y monitoreo (1970/1980-2015). Comisión Nacional para el Conocimiento y Uso de la Biodiversidad. México, D.F; 2016. 16p

[51] Tropicos.org. Missouri Botanical Garden. 2017. http://tropicos.org/Name/33701106

[52] Tropicos.org. Missouri Botanical Garden. 2017. http://tropicos.org/Name/33702454

[53] Tropicos.org. Missouri Botanical Garden. 2017. http://tropicos.org/Name/21600010

[54] Tropicos.org. Missouri Botanical Garden. 2017. http://tropicos.org/Name/8200192

[55] Tropicos.org. Missouri Botanical Garden. 2017. http://tropicos.org/Name/8200077

[56] Tropicos.org. Missouri Botanical Garden. 2017. http://tropicos.org/Name/8200193

[57] Tropicos.org. Missouri Botanical Garden. 2017. http://tropicos.org/Name/27600064

[58] Tropicos.org. Missouri Botanical Garden. 2017. Flora Mesoamericana. http://www.tropicos.org/Name/27600064?projectid=3

[59] Tropicos.org. Missouri Botanical Garden. 2017. http://www.tropicos.org/Name/27600058

[60] Valderrama L, Troche C, Rodríguez MT, Márquez D, Vázquez B, Velázquez S, Vázquez A, Cruz MI, Ressl R. Evaluation of mangrove cover changes in Mexico during 1970-2005 period. Wetlands. 2014;**34**:747-758

[61] Valderrama-Landeros LH, Rodríguez-Zúñiga MT, Troche-Souza C, Velázquez-Salazar S, Villeda-Chávez E, Alcántara-Maya JA, Vázquez-Balderas B, Cruz-López MI, Ressl R. Manglares de México: actualización y exploración de los datos del sistema de monitoreo 1970/1980-2015. Comisión Nacional para el Conocimiento y Uso de la Biodiversidad. Ciudad de México. p. 128

[62] Vázquez-Yanes C, Batis Muñoz AI, Alcocer Silva MI, Gual Díaz M, Sánchez Dirzo C. Árboles y arbustos potencialmente valiosos para la restauración ecológica y la reforestación. Reporte técnico del proyecto J084. Instituto de Ecología, Universidad Nacional Autónoma de México; 1999. p. 262

[63] Von Linsingen LC, Armando C. Conocarpus erectus Linnaeus, nova ocurrencia para a flora do Sul do Brasil. Adumbrationes ad SummÆ editionem 26: 1-6. Madrid, 12-V-2007

[64] Wunderlin RP, Hansen BF, Franck AR, Essig FB. Atlas of Florida Plants. http://florida.plantatlas.usf.edu/. [S. M. Landry and K. N. Campbell (application development), USF Water Institute.] Institute for Systematic Botany, University of South Florida, Tampa; 2017

8

Biological Conservation and Nature Protection Strategies in Spanish Atlantic Region

Javier Ferreiro da Costa and Pablo Ramil-Rego

Abstract

Despite the fact that the loss of biodiversity continues to be an unresolved problem at a global level, it is possible to identify new alternatives and initiatives in biodiversity conservation. From the environmental strategic framework created worldwide, an important World Network of Biosphere Reserves has been implemented under UNESCO's Man and the Biosphere (MAB) Programme, which in the European Union in general, and in Spain in particular, is strongly interrelated with Natura 2000 network that is created under Directive 92/43/EEC. In the present work, the importance of the Spanish Atlantic region for biodiversity is assessed, contextualizing the networks of protected areas created in this territory and valuing the created synergies that have favored the start-up of projects and agreements aimed at reducing the loss of biodiversity and achieving sustainability.

Keywords: biodiversity conservation, habitats, species, Natura 2000 network, Biosphere Reserves, Atlantic biogeographical region, LIFE Programme

1. Introduction

After the declaration of the first protected areas in the mid-nineteenth century [1], the American model of national parks and monuments moved progressively to other continents and countries [2] until reaching an important development and covering important milestones in the conservation of world biodiversity and natural heritage. However, the model of protected areas generated frequent problems with the local population, especially at the time of reconciling the conservation objectives with the exploitation of the resources that had been carried out in the territory or that are incorporated after the protected area's declaration. This determined that during the mid-twentieth century the increase of protected areas stopped in many countries and even worldwide [3]. This situation changed during the 1970s through the incorporation of two new protected figures: the Wetlands of International Importance especially as Waterfowl Habitat (Ramsar, Iran, 02/02/1971) and the Biosphere Reserves created by the UNESCO in 1974 within the Man and the Biosphere (MAB) Programme.

In 1992 during the United Nations Conference on Environment and Development (Rio de Janeiro, Brazil), known as the "Earth Summit," the Convention on Biological Diversity (CBD) was signed. Months before, the Council of European Communities promulgated Directive 92/43/EEC, creating Natura 2000 network and establishing a new nature conservation policy in the European Economic Community. CBD endorsed a really ambitious goal when the parties approved "to achieve by 2010 a significant reduction in the current loss rate of biological diversity, globally, regionally and nationally, as a contribution to the mitigation of poverty and for the benefit of all forms of life on earth" [4].

Despite world and European efforts, it was not possible to stop the loss of biodiversity before 2010. This failure led to the elaboration and adoption by the CBD of the "Strategic Plan for Biodiversity," for the 2011–2020 period, which includes 20 operational targets (known as the Aichi Biodiversity Targets) in terms of conservation and sustainable use of biodiversity [5] to be met at the end of the programming period. Notable efforts have been made since then. The European Union has been one of the organizations, which has promoted the most activity from its competences, with the promulgation of different environmental regulations, the creation of an extensive network of protected areas, Natura 2000 network, and the adoption in 2011 of a new Biodiversity Strategy to protect and improve the state of biodiversity during the next decade [6].

2. Two different and complementary models of protected areas

Two models of networks of protected areas, the Biosphere Reserves Network and the Natura 2000 network, currently coexist in the territory of the European Union. Although they have a clearly different configuration, organization, and functioning, since their inception, the two models have shared common objectives oriented to biodiversity conservation and sustainable development. This coincidence determines that in many cases there is a strong complementarity between both figures, resulting in greater effectiveness of their objectives and especially in a greater interrelation of environmental policies with the society that lives in these territories.

The Biosphere Reserves have been consolidated as a long-term conservation model, with 45 years of experience. The Earth Summit of Rio, the Convention on Biological Diversity, and the European Biodiversity Strategy have had a wide impact on the evolution of the Man and Biosphere Programme, reflected in its Seville Strategy (worldwide strategy of working and sustainable development) and the approval of the Statutory Framework of the World Network of Biosphere Reserves in 1995 [7]. Biosphere Reserves have become a successful example of adaptive management, incorporating the latest trends and the best available scientific-technical knowledge through the implementation of the successive Action Plans for the World Network of Biosphere Reserves until the last one approved recently in Lima [8].

In 1992, the European environmental policy was reinforced with the approval of the Directive 92/43/EEC and the creation of Natura 2000 network, made up of various types of protected areas (including Sites of Community Importance and Special Areas of Conservation and incorporating Special Protected Areas according to Directive 79/409/EEC, now replaced by Directive 2009/147/EC), and destined to the conservation of biodiversity. Up to now, Natura 2000 network has been established as the largest ecological network worldwide [9, 10], with a total area of 1,147,956 km^2, which represents 18.12% of the European Union territory. Therefore, Natura 2000 network accomplishes and surpasses the 11th target established in the Strategic Plan for Biodiversity for the 2011–2020 period [5]. In addition, Natura 2000 areas are estimated to provide ecosystem services valuated between 223,000 and 314,000 million euros a year; to retent 35,000 million tons of carbon dioxide (CO2); and to generate about 12 million jobs every year in the European Union [11].

Both in the model of Biosphere Reserves Network and Natura 2000 network, the protected areas must meet a set of environmental criteria, evaluated in the context of a biogeographical sectorization with an impact on the organization and management of the sites.

Designated Biosphere Reserves must be able to fulfill three basic functions: conservation of natural resources, economic and social development according to sustainability principles, and creation and participation in networks [7]. To achieve this, a zoning model is established, which includes one or several separate core areas within the Biosphere Reserve, aimed to ensure the biodiversity conservation function. On the contrary, Natura 2000 network areas (SCI, SAC, SPA) have been designed by adjusting their morphology to the spatial distribution of certain types of habitats or species, so they are identified with areas destined primarily for conservation. These are comparable with the core areas of the Biosphere Reserves but lacking a clear designation of buffer and transition zones. This absence generates serious problems when Natura 2000 areas are designated into the member states, but it is being addressed with the brand new Green Infrastructure Strategy [12].

Natura 2000 network has an important level of regulatory development, supported by both Directives 92/43/CEE and 2009/147/CE. This regulatory complexity of Natura 2000 network clash against the maintenance of an unsustainable use generating a widespread rejection among sectors that maintain the hopes of achieving a profit based on an irrational exploitation.

Biosphere Reserves lack specific regulations, or rather, do not generate their own regulations, although their establishment and development implies the incorporation of European Union regulations, including reference to the Natura 2000 network. Without these protected areas, it would be impossible to comply the objectives established by the Man and Biosphere Programme.

The model management shows important differences between Biosphere Reserves and Natura 2000. The ordinary management of Biosphere Reserves is carried out from the own territory, through a set of management bodies, including a manager, as well as public participation and advisory bodies. Regarding to Natura 2000 network, the national or regional authorities directly manage the protected areas, but there are no participation bodies, and it's not possible to identify the manager of each of them into the territory. Therefore, Natura 2000 management is more detached from the territory, which in many cases supposes malfunctioning of participation channels and the lack of collaboration with local society and entities during the implementation of specific plans or programs or the search for the simple implication of citizenship. At this point, the Biosphere Reserves have demonstrated a greater capacity for management and involvement with real problems.

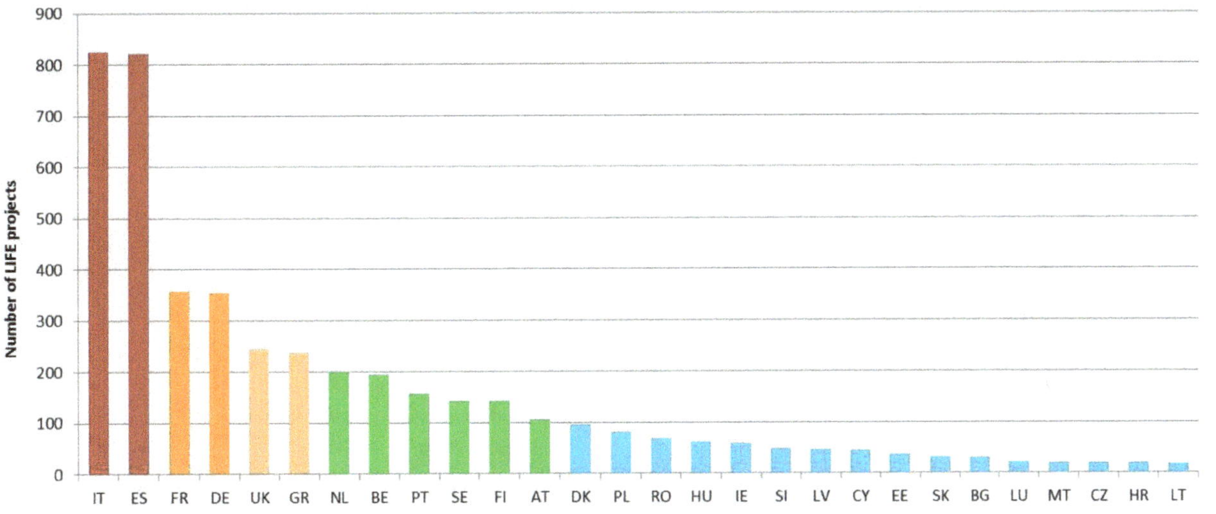

Figure 1. Number of LIFE projects led by the EU member states (prepared from [13]).

Funding is another key aspect in both models of protected areas. The MAB Programme does not provide any kind of financial aid or credit to the Biosphere Reserves or to the countries that have designated them. On the contrary, the countries are responsible to bear the expenses of the UNESCO and United Nations. Even in most European countries, there is also no specific financial program for Biosphere Reserves.

The European Union financing for environment is carried out globally through the ERDF, ESF, and ECF funds. In addition, the EU maintains a successful program for support environmental conservation and restoration activities, the LIFE Programme (acronyms that come from the French l'instrument financier pour l'environnement). This program, mainly focused on the Natura 2000 network, has cofinanced more than 4,600 projects dedicated exclusively to the environment and climate change [13] across the European Union and beyond (**Figure 1**).

3. Natura 2000 network and diversity of habitats in Spain

Undoubtedly, the most relevant tool in the promotion of biodiversity conservation in Spain has been the approval of the Directive 92/43/CEE (Habitats Directive) and the implementation of the Natura 2000 network. In fact, Spain has been one of the EU member states that has put more efforts on this figure, since it is the country that holds the largest area occupied by Natura 2000 sites (**Figure 2**), with a total amount of 222,276 km^2 and 27.3% of its continental area (**Figure 3**). Natura 2000 network in Spain meets, therefore, the 11th Aichi Biodiversity Targets [5].

Currently, the Spanish Natura 2000 network includes a set of 1,863 sites, consisting of 1,467 SCI/SAC and 644 SPA, which are distributed among the biogeographical regions represented in Spain, i.e., Atlantic, Alpine, Mediterranean, and Macaronesian (**Figure 4**). 85.7% of the Spanish continental area is included in the Mediterranean region, divided between the Iberian Peninsula, the Balearic Islands, Ceuta, and Melilla. The following region by occupation order is the Atlantic region, with 11.0% of the total, present in the NW Spanish mainland. Finally,

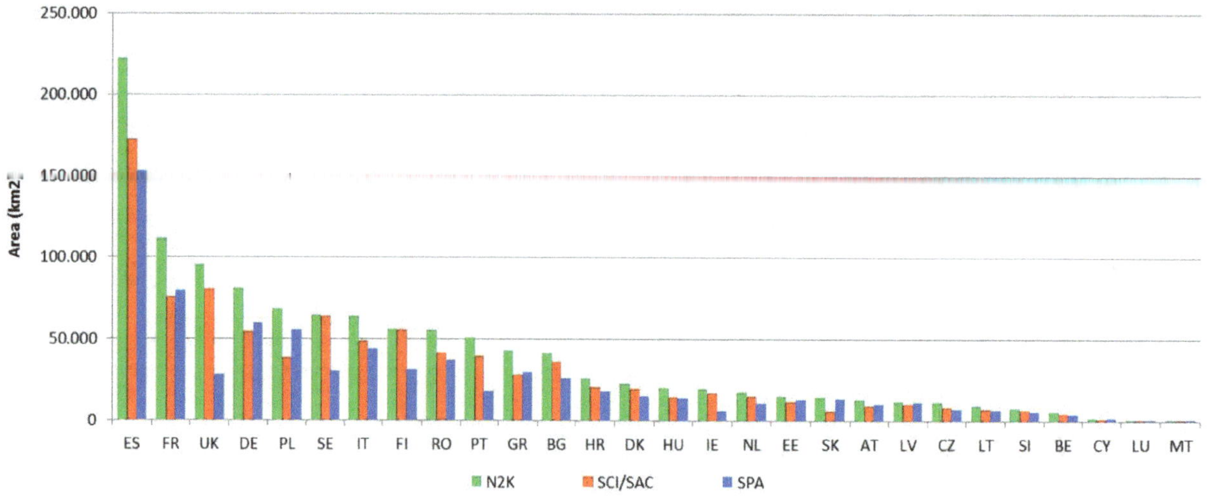

Figure 2. Total area occupied by Natura 2000 sites in the EU member states (prepared from [14]).

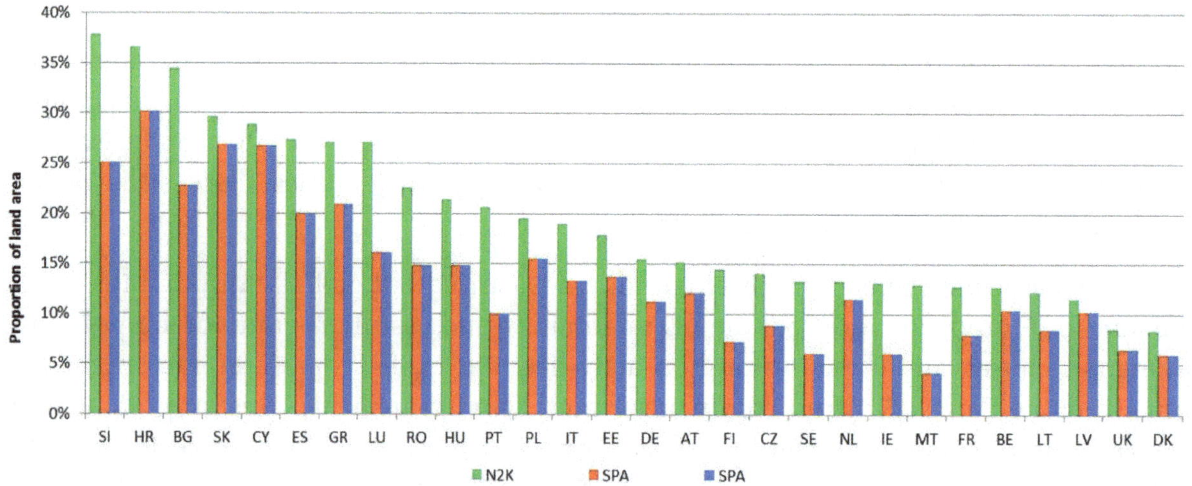

Figure 3. Proportion of the terrestrial and inland water areas occupied by Natura 2000 sites in the EU member states (prepared from [14]).

it should be noted that 1.9% of Spain is included in the Alpine region, delimited around the Pyrenees mountain range, while the remaining 1.4% corresponds to the Macaronesian region represented in the Canary Islands.

According to distribution analysis of habitat types in the Iberian Peninsula [17], the areas of the highest diversity are concentrated in the Atlantic portion (**Figure 5**). The 10×10 km^2 UTM grids with a higher number of habitats are placed mainly in the northern third of the peninsula, highlighting those that host the maximum (more than 25 types) in the NW coastal area (Galician region) and along the mountain foothills of the Cantabrian Mountains.

The Spanish Atlantic biogeographical region includes the most diverse areas in Spain and even several of the most diverse habitat areas of the whole Atlantic biogeographical region or of the entire European Union, according to reports on conservation status and trends of habitat types and species, which are developed every 6 years under Article 17 of Directive 92/43/EEC and Article 12 of Directive 2009/147/EC. The report regarding the last evaluation period

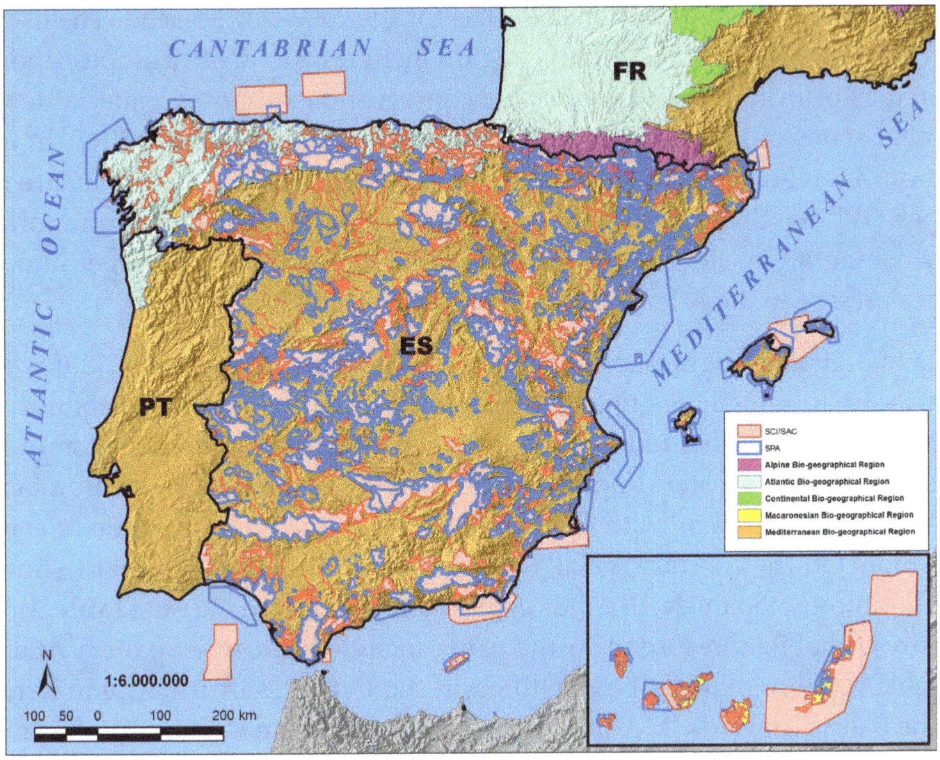

Figure 4. Distribution by biogeographic regions of Spanish Natura 2000 network (prepared from [15, 16]).

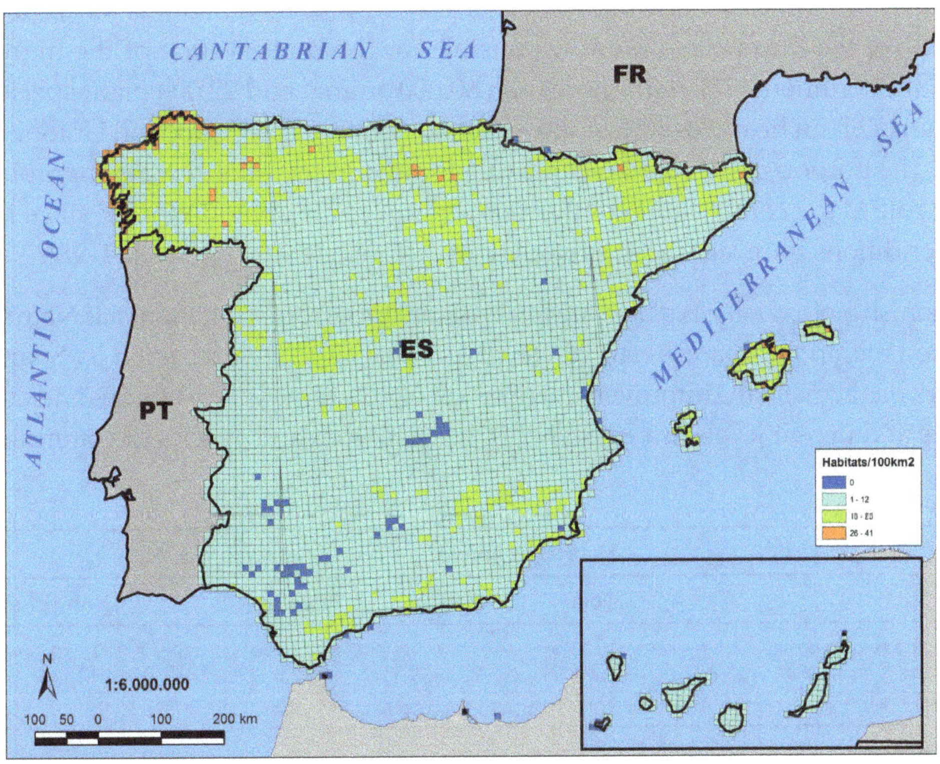

Figure 5. Number of habitats per 10 × 10 km^2 UTM grid in Spain (prepared from [17]).

available [18] includes the information collected for the 2007–2012 period. The list of habitats provided in this report [19] for the Atlantic region includes a total of 121 types, and 30 of them are considered as priority types (those whose conservation supposes a special responsibility for the member states of European Union) in Annex I of Directive 92/43/EEC. In the Spanish Atlantic region, a total of 89 habitat types are considered present, of which 19 are considered priority types [19]. So, Spanish Atlantic biogeographical region plays an important role in biodiversity conservation, since it houses 72% of the variety of habitat types and 63% of the priority types present in the Atlantic biogeographical region (**Table 1**).

The analysis of digital cartographic information housed in web servers of [20] and classified in LAEA grids 10×10 km^2 allows to assess the territorial distribution of habitat types in different intervals by the number of habitat types (**Figure 6**). The Atlantic region of the Iberian Peninsula concentrates a greater proportion of the top three intervals of the number of habitat types by LAEA grids 10×10 km^2 with respect to the rest of the Atlantic region or even from the European Union. On the contrary, grids with the lowest number of habitats show a scarcer representation into the Spanish Atlantic region than in the other areas (**Table 2**). The levels of biodiversity of habitats recorded in the grids included into the Spanish Atlantic region are comparable to those that can be identified in the Pyrenees or the Alpine Range (Alpine region), in the Balearic Islands (Mediterranean region), or in small points of the Scandinavian coast (Boreal region). These results are consistent with those obtained by analysis made for the period 2001–2006 [21].

Regarding to the species across the territory, the number of taxa present in the Iberian Peninsula is estimated to amount to a value above 80,000 [22]. In this way, Spain is attributed to be the member state of the European Union responsible for the conservation of the highest species diversity of the continent [23], with more than 60,000 animal and 22,000 plant species. In some of these groups, Spain hosts 80–90% of the total taxa found in the European Union. Within this framework, the portion of the Iberian Peninsula located in the Atlantic biogeographical region plays a key role in the conservation of the main components of biodiversity, since it meets the optimum conditions for a significant diversity of different taxon groups [24–28].

The high Spanish biodiversity explains the relevant territorial representation that Natura 2000 sites have achieved within the Spanish national territory. This large area occupied by Spanish Natura 2000 network has allowed the improvement of biodiversity conservation policies and environmental measures at different levels. Inventorying of habitat types and species of community interest

	Habitat number		
	Total	**Priority**	**Non-priority**
European Atlantic region	121	30	91
Spanish Atlantic region	89	19	70

Table 1. Comparison of the number of habitats (priority and non-priority) of Annex I of the DC 92/43/EEC present in the European and Spanish Atlantic regions (taken from [19]).

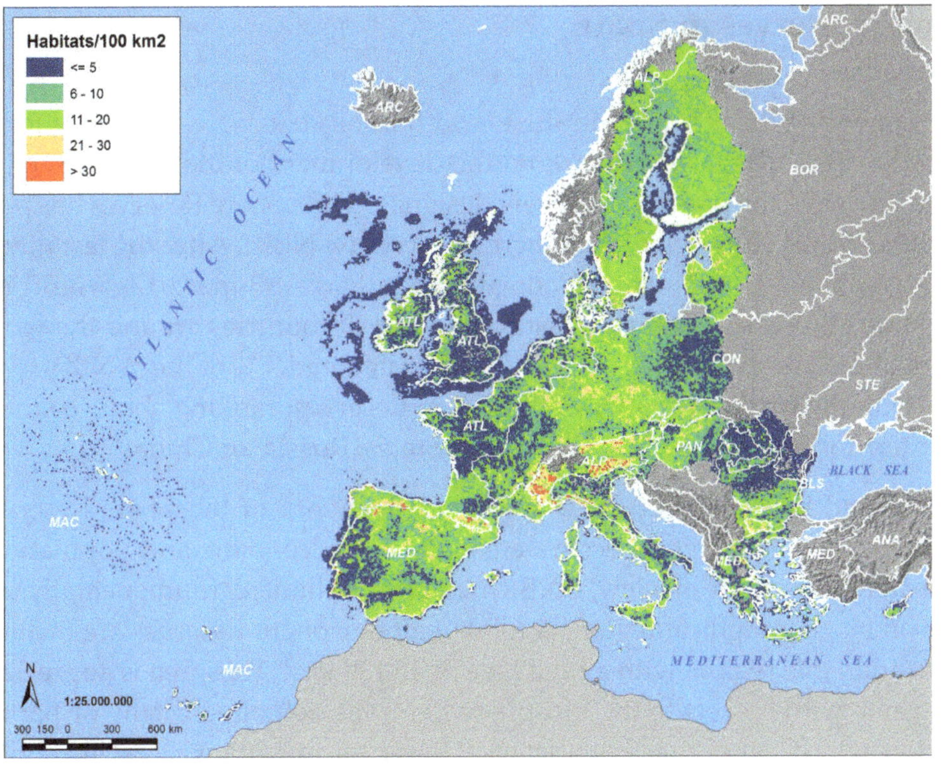

Figure 6. Number of habitats per LAEA grid 10 × 10 km² in the European Union, with respect to biogeographical regions (prepared from [15, 20]).

	European Union		European Atlantic region		Spanish Atlantic region		Spanish Atlantic Biosphere Reserves	
Nº hab	N LAEA	%	N LAEA	%	N LAEA	%	N UTM	%
≤5	17.376	32.5	8.223	59.6	113	15.4	1	0.4
6–10	15.004	28.1	2.803	20.3	106	14.4	24	10.6
11–20	18.093	33.8	2.445	17.7	338	46.1	155	68.3
21–30	2.764	5.2	299	2.2	150	20.4	43	18.9
>30	207	0.4	35	0.2	27	3.7	4	1.8

Table 2. Comparison of the number of habitats per grid 10 × 10 km² and percentages in the European Union, the Atlantic region, the Spanish Atlantic portion, and the Spanish Atlantic Biosphere Reserves (prepared from [15, 16, 20]).

and delimitation of the Natura 2000 areas have allowed increasing the knowledge on Spanish interesting areas for biodiversity conservation, strengthening, and consolidating the national and regional networks of natural protected areas [29]. On the other hand, the provision of an extensive Natura 2000 network has allowed Spain to lead 821 projects and to participate in 903 projects of the LIFE Programme in 25 years [13]. This means an average participation in 34 LIFE projects per year, making Spain one of the leader countries in terms of the implementation of these initiatives for nature conservation. Finally, it is worth mentioning the establishment of strong synergies with other protection figures, especially the Spanish Network of Biosphere Reserves, whose core areas have been delimited fundamentally around the Spanish Natura 2000 network [30].

4. Biosphere Reserves in Spain

Spain has been one of the countries in the world that has bet more decisively for Biosphere Reserves model, becoming a benchmark on the development of the objectives of the Seville Strategy [31, 32], while the promotion of new designated Biosphere Reserves has placed Spain at the top of the World Network of Biosphere Reserves (WNBR). After the last meeting of the UNESCO International Coordination Council held in Paris in June 2017, WNBR is made up of 669 designated Biosphere Reserves distributed in 120 countries around the world [33]. In Spain, a total of 48 Biosphere Reserves have been designated, 7% of the WNBR, which makes the country with the highest number of Biosphere Reserves, ranking above countries with a much larger territorial scope, such as the United States, Russia, or China.

The numerous and large set (**Figure 7**) of Spanish Network of Biosphere Reserves (SNBR) houses a wide variety of habitats, ecosystems, and socioeconomic and population realities. From a geo-statistical point of view, SNBR has been delimited fundamentally around the Mediterranean biogeographical region. A total of 29 Biosphere Reserves are included totally or partially within this region, with a total area of 40,139 km^2. This area is appreciably higher in absolute value with respect to the Biosphere Reserves delimited totally or partially in the Atlantic region, which brings together a set of 18 Reserves, including a total area of 12,446 km^2. These differences are even greater with respect to the Biosphere Reserves in the Spanish Macaronesian or Alpine regions. However, if the analysis is relativized from a biogeographical point of view, the values indicate a completely different reality (**Table 3**).

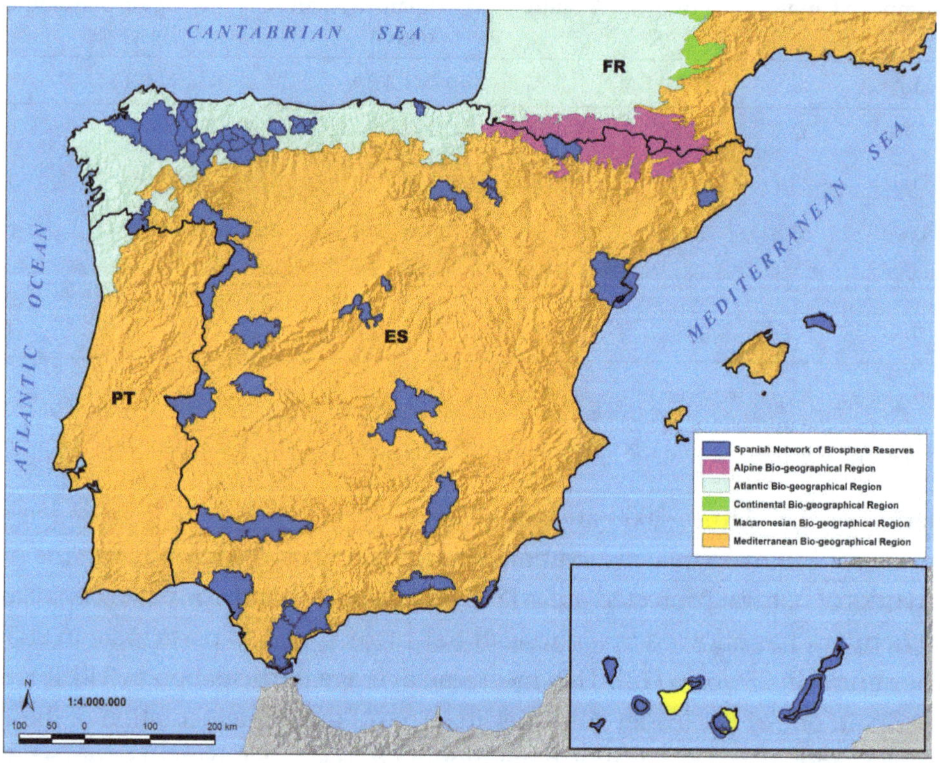

Figure 7. Distribution of the Spanish Network of Biosphere Reserves with respect to biogeographical regions (prepared from [15, 16]).

Region	Spain (%)	No. of reserves	Biosphere Reserves area (km^2)			Region (%)
			Total	Marine	Continental	
Mediterranean	85.7	29	40.139	1.185	38.954	8.9
Atlantic	11.0	18	12.446	33	12.413	22.3
Alpine	1.9	1	992	–	992	10.2
Macaronesian	1.4	7	8.400	4.721	3.679	52.4

Table 3. Surface distribution by biogeographical region of the Biosphere Reserves in Spain (prepared from [15, 16]).

Although Spanish Mediterranean Biosphere Reserves include 38,954 km^2 of terrestrial areas, they only represent 8.9% of the Spanish Mediterranean continental areas. Both in the Atlantic and Macaronesian regions, the percentage included in the SNBR is significantly higher (**Table 3**), and they also exceed the index values of 11th Aichi Biodiversity Targets [5].

The high percentage of the Spanish Atlantic region (22.3%) that has been included in the Biosphere Reserves is consistent with the territorial distribution of habitat diversity in the Iberian Peninsula (**Figure 5**). In the Spanish Atlantic region, there is a high presence of areas with the highest abundance of habitats in the Iberian Peninsula [17] and even in the European Union [18, 20]. This is indicative of the biodiversity harbored by the Iberian Atlantic area and of the sustainability that has governed the management and conservation of this territory, and therefore it has allowed the establishment of a large set of Biosphere Reserves into the Spanish Atlantic region.

The reason for the harbored biodiversity by the Iberian Atlantic region combines climatic and geographical factors [34–36]. Rainfall occurs abundantly and is well distributed throughout the year, so that the summer drought does not exist or is very low, because even in summer the winds of component N loaded with moisture provide effective and hidden precipitation. The distance from the coast to the continent does not exceed 120 km from the shore of the Atlantic Ocean, so a low continentality and a reduced annual thermal amplitude are typical, especially in comparison with Mediterranean territories further away from the sea, in which continentality produces a greater oscillation between winter and summer temperatures. Everything causes the optimal conditions for Atlantic vegetation to converge in NW Spain [25], which allows a higher variety of ecosystems [24, 37, 38], and consequently a greater richness of habitat types [17] and species [39].

5. Biosphere Reserves in the Spanish Atlantic region

The Biosphere Reserves delimited totally or partially in the Spanish Atlantic region form a set of 18 Biosphere Reserves (**Figure 8**) whose total area amounts to 12,446 km^2 of which 33 km^2 (0.3% of the total) correspond to marine waters, while the remaining 12,413 km^2 (99.7% of the total) correspond to terrestrial and inland water areas.

In this set of Biosphere Reserves, a wide ecological range is represented, being possible to identify reserves in which marine, littoral, and coastal environments are represented

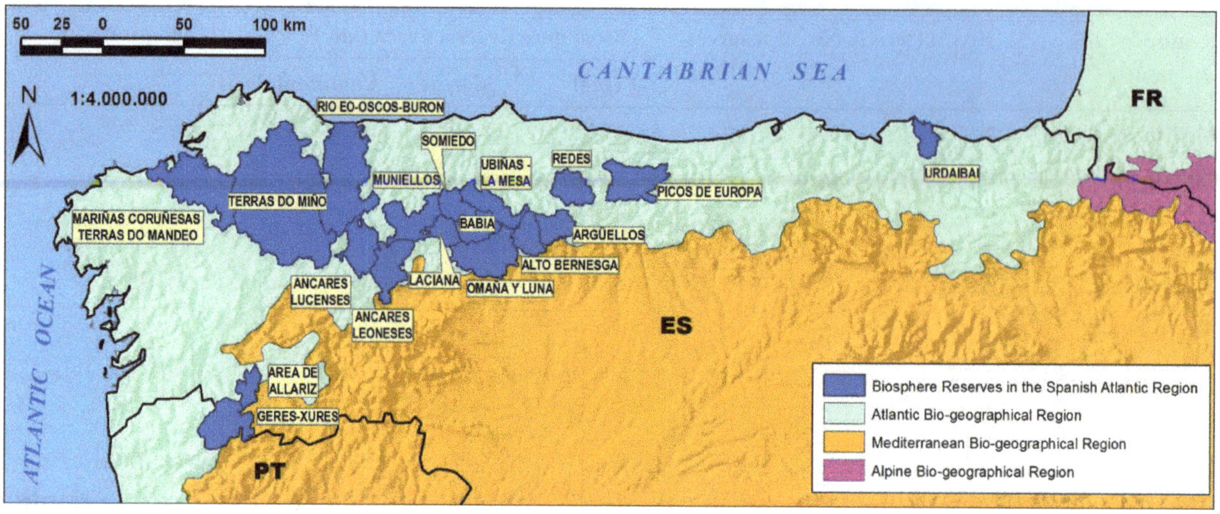

Figure 8. Territorial distribution of the Biosphere Reserves in the Spanish Atlantic region (prepared from [15, 16]).

(Urdaibai, Mariñas-Mandeo, Rio Eo-Oscos-Burón), whereas in other cases, wetlands are the main component (Terras do Miño). There are also examples in which traditional agro-systems are the main component (Allariz), although most include inland valley-range systems (Muniellos, Laciana, Babia, Omaña-Luna, Alto Bernesga, Los Argüellos, Gerês-Xurés, Redes) or high mountain ecosystems (Somiedo, Picos de Europa, Ancares Lucenses, Ancares Leoneses, Ubiñas-La Mesa).

The 18 Biosphere Reserves in the Spanish Atlantic region respond to different governance models (**Table 4**). Eight of them are managed by the regional governments, differentiating those that coincide territorially with the limits of a national park (NP) and that are managed jointly with this protected area, of which they are directed directly by the regional government. As for the rest, it is possible to differentiate those Biosphere Reserves linked to provincial and local entities, from those that are directed by foundations or associations of various kinds. This variety in the management bodies is representative of the success example of adaptive management that Biosphere Reserves of the Spanish Atlantic region have become, so their management is designed from the development needs of local communities to regional priorities.

The set of Biosphere Reserves in the Spanish Atlantic region, given their proximity to the biogeographical boundary and as transition to the Mediterranean territories, has a large number of unique components of the natural heritage and biodiversity. They harbor a remarkable richness of natural and seminatural habitats that are threatened with disappearance in the whole of the European Union or that have a very small territorial representation (marshes, dune ecosystems, rivers and lagoons, peat bogs, wet heathland, dry heathland, orophilic scrubs, hay meadows, deciduous and evergreen forests, rocky habitats, caves not exploited by tourism, etc.), as well as an important group of endemic or threatened wild species or subspecies.

According to distribution analysis of habitat types in the Iberian Peninsula [17], the Biosphere Reserves in the Spanish Atlantic region have been delimited over the areas that harbor the highest diversity of habitat types (**Figure 9**), so the distribution of the number of habitats per 10×10 km^2 grid is similar to the rest of the Spanish Atlantic region (**Table 2**). It is worth highlighting the maximum interval of habitat types (>30 types), which reaches a proportion that far exceeds

	NP	Managing entity	Habitat number		
			Total	Priority	Non-priority
Mariñas-Mandeo		Foundation/association	49	12	37
Río Eo-Oscos-Burón		Regional government	48	12	36
Picos de Europa		Regional government	41	10	31
Somiedo	✷	Regional government	37	8	29
V. Omaña y Luna		Local entity/consortium	36	10	26
Terras do Miño		Provincial Deputation	35	11	24
Ancares Lucenses		Provincial Deputation	35	11	24
Las Ubiñas-La Mesa	✷	Regional government	35	8	27
Redes	✷	Regional government	34	8	26
Ancares Leoneses		Local entity/consortium	34	8	26
Muniellos	✷	Regional government	33	7	26
Babia		Local entity/consortium	33	9	24
Valle Laciana		Foundation/association	32	8	24
Urdaibai		Regional government	32	6	26
Alto Bernesga		Local entity/consortium	30	8	22
Argüellos		Local entity/consortium	29	7	22
Gerês-Xurés	✷	Regional government	28	8	20
Área de Allariz		Foundation/association	13	2	11
Total			78	19	59

NP: presence of National/Natural Park.

Table 4. Typology of the management bodies and richness of habitat types in Biosphere Reserves of the Spanish Atlantic region.

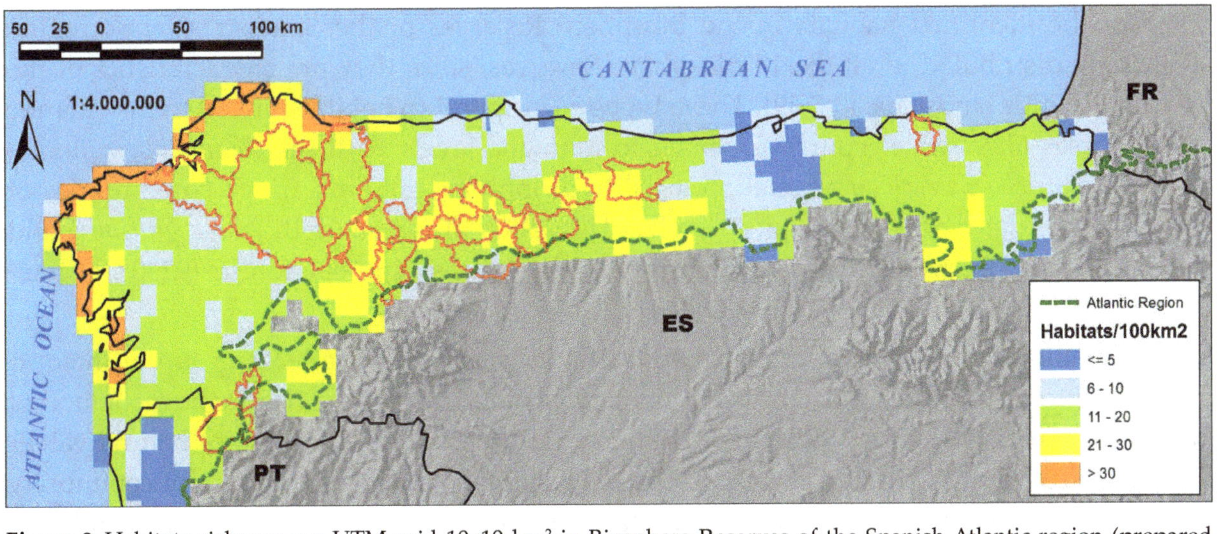

Figure 9. Habitats richness per UTM grid 10×10 km^2 in Biosphere Reserves of the Spanish Atlantic region (prepared from [17]).

(between 4 and 9 times) the value reached in the entire Atlantic Biogeographic region and even in the European Union (**Table 2**). The four UTM 10×10 km^2 grids with more than 30 types of habitat are located in littoral Biosphere Reserves (Mariñas-Mandeo, Rio Eo-Oscos-Burón), due to the great diversity of the Atlantic coastal environments. In contrast, Urdaibai, Allariz, and Gerês-Xurés Biosphere Reserves do not include grids of more than 20 habitat types. The remaining 13 Biosphere Reserves are in an intermediate situation, housing grids of up to 30 habitat types.

The total amount of habitat types housed by each Biosphere Reserve of the Spanish Atlantic region (**Table 4**) is directly proportional to their territorial distribution over the number of habitat types by UTM 10×10 km^2 grids (**Figure 9**). So Biosphere Reserves with a greater number of habitat types are Mariñas Coruñesas, Terras do Mandeo (49 habitats), Río Eo, and Oscos e Terras de Burón (48 habitats). The high diversity of habitat types housed in these two is due to the fact that they include coastal, sublittoral, and continental areas. Excluding the coastal areas, Picos de Europa would be the Biosphere Reserve with the largest number of habitats of community interest (41 habitats), followed by the Somiedo Biosphere Reserve (37 habitats). On the contrary, Allariz Biosphere Reserve is the one with the smallest number of habitat types included in Annex I of the DC 92/43/CEE, with 13 types.

Similarly, Mariñas Coruñesas y Terras do Mandeo and Río Eo, Oscos e Terras de Burón Biosphere Reserves are the ones that hold the largest set of priority habitat types, with a total amount of 12 types, due to the presence of coastal priority types that are not able to be found in the rest of the Biosphere Reserves in the Spanish Atlantic region. However, it is also worth highlighting the presence of 11 priority types in Terras do Miño and Os Ancares Lucenses Biosphere Reserves or of 10 types in Picos de Europa and Valles de Omaña and Luna Biosphere Reserves. Biosphere Reserves of the Spanish Atlantic Region host an amount of 78 types of habitats, of which 19 are considered priority (**Table 4**). That is, the Spanish Atlantic Biosphere Reserves contain 88% of the diversity of habitat types identified in the Spanish Atlantic region and 64% of the diversity of habitats in the European Atlantic region (compared with **Table 1**). Regarding to priority habitats, the Biosphere Reserves of the Spanish Atlantic region still play a more important role: they include 100% of priority types in Spanish Atlantic region and 63% of priority habitats identified in the whole European Atlantic biogeographical region (compared with **Table 1**).

Some of the identified habitats in the Biosphere Reserves of the Spanish Atlantic region are widely distributed in all the 18 Biosphere Reserves, since they are characteristics of the Spanish Atlantic landscape [38, 39]. These types correspond to habitats linked to the Atlantic river courses (3260 water courses of plain to montane levels with the *Ranunculion fluitantis* and *Callitricho-Batrachion* vegetation, 6430 hydrophilous tall herb fringe communities of plains and of the montane to alpine levels, 91E0* alluvial forests with *Alnus glutinosa* and *Fraxinus excelsior*), dry heathland (4030 European dry heaths), and oak forests (9230 Galician-Portuguese oak woods with *Quercus robur* and *Quercus pyrenaica*).

The set of Biosphere Reserves in the Spanish Atlantic region hosts a large group of interesting species for nature conservation (protected and catalogued species). The implementation of Natura 2000 network has led to standardization regarding the inventorying of natural values through the Standard Data Forms (SDFs). These register the habitats and species of community interest in each Natura 2000 area, allowing a homogeneous treatment of the data hosted in the database of the European Topic Centre on Biological Diversity [40], dependent on the European

	Plant	Inverteb.	Fish	Herpet.	Birds	Mamm.	Total
Urdaibai	3	7	1	3	153	6	173
Gerês-Xurés	11	7	7	5	93	9	132
Río Eo-Oscos-Burón	2	9	4	4	85	11	115
Mariñas-Mandeo	6	6	4	4	72	7	99
Terras do Miño	6	6	4	4	61	6	87
Picos de Europa	7	13	2	4	35	13	74
Redes	2	8	0	4	45	9	68
Ancares Lucenses	7	6	1	4	42	7	67
Babia	4	3	2	3	36	8	56
Las Ubiñas-La Mesa	3	4	1	4	35	8	55
V. Omaña y Luna	4	3	2	3	33	8	53
Ancares Leoneses	4	6	2	5	30	5	52
Somiedo	4	3	0	3	35	7	52
Valle Laciana	5	4	2	2	33	6	52
Muniellos	2	6	0	3	32	7	50
Área de Allariz	1	1	1	0	32	5	40
Argüellos	5	3	2	4	6	7	27
Alto Bernesga	3	1	0	3	6	5	18

Table 5. Richness of species in Biosphere Reserves of the Spanish Atlantic region.

Environment Agency (EEA). The core areas of the Biosphere Reserves in the Spanish Atlantic region have been included almost completely within the Natura 2000 areas, so an analysis of their SDFs has been carried out to compare diversity of the species of community interest present in each of the Biosphere Reserves of the Spanish Atlantic region has been carried out (**Table 5**).

The total number of present species in the Biosphere Reserves of the Spanish Atlantic region is high as a whole and depends to a large extent on the diversity of birds, since they are the most numerous group by far. In this way, Biosphere Reserves with the largest number of species (>80 species) are Urdaibai, Gerês-Xurés, Río Eo-Oscos-Burón, Mariñas-Mandeo, and Terras do Miño, which in turn are those that have a greater number of bird species (>60 species). Also, noteworthy are a set of flora (11) and fish (7) species of Gerês-Xurés, invertebrates (13) in Picos de Europa, or mammals in Picos de Europa (13) and Rio Eo-Oscos-Burón (11).

6. The seed for new projects and initiatives

The Spanish Atlantic region is one of the European areas with greater value for biodiversity conservation of the European territory, including a large group of habitat types and species considered of interest for conservation under Directives 92/43/EEC and 2009/147/CE. This has enabled the delimitation of a profuse set of Natura 2000 areas, which have served as support for the core areas of the set of Biosphere Reserves established in the Spanish Atlantic region. The synergy created between the importance of the hosted values and these new instruments for the protection, planning, and management of the territory have made possible the start-up of new projects and initiatives for nature conservation and sustainable development in this territory, fundamentally through the LIFE Programme of the European Union [13].

The peatland ecosystems, present throughout the entire Iberian Atlantic region, have been the subject of several restoration projects in Natura 2000 areas. It should be highlighted LIFE Parga-Ladra-Támoga project (LIFE00 NAT/E/007330), that has been complemented by LIFE Tremedal project (LIFE11 NAT/ES/000707), should be highlighted. In both cases, actions have been carried out to restore peatland habitat types (7110*, 7130*, 7140, 7150, 7210*, 7230) and other types of wetlands (3110, 3190, 6410, 6510, 91E0*), as well as the taxa of interest for the conservation that they harbor, with special attention to the priority species *Eryngium viviparum** and others such as *Spiranthes aestivalis* or *Narcissus pseudonarcissus nobilis*. This kind of projects is of great interest not only for the environmental benefits they generate but also because they allow for the increase of knowledge as they generate up-to-date and homogeneous information on the distribution and conservation status of the key elements for biodiversity, gathering experiences for its better management and restoration [41]. In a complementary way, the generated information allows improving and increasing the communication and awareness of society about the values and environmental services that provide the ecosystems for biodiversity conservation and specifically the types of habitats and species of interest for conservation. The forest ecosystems have also been the subject of several LIFE projects, highlighting the LIFE BACCATA project, which aims to improve the conservation status of the yew forests (*Taxus baccata*), considered a priority habitat type (9580*) that has a scarce distribution at the European Union level, and which is present in the Iberian Atlantic region [42]. The relevance of this project is high, since it includes the increase of the area occupied by this habitat type by 7% and an improvement in the conservation status by 26%, complemented by a genetic characterization of *Taxus baccata* in the Spanish Atlantic region, in order to determine the variability and degree of genetic connectivity between populations and the kinship structure. The dissemination, diffusion, and awareness of society about the conservation of forests in general, and of yew forests in particular, are also included in LIFE BACCATA project, together with the transfer of its results for replication to the different agents involved throughout Spanish and European territory.

Other conservation projects have focused on species of interest for conservation, with special attention to those considered in danger of extinction. In this case, there is no doubt that in the Cantabrian Mountains two species have received the greatest investigative and conservative effort: the brown bear (*Ursus arctos**) and the capercaillie (*Tetrao urogallus*).

The territorial area formed for the most of the Biosphere Reserves of the Spanish Atlantic region (Os Ancares Lucenses, Los Ancares Leoneses, Muniellos, Somiedo, Valle de Laciana, Babia, Valles de Omaña y Luna, Las Ubiñas-La Mesa, Alto Bernesga, Los Argüellos, Redes, and Picos de Europa) forms a continuum in which various instruments and actions for the conservation of *Ursus arctos** have been set in motion. These instruments and initiatives include the monitoring of the Cantabrian brown bear population and brown bear habitat improvement actions, through several LIFE projects that have been developed in recent decades (Corredores Oso, Oso/Cantabria, Oso/Galicia, Oso en Asturias, LIFE Bear Defragmentation, etc.). The initiative for the implementation of these projects arose from the provision of the Spanish Strategy for the Conservation of the Cantabrian Brown Bear [43]. However, new projects were also initiated seeking synergy between brown bear conservation and other aspects such as climate change ("Osos CO2") or insect conservation ("Abejas en Acción"). In a complementary way, support is available for the maintenance of mountain livestock by the different regional governments where brown bear is present.

Regarding the capercaillie (*Tetrao urogallus*), this species has been the object of a conservation strategy parallel to the brown bear's, since its presence is distributed in the same Biosphere Reserves as the plantigrade, reaching in some cases to share programs and actions of conservation (LIFE Ancares/Galicia and Ancares/Castilla y León) that were designed in a coordinated and complementary way. Nevertheless, capercaillie has had its own projects (LIFE + Urogallo Cantabrico). All these projects have pursued the improvement of the habitat for the capercaillie in the Cantabrian Mountains and the reintroduction of the species in areas where it had disappeared, and they have been supplemented by the aids for the maintenance of mountain livestock.

In addition to the previous habitats and species, other examples have been the subject of LIFE projects in the Spanish Atlantic region in response to its high variety of ecosystems and because of its extraordinary biodiversity. In this way, the dune ecosystems (2110, 2120, 2130*) have also been the subject of LIFE projects (LIFE + ARCOS, LIFE Dunas Laida), while several taxa of interest for conservation have also been the main beneficiaries of LIFE projects, as is the case of *Galemys pyrenaicus* (LIFE Desmania), *Margaritifera margaritifera* (LIFE Margal Ulla), or *Woodwardia radicans* (LIFE Miera), among many others.

7. Final considerations

The Atlantic portion of Spain is one of the most biodiverse areas of the entire Atlantic biogeographical region and even of the entire European Union, since it includes a significant proportion of the whole variety of habitat types in the European Atlantic region. This has motivated Spain to be one of the countries that has bet more strongly on the Natura 2000 network, mainly through the SCI/SAC and SPA delimited in the Atlantic region.

The importance of the housed biodiversity values and the territorial representativeness of the Natura 2000 network have favored the establishment of a profuse and large set of Biosphere Reserves in the Spanish Atlantic region, which constitute territories that develop initiatives for the biodiversity conservation, the sustainable development, and the improvement of knowledge and capacities of the population.

The situation generated in the Spanish Atlantic region has allowed the development and implementation of LIFE projects for the conservation and restoration of habitats and populations of species of interest for conservation. Consequently, the Spanish Atlantic region has become a reference area when launching new strategies, functioning as a living laboratory to develop and test new experiences of biodiversity conservation, promotion of sustainable development, and logistical support for research, monitoring, education, and information exchange.

Author details

Javier Ferreiro da Costa* and Pablo Ramil-Rego

*Address all correspondence to: ibader.bio@gmail.com

IBADER (Instituto de Biodiversidade Agraria e Desenvolvemento Rural)-GI-1934 TB, Universidade de Santiago de Compostela (USC), Lugo, Spain

References

[1] United States Department of the Interior. The National Parks: Shaping the system. Washington, DC: National Park Service, US Department of the Interior, US Government Printing Office; 2005. 45 p.

[2] Hernández-Pacheco E. Comunicación respecto a los Parques nacionales y a los Monumentos naturales de España. Boletín de la Real Sociedad Española de Historia Natural. 1920; **20**:267-282

[3] Chape S, Harrison J, Spalding M, Lysenko I. Measuring the extent and effectiveness of protected areas as an indicator for meeting global biodiversity targets. Philosophical Transactions of the Royal Society B: Biological Sciences. 2005;**360**(1454):443-455

[4] Convention on Biological Biodiversity. Proposed Biodiversity Indicators Relevant to the 2010 Target. Montreal: United Nations, UNEP/CBD/SBSTTA/9/INF/26; 2003. 31 p

[5] Convention on Biological Biodiversity. The Strategic Plan for Biodiversity 2011-2020 and the Aichi Biodiversity Targets. Decision Adopted by the Conference of the Parties to the Convention on Biological Diversity at its Tenth Meeting, X/2. Nagoya: United Nations, UNEP/CBD/COP/DEC/X/2; 2010. 13 p.

[6] European Commission. Communication from the Commission to the European Parliament, the Council, the Economic and Social Committee and the Committee of the Regions. Our Life Insurance, Our Natural Capital: An EU Biodiversity Strategy to 2020. Brussels: European Commission, COM (2011) 244 final; 2011. 16 p.

[7] UNESCO. Biosphere Reserves. The Seville Strategy and the Statutory Framework of the World Network. París: UNESCO; 1996. 21 p

[8] UNESCO. A New Roadmap for the Man and the Biosphere (MaB) Programme and its World Network of Biosphere Reserves. MAB Strategy (2015-2025). Lima Action Plan (2016-2025). Lima Declaration. Paris: Man and the Biosphere Programme, United Nations Educational, Scientific and Cultural Organization; 2017. 54 p

[9] Sundseth K, Creed P. Natura 2000: Protecting Europe's Biodiversity. Luxembourg: European Commission, Office for Official Publications of the European Communities; 2008. 296 p

[10] Orlikowska EH, Roberge JM, Blicharska M, Mikusinski G. Gaps in ecological research on the world's largest internationally coordinated network of protected areas: A review of Natura 2000. Biological Conservation. 2016;**200**:216-227. DOI: 10.1016/j.biocon.2016. 06.015

[11] European Commission. The Economic Benefits of the Environment Natura 2000 Network. Synthesis Report. Luxembourg: European Commission, Publications Office of the European Union; 2013. 76 p

[12] European Commission. Communication from the Commission to the European Parliament, the Council, the European Economic the Social Committee and the Committee of the Regions. Green Infrastructure (GI)—Enhancing Europe's Natural Capital. Brussels: European Commission, COM (2013) 249 final; 2013. 11 p.

[13] European Commission. LIFE Programme. [Internet]. 2017. Available from: http://ec.europa.eu/environment/life/index.htm [Accessed: 2017-09-01]

[14] European Commission. Natura 2000 barometer—Update 2017. Nature and Biodiversity Newsletter. 2017;**41**:8-9

[15] European Environmental Agency. Biogeographical Regions, Europe 2016. [Internet]. 2016. Available from: https://www.eea.europa.eu/data-and-maps/data/biogeographical-regions-europe-3 [Accessed: 2017-09-01]

[16] Ministerio de Agricultura, Pesca, Alimentación y Medio Ambiente. Banco de Datos de la Naturaleza. [Internet]. 2017. Available from: http://www.mapama.gob.es/en/biodiversidad/servicios/banco-datos-naturaleza/ [Accessed: 2017-09-01]

[17] Ramil Rego P, Rodríguez Guitián MA, Hinojo Sánchez BA, Rodríguez González PM, Ferreiro da Costa J, Rubinos Román M, Gómez-Orellana L, de Nóvoa Fernández B, Díaz Varela RA, Martínez Sánchez S, Cillero Castro C. Os Hábitats de Interese Comunitario en Galicia. Descrición e Valoración Territorial. Lugo: Monografías do IBADER, Universidade de Santiago de Compostela. 189 p.

[18] European Commission (EC). The State of Nature in the EU. Reporting under the EU Habitats and Birds Directives 2007-2012. Luxembourg: Office for Official Publications of the European Union; 2015

[19] European Topic Centre on Biological Diversity. Checklists for Habitats Directive Article 17 reporting. [Internet]. 2015. Available from: http://biodiversity.eionet.europa.eu/activities/Natura_2000/Folder_Reference_Portal/Art17_habitats_check_list.xls [Accessed: 2017-09-01]

[20] European Environmental Agency. Conservation Status of Habitat Types and Species (Article 17, Habitats Directive 92/43/EEC). [Internet]. 2015. Available from: https://www.eea.europa.eu/data-and-maps/data/article-17-database-habitats-directive-92-43-eec-1#tab-gis-data [Accessed: 2017-09-01]

[21] Martínez Sánchez S, Ramil Rego P, Hinojo Sánchez B, Chuvieco Salinero E. Assessing loss of biodiversity in Europe through remote sensing: The necessity of new methodologies. In: Grillo O, Venora G, editors. Biodiversity Loss in a Changing Planet. Rijeka: Intech; 2011. pp. 19-48

[22] Ministerio de Medio Ambiente. Estrategia Española para la Conservación y el uso Sostenible de la Diversidad Biológica. Madrid: Secretaría General de Medio Ambiente, Dirección General de Conservación de la Naturaleza, Ministerio de Medio Ambiente; 1999. 160 p.

[23] Pineda FD, de Miguel JM, Casado MA, Montalvo J, editors. La diversidad biológica de España. Madrid: Ed. Prentice Hall; 2002. 432 p.

[24] Loidi Arregui JJ, Díaz González TE, Herrera Gallástegui M. El paisaje vegetal del Norte-Centro de España. Itinera Geobotánica. 1997;**9**:5-160

[25] Loidi J, Biurrun I, Campos JA, García-Mijangos I, Herrera M. A biogeographical analysis of the European Atlantic lowland heathlands. Journal of Vegetation Science. 2010; **21**:832-842

[26] Rey Benayas JM, de la Montaña E. Identifying areas of high-value vertebrate diversity for strengthening conservation. Biological Conservation. 2003;**114**:357-370

[27] Assunçao-Albuquerque MJT, Rey Benayas JM, Albuquerque FS, Rodríguez MA. The geography of high-value biodiversity areas for terrestrial vertebrates in Western Europe and their coverage by protected area networks. Web Ecology. 2012;**12**:65-73

[28] Gómez A, Lunt DH. Refugia within refugia: Patterns of phylogeographic concordance in the Iberian Peninsula. In: Weiss S, Ferrand N, editors. Phylogeography of Southern European Refugia. Dordrecht: Springer; 2006, 2006. pp. 155-188

[29] Ferreiro da Costa J, Ramil-Rego P. Implicacións da propiedade da terra na planificación e xestión dos espazos protexidos en Galicia. Recursos Rurais. 2015;**11**:19-36

[30] Ramil-Rego P, Ferreiro da Costa J. As Reservas de Biosfera de Galicia (NW España). Recursos Rurais Serie Cursos. 2014;**7**:7-24

[31] Bonnin M, Jardin M. Le concept de réserve de biosphère en droit national. Paris: UNESCO, SC-09/CONF.207/INF.4; 2009. 60 p.

[32] Comba D. Le Régime Légal des Réserves MaB en France. Revue juridique de l'environnement. 2011;**3**:297-304

[33] UNESCO. Directory of the World Network of Biosphere Reserves (WNBR). [Internet]. 2017. Available from: http://www.unesco.org/new/en/natural-sciences/environment/ecological-sciences/biosphere-reserves/world-network-wnbr/wnbr/ [Accessed: 2017-09-01]

[34] Rodríguez Guitián MA, Ramil-Rego P. Revisión de las clasificaciones climáticas aplicadas al territorio gallego desde una perspectiva biogeográfica. Recursos Rurais. 2007;**3**:31-55

[35] Rodríguez Guitián MA, Ramil-Rego P. Fitogeografía de Galicia (NW Ibérico): análisis histórico y nueva propuesta corológica. Recursos Rurais. 2008;**4**:19-50

[36] Ninyerola M, Pons X, Roure JM. Atlas Climático Digital de la Península Ibérica. Metodología y aplicaciones en bioclimatología y geobotánica. Bellaterra: Universidad Autónoma de Barcelona; 2005. 45 p

[37] Rivas-Martínez S, Fernández-González F, Loidi J, Lousa M, Penas A. Syntaxonomical checklist of the vascular plant communities of Spain and Portugal to association level. Itinera Geobotanica. 2001;**14**:5-341

[38] Díaz González TE, Fernández Prieto JA. El paisaje vegetal de Asturias. Itinera Geobotanica. 1994;**8**:5-242

[39] Loidi J, Biurrun I, Campos JA, García-Mijangos I, Herrera M. A survey of heath vegetation of the Iberian Peninsula and Northern Morocco: A biogeographic and bioclimatic approach. Phytocoenologia. 2007;**37**(3-4):341-370

[40] European Topic Centre on Biological Diversity. Reference Portal for Natura 2000. [Internet]. 2016. Available from: https://bd.eionet.europa.eu/activities/Natura_2000/reference_portal [Accessed: 2017-09-01]

[41] Berastegi Gartziandia A, Zaldua Esteban A, Ibarrola Manterola I, Larumbe Arricibita J, Perez Perez J, Zulaika J, Carreras J, Valderrábano Luque J, Díaz González TE, Bueno Sánchez Á, Mora Cabello de Alba A, Férnandez Pascual E, Rubinos M, Hinojo B, Ramil P, HAZI, Diputación Foral de Álava (DFA), Diputación Foral de Gipuzkoa (DFG). Manual de buenas prácticas en la gestión de turberas y humedales. Pamplona: Equipo de LIFE TREMEDAL. Navaprint; 2016. 222 p.

[42] European Topic Centre on Biological Diversity. Article 17 Web Tool on Biogeographical Assessments of Conservation Status of Species and Habitats under Article 17 of the Habitats Directive. [Internet]. 2016. Available from: https://bd.eionet.europa.eu/article17/reports2012/ [Accessed: 2017-09-01]

[43] Ministerio de Medio Ambiente. Estrategia para la Conservación del Oso Pardo Cantábrico (Ursus arctos) en España. Criterio Orientadores. Madrid: Comisión Nacional de Protección de la Naturaleza; 1999. 37 p.

9

Methods for Biodiversity Assessment: Case Study in an Area of Atlantic Forest in Southern Brazil

Maria Raquel Kanieski, Solon Jonas Longhi and Philipe Ricardo Casemiro Soares

Abstract

Populations and species are disappearing due to disturbances in the environment caused by human activities. Given, the obvious risk of loss of diversity, it is increasingly necessary to take actions concerning preservation, in which safety features are necessary for measuring the variation of diversity in space and time. The aim of this study was to evaluate the structure and diversity in the arboreal component and natural regeneration in an area of Araucaria Forest in Southern Brazil. The vegetation sampling was performed by analyzing 180 subunits of 10 × 10 m, where all the arboreal individuals and natural regeneration were inventoried. Different alpha and beta indexes of diversity were calculated. The Margalef, Shannon, and the Beta indexes were underestimated, possibly influenced by the size of sample unit. Index Menhinick represented the diversity in a very real form, even in small sampling units. The indexes of Simpson and MacIntosh denote low dominance and the equity indexes showed high uniformity in species.

Keywords: Mixed Rainy Forest, alpha indexes, beta indexes, biodiversity evaluation, natural regeneration, arboreal component

1. Introduction

Biodiversity is one of the fundamental properties of nature and source of immense potential for economic use. It is the basis of agricultural activities, livestock, forestry, and fishing, and also for the strategic biotechnology industry. The ecological functions performed by biological diversity are still little understood, although we consider it to be responsible by natural processes and products provided by ecosystems and species that sustain other life forms and change the biosphere, making it suitable and safe for life [1].

Biological diversity or biodiversity are expressions that refer to the variety of life on the planet, or to the property of living systems to be distinct. It includes plants, animals, microorganisms, ecosystems, and ecological processes in a functional unit [2].

The diversity is regarded as an indication of the ecosystem well-being. It indicates directly the protection of certain place, the higher the value, the larger the biodiversity of the ecosystem in question [3].

Because it is an extremely complex structure, there is a problem to evaluate biodiversity, so we need to find simplified variables for its determination [4]. Some of them—as the composition, type of regeneration, introduced species, the presence of dead trees and the landscape—have been defined according to the 4th Ministerial Conference on the protection of forests, in Vienna, 2002. Especially biodiversity is important, since it represents the state of conservation of ecosystems, including forests, where it can help evaluate the sustainability of the resources are managed [1].

Biological diversity is a central theme of ecological theory and has been the subject of many discussions. Currently, researchers have developed a large number of parameters for the measurement of biodiversity as an indicator of the state of ecological systems, with practical applicability for purposes of conservation, management, and environmental monitoring [5].

Biodiversity can be expressed in four levels: diversity of genes, species, and ecosystems settlement. It can be appreciated by the number (wealth) of different biological categories and relative abundance (evenness) of these categories; and also for variability in local level (alpha diversity), the complementarity between biological habitats (beta diversity) and variability between landscapes (gamma diversity) [6]. It includes, therefore, all the living or biological resources, genetic resources, and their components.

Any environmental protection strategy must ensure the maintenance of biodiversity. All the living beings that inhabit a country constitute an irreplaceable heritage, because each species, as well as each population, has in its genetic composition information from millions of years of evolutionary adaptations. However, in order to exercise management plans and protection of nature reserves, trusted tools able to measure its variation in space and time are required.

Assessing the biodiversity of an ecosystem by its forestry component assumes that the arboreal component is what sustains a forest ecosystem [7]. Already the quantification of diversity through natural regeneration allows a complete characterization of the forest as an ecosystem, and not only by its arboreal component. This approximate evaluation of possible interventions made in settlements, as well as the intensity and provides a rough overview of the volume of the existing biomass in the forest, whose presence can be very relevant with regard to fire prevention, energy potential, and characterization of biodiversity.

Considering the importance of the subject, the objective of this work is to evaluate the arboreal component and natural regeneration biodiversity in an Atlantic Rainy Forest area in Southern Brazil, through different diversity indices in order to find the variables that better represent the current status of diversity in study environment.

2. Material and methods

2.1. Characterization of the study area

This study was conducted on National Forest (FLONA) of São Francisco de Paula, administered by the Chico Mendes Institute for Biodiversity Conservation (ICMBio), constituting a conservation unit of sustainable use. The aim of this type of conservation unit (UC) is to make nature conservation with the sustainable multiple uses of its natural resources and stimulate scientific research, with emphasis on methods for sustainable forest management [8].

The FLONA is located in the northeast of the State of Rio Grande do Sul, in southern Brazil, in the city of San Francisco de Paula. It has a total area of 1606.69 ha, which 9019 ha are occupied by native forests (Mixed Rainy Forest and Dense Rainy Forest), over 600 ha by planted forests (*Pinus* spp., *Araucaria angustifolia*, and *Eucalyptus* spp.) and the rest by other areas such as fields, lakes, infrastructure among others.

The region is one of the wetter of the State, with more than 2000 mm rainfall per year and with annual average temperature of approximately 14.5°C. According to the Köppen's climate classification, the climate is classified as "Cfb" mesothermic medium [9]. The average relative humidity is 83.9%. Summer is characterized by mild temperatures and winter by constant and intense cold, where the average temperature is close to 0°C. Frosts are common in the cooler half and may snow in winters.

The types of soil found in the FLONA are Cambisols, Chernozems, and Newsoils [10]. Geomorphology is marked by a strongly wavy relief in the northern part, with an altitude of 930 m and rugged in the South, forming canyons with more than 100 m in depth [11] (**Figure 1**).

The predominant forest vegetation belongs to the Mixed Rainy Forest. In addition to *Araucaria angustifolia* (Bertol.) Kuntze, which prints a physiognomic character in this vegetation, it is common to find other tree species such as *Sebastiania commersoniana* (Baill.) L. b. SM. & Downs, *Cedrela fissilis* Vell., *Podocarpus lambertii* Klotzsch ex Endl., among others. It has the peculiarity to be a transition zone between the Mixed Rainy Forest and Dense Rainy Forest, with endemic species as *Oreopanax fulvum* Marchal, typical of the Atlantic forest [12]. The author even describes the endemism does not get more accentuated because there is a physical barrier to the species by the existence of the Canyons.

2.2. Sampling

The work was performed in six permanent plots of 100 × 100 m (10,000 m^2), which were divided into 10 tracks of 10 × 100 m (1000 m^2) and these subdivided into 10 subunits of 10 × 10 m (100 m^2). Within each plot, 30 subunits were drawn, being three subunits by track from 10 total tracks in each unit of 100 × 100 m, where vegetation surveys and environmental parameters were made. The distribution of selected subplots in each parcel can be observed in **Figure 2**.

Figure 1. View from the FLONA of São Francisco de Paula, Rio Grande do Sul, Brazil.

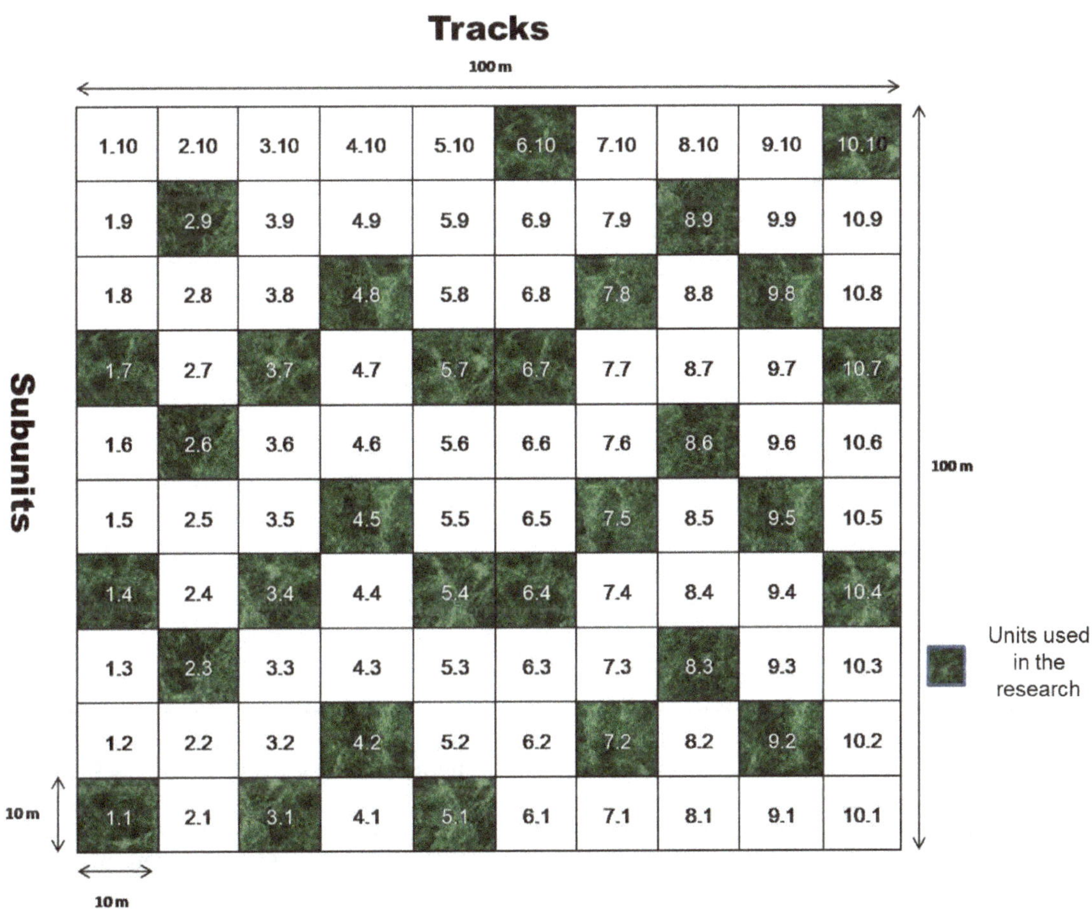

Figure 2. Scheme of subplots in each sampling unit used in this study in southern Brazil.

2.3. Data collection

All individuals over 1.30 m and with a circumference at breast height (CBH) from 3 cm, minimum, and up to the limit of 29.9 cm are considered as natural regeneration. As an arboreal component, we considered all individuals with CBH greater than or equal to 30 cm. Individuals sampled were numbered and identified botanically, collecting botanical material, in case of doubt, registering the number of plants and the common names. The identification in level of family, genus, and species, was carried out consulting the Herbarium of the Department of Forest Sciences at de Santa Maria Federal University, according to the APG II system [13].

2.4. Biodiversity indices

Biodiversity indices (**Table 1**) that provided the evaluation for this work were presented by [14, 15]. These were calculated for each sampling unit leased in the study area, both for natural regeneration and for the arboreal component.

2.4.1. *Alpha diversity*

The majority of the proposed methods to quantify the biodiversity of species refer to the diversity within communities, namely as alpha diversity. Within this, there are methods based

Species richness	Margalef index	$R_1 = \frac{(S-1)}{\ln n}$
	Menhinick index	$R_2 = \frac{S}{\sqrt{n}}$
Dominance	Simpson index	$D = \sum p_i^2$
	MacIntosh index	$D = \frac{n-U}{n-\sqrt{n}}$
Information	Shannon index	$H' = -\sum_{i=1}^{s}(p_i)(\ln p_i)$
Equity	Pielou index	$J' = \frac{H'}{H'_{max}}$
	Alatalo index	$F = \frac{N2-1}{N1-1}$
	Hill index	$E' = \frac{N2}{N1}$
Similarity	Jaccard index	$I_j = \frac{c}{a+b+c} = \frac{c}{A+B-c}$
	Sorensen index	$I_s = \frac{2c}{a+b} = \frac{2c}{A+B-c}$

R_1 = Margalef index; R_2 = Menhinick index; S = total number of species; n = total number of individuals; D = Simpson index; p_i = proportion of species in a Community ($p_i = n_i/n$); d = MacIntosh index (i = 1, 2, 3, ..., S); H' = Shannon index; J = Pielou index; H'max = ln(S) = Alatalo index; N = number of abundant species = $e^{H'}$; $N2$ = number of very abundant specie = $1/D$; E = Hill index. I_j = Jaccard index; I_s = Sorensen index; A = number of species present on the site A; B = number of species present on the site B; a = number of species unique to the site A; b = number of species unique to the site B; and c = number of species present in both places (A and B).

Table 1. Alpha and Beta diversity indices used in the work evaluation.

on quantification of the number of species (species richness) and those based on community structure, in the other hands, in proportion of the value of importance of each species. These can still be based on the information, on dominance or equity of the Community [15].

2.4.1.1. *Species richness*

The measures of species richness provide an understandable and instantaneous expression of diversity [5]. Within the species richness indices, we can cite the Margalef and Menhinick indices (**Table 1**). These indices do a relationship between the number of species and the number of individuals, and the larger the area of the sampling unit. The greater the number of species entered, the greater the value of the index [5, 15, 16]. One of the great advantages of Margalef and Menhinick index is its ease calculation, being used successfully in several scientific papers [5].

2.4.1.2. *Information*

The most commonly indices used are those based on information theory. These indices are based on the logic that the diversity, or information, in a natural system, can be measured in a similar way to information contained in a code or message [5]. Shannon and Wiener obtained the function known as Shannon-Wiener index (**Table 1**). In many works, researches wrongly named it as "Shannon-Weaver" [17, 18]. This index considers that individuals are sampled at random from a population "indefinitely large" [5, 16, 19].

The Shannon index assumes zero value when there is only one species and the logarithm of the number of species, when the same number of individuals represents all species [5]. Shannon values generally vary between 1.3 and 3.5, and may exceed 4.0, and reach around 4.5 in tropical forest environments [20]. The authors even claim that this index assigns greater value to rare species and is one of the best indices to be used in comparisons, if there is no objective in separating abundance to rarity.

2.4.1.3. *Dominance*

The indexes based on dominance are inverse parameters to the concepts of uniformity or equity of the community. These indices take into account the representativeness of the species with the highest value of importance, without evaluating the contribution of other species [15]. Among these, we can cite the Simpson and MacIntosh indices (**Table 1**). The Simpson index shows the probability that two individuals taken at random from a given community are of different species. This index is heavily influenced by the most abundant species of sampling unit while it is least sensitive to species richness [5, 16]. The Simpson index varies from 0 to 1, and the closer to 1 the more dominance. On the other hand, the MacIntosh will introduce greater dominance when their values are closer to 0. The index of MacIntosh is not a dominance index; however, it can be calculated as a measure of diversity or dominance, which is independent of the total number of individuals [5].

2.4.1.4. *Equity or uniformity*

The equity or uniformity indices show how the number of individuals are distributed among the species, denoting less or greater uniformity in the composition of the parcels. Among these, we can cite the Pielou, Hill, and Alatalo (**Table 1**). Pielou index is the most commonly used within the category equity, and measures the proportion of diversity observed with respect to the maximum expected diversity. The value of this index varies from 0 to 1, and when it reaches the value 1, it means that all species are equally abundant [15, 16].

The index of Hill is less used for vegetation, with greater use in wildlife studies. This index refers to the distribution of the abundance of the species, or the manner in which the abundance is distributed between the species in a community. When all species in a sample are also abundant, the equity index must take the maximum value, and decreases, tending to zero, as the relative abundances of species diverge from its equality. It has as parameters Simpson and Shannon indices. The Hill index can cause some misunderstandings in some particular cases: reaches high value when equity is high or when a species dominate the community [15].

Alatalo index is known as Hill modified index. This index will approaching zero as the smaller gender equality in the composition of species in different plots [16]. This index is not recommended because, when used in comparisons, tends to overvalue the equity, having a non-linear relationship with this [15].

2.4.2. Beta diversity

Beta diversity or among habitats represents the degree of change of species, as well as biotic change through environmental gradients [6]. It is based in proportions or differences, that can be quantified based on indices or similarity coefficients of dissimilarity or distance between the plots, from quantitative or qualitative data, as well as by beta diversity indices themselves.

The easiest system to measure beta diversity between pairs of localities is through the use of similarity coefficients [5]. The most used is the Jaccard and Sorensen indices (**Table 1**). These indices are designed to be equal to 1 if there are complete similarity cases (when the two species are identical) and equal to zero if the plots are dissimilar and have no species in common. One of the great advantages on these measures is your simplicity. However, this can also be a disadvantage, since the coefficients do not consider the abundance of species. All species have an equal weight in the equation, whether abundant or rare [5].

3. Results and discussion

3.1. Natural regeneration

The diversity indices calculated at work (**Table 2**) included the contents of Margalef and Menhinick (species richness); Simpson and MacIntosh (dominance); Shannon (information); Pielou, Alatalo and Hill (equity) and Jaccard and Sorensen (similarity).

The value found by index of Margalef (3.81) featuring low diversity when compared to other regeneration areas. In a study of similarity in floristic composition between an area of natural regeneration of Mixed Rainy Forest and natural regeneration in an adult Brazilian pine plantation, in Irati, Paraná [21], the authors found a value of 10.68 floristic diversity in Margalef index for the Mixed Rainy Forest, denoting high floristic diversity. The low diversity in the present study may be explained by the fact that the index does not consider the size of the sampled area, which in this case is less than the commonly found in the works.

Diversity indices		Values	Classification*
Species richness	Margalef index	3.81	Low diversity
	Menhinick index	2.05	High diversity
Dominance	Simpson index	0.21	Low dominance
	MacIntosh index	0.64	Low dominance
Information	Shannon index	2.13	Medium diversity
Equity	Pielou index	0.76	High uniformity
	Alatalo index	0.65	High uniformity
	Hill index	0.71	High uniformity
Similarity	Jaccard index	0.24	Low similarity
	Sorensen index	0.38	Low similarity

*Sorting by comparing values found in similar areas in other works.

Table 2. Alpha and Beta diversity indices for the natural regeneration.

The Menhinick index denoted high diversity through the value of 2.05. Even without taking into consideration the size of the sampling area, different of Margalef index, represents, in this case, the diversity of real way in the area, even a small sampling area.

The indices of dominance express if there is dominance by one or a few species in the sampled area. In species dominance, the two indices, Simpson and MacIntosh, presented low dominance by one or a few species on the plots with values of 0.21 and 0.64, respectively.

The most commonly used indices are those based on information theory. These indexes express directly the value of diversity at that location; the most used is the Shannon index. For this index, we found an average of 2.13, which is characteristic of species diversity for the average site. Similar values were found using the same size of sampling units in the FLONA of São Francisco de Paula [18], with a value by the Shannon index of 2.22. Due to the high density of individuals that occur in natural regeneration, and these are heavily clustered, the Shannon index have a good representation of the diversity of natural regeneration even in smaller sampling units.

The equity indexes show how the number of individuals are distributed among the species, denoting less or greater uniformity in the composition of the parcels. For the area in question, the three indices used, Pielou, Alatalo, and Hill showed the same trend in each sub-plot evaluated by means of the average values of 0.76, 0.65, and 0.71, respectively, which denotes high uniformity in composition of species on the plots.

On natural regeneration, the Jaccard index varied from 0 to parcels with less interaction, up 0.75 to parcels with greater interaction, with an average of 0.24, which denotes low floristic similarity between the plots. Already the Sorensen index reached values from zero to the natural regeneration in smaller plots, interaction to 0.86 to portions of greater interaction, with an average of 0.38, featuring low floristic similarity. In the FLONA of São Francisco de Paula, in portions of 100 × 100 m [11], the authors found high similarity between floristic groups formed within the area through Sorensen index, with values ranging from 0.66 to 0.75.

3.2. Arboreal component

The studied indices for the arboreal component (**Table 3**) included the contents of Margalef and Menhinick (species richness); Simpson and MacIntosh (dominance); Shannon (information); Pielou, Alatalo and Hill (equity) and Jaccard and Sorensen (Similarity).

The value found for Margalef index denoted low diversity when compared to other areas of Mixed Rainy Forest. This low diversity can be explained by the fact that the index does not consider the size of sampled area, which in this case is less than the commonly found in other researches. In a study on FLONA of São Francisco de Paula [22], the authors found values for Margalef index, through 10 parcels of 100 × 100 m, which ranged from 6.24 to the lower portion to 10.05 for greater diversity, with an average of 8.38, denoting high diversity in the arboreal component. It allows us to conclude that this index is not ideal for assessing the diversity in small areas, such as those used in this work (10 × 10 m).

In a research of floristic and structural changes that have occurred in a Mixed Rainy Forest, located in São João do Triunfo, PR [23], the authors found a value in the year 1979, of 6.52 and in the year 2000, of 7.02 for Margalef index, featuring of medium to high biodiversity.

The Menhinick index presented a 1.96 value of species richness. This value represented medium to high diversity in the study area, when compared to other areas of the same type. Also in the FLONA of São Francisco de Paula [22], however, with plots of 100 × 100 m, the authors found an average of 1.96 for Menhinick index, denoting high species richness.

Diversity indices		Values	Classification*
Species richness	Margalef index	2.20	Low diversity
	Menhinick index	1.96	Medium to high diversity
Dominance	Simpson index	0.26	Low dominance
	MacIntosh index	0.80	Low dominance
Information	Shannon index	1.54	Low diversity
Equity	Pielou index	0.93	High uniformity
	Alatalo index	0.88	High uniformity
	Hill index	0.91	High uniformity
Similarity	Jaccard index	0.12	Low similarity
	Sorensen index	0.21	Low similarity

*Sorting by comparing values found in similar areas in other works.

Table 3. Alpha and Beta diversity indices for the arboreal component.

In the same study of floristic and structural changes that have occurred in a Mixed Rainy Forest, located in Paraná [23], the authors found a value for Menhinick index of 1.10, in 1979, and 1.17 in 2000, characterized by medium diversity.

Even without considering the size of the sampling area, the Menhinick index, different of Margalef index, represents, in this case, the real diversity of the area, even a small sampling area. The suggestion would be in smaller sampling units, use the Menhinick index, and, in larger sampling units, Margalef index, both to assess the species richness within the area.

The dominance of species by the two indices, Simpson and MacIntosh, presented values of 0.26 and 0.8, respectively. The contents of Simpson and MacIntosh presented low dominance by one or a few species on the plots. Other research found very similar values in Mixed Rainy Forest area in São João do Triunfo, PR [23], with 0.21, in the year of 1979, and 0.26 in 2000, for the Simpson index, and 0.97 (1979) and 0.96 (2000), for the MacIntosh index, characterizing low dominance by one or a few species in the area.

It is common to find misinterpretation in Simpson index, in which the authors use this index as a parameter of the diversity of local information, when in fact he expressed only the dominance of some species in the plot. Other researchers found values of 0.15 [24] and 0.04 [25] for the Simpson index in areas of Mixed Rainy Forest and interpreted as areas of middle and low diversity, when in fact these values indicate only that there are low dominance of one or a few species in the area, it means that there are, approximately, the same number of species in each sample unit.

We obtained a value for Shannon index of 1.54, denoting low diversity when compared to other studies. The low floristic diversity is common in Mixed Rainy Forest remnants located in areas of higher altitude, where the climatic rigors selective pressure on plant diversity [26].

These low values of diversity can also be a result of sample unit size, as well as the index of species richness of Margalef. In the same FLONA of São Francisco de Paula, in sampling units of 100 × 100 m [22], the authors found values for the arboreal component between 2.78 to plots of less diversity, to 3.40 to the greater diversity, with an average of 3.19, which features high diversity. Also in the FLONA of São Francisco de Paula, others researches found high diversity in the areas sampled by the Shannon index, with values of 3.53 [27] and 2.95 [28]. Highest values of diversity were found in several works with Mixed Rainy Forest [17, 25, 29–32], which characterized the diversity in these areas of medium to high, according to the Shannon index.

Pielou, Alatalo, and Hill indices presented values of 0.93, 0.88, and 0.91, respectively, demonstrating high uniformity in the composition of the parcels. Other research [28], in the FLONA of São Francisco de Paula, also found high uniformity in the composition of the parcels, with value of 0.80 for the Pielou index. Similar values were also found in other areas of Mixed Rainy Forest, like [17] (Pielou = 0.90) and [32] (Pielou = 0.83).

Other research also found similar results by evaluating the arboreal component in National Forest of São Francisco de Paula, in sampling units of 100 × 100 m, with a value of 0.60 at Alatalo index and 0.63 for Hill index, featuring high uniformity.

For the arboreal component, the Jaccard similarity index values varied between zero, for parcels with no interaction, which means that no species coincides in the composition of the two parcels, up to 1.00 for portions of greater interaction, meaning that all species present in a portion are equal to those in another. The average Jaccard index in arboreal component was 0.12, what characterizes low similarity in species composition between different plots. In 20 × 100 m plot in a remnant of Mixed Rainy Forest in Santa Catarina [31], the authors also found low floristic similarity in composition of plots according to the Jaccard index, with values rarely exceeding 0.5.

The coefficient of similarity of Sorensen reached values between 0, to the parcels with less interaction, to 1 for parcels with greater interaction, with an average value of 0.21, which denotes, as well as the Jaccard index, low similarity in species composition for different plots. These low values of similarity found for the two indices can be related to the size of the sampling unit.

4. Conclusions

The specific richness of Margalef index was highly influenced by the size of the sampling unit, obtaining values of diversity, both in arboreal component and in natural regeneration. As a result, this index is not suitable for assessing the diversity of Mixed Rainy Forest in small sampling units, as adopted in this methodology.

Even without considering the size of the sampling area, Menhinick index represented the diversity in a real way of the area, fact that indicates this index for diversity evaluation in small sampling units of Mixed Rainy Forest.

The Simpson and MacIntosh indices showed similar patterns, both to the arboreal component and natural regeneration, denoting low dominance of one or a few species. These values are similar to those found in other studies in Mixed Rainy Forest, which denotes that the two indices show the real dominance of species in these areas, in accordance with the methodology adopted.

The Shannon-Wiener index, which reports directly to the value of diversity at the site, showed an average value of 1.54 for the arboreal component, which denotes low diversity in place. This low value of diversity can also be a result of sample unit size, as well as the Margalef index.

On natural regeneration, the Shannon index was not influenced by the size of the sampling unit, showing an average of 2.13, which denotes medium diversity in place.

The three equity indices used, Pielou, Alatalo, and Hill, showed the same pattern in each subplot evaluated, denoting high uniformity in the composition of the parcels. The indices presented the uniformity in composition of species on the plots, being suitable for the assessment of equity in Mixed Rainy Forest areas, according to the adopted methodology.

Beta diversity indices presented low similarity in the floristic composition of different samples with the low value may also be a consequence of the small area of sampled units.

Acknowledgements

To FLONA of São Francisco de Paula, CAPES, and CNPQ.

Author details

Maria Raquel Kanieski[1*], Solon Jonas Longhi[2] and Philipe Ricardo Casemiro Soares[1]

*Address all correspondence to: raquel.kanieski@udesc.br

1 Santa Catarina University State, Lages, Santa Catarina, Brazil

2 Santa Maria Federal University, Santa Maria, Rio Grande do Sul, Brazil

References

[1] Salas GR, Fidalgo B, Torres R. Relationship between biodiversity and structure indices in maritime pine forests: Usefulness in forest management. In: IUFRO, editor. Conference of the Biodiversity and Conservation Biology in Plantation Forests 2005; April, 2005; Bordeaux, France; 2005

[2] Dias BFS, Coradin L. Diversidade biológica e sua importância para o Brasil: Programa de Dados para a Conservação da Biodiversidade. PROBIO/SP: São Paulo, SP; 1999

[3] Ricklefs RE. Ecology. 3rd ed. New York, NY: Freeman; 1990. p. 456

[4] Noss RF. Assessing and monitoring forest biodiversity: A suggested framework and indicators. Forest Ecology and Management. 1999;**115**

[5] Magurran AE. Ecological Diversity and its Measurements. New Jersey, NJ: Princeton University Press; 1988. p. 179

[6] Whittaker RH. Evolution and measurement of species diversity. Taxon; 1972;**21**(2/3): 213-251

[7] Couto HTZ. Métodos de inventário da biodiversidade de espécies arbóreas. ESALQ/FAPESP: Piracicaba, SP; 2005. p. 108

[8] Brasil. Lei 9.985/2000 [Internet]. 2000 [Updated: September 10, 2017]. Available from: http://www.planalto.gov.br/ccivil_03/leis/L9985.htm [Accessed: September 16, 2017]

[9] Alvares CA, Stape JL, Sentelhas PC, Gonçalves LM, Sparovek G. Köppen's climate classification map for Brazil. Meteorologische Zeitschrift. 2013;**22**(6)

[10] Embrapa. Sistema brasileiro de classificação de solos. Rio de Janeiro, RJ: EMBRAPA Solos; 1999. p. 412

[11] Ribeiro SB, Longhi SJ, Brena DA, Nascimento AT. Diversidade e classificação da comunidade arbórea da Floresta Ombrófila Mista da Flona de São Francisco de Paula RS. Ciência Florestal. 2007;**17**(2)

[12] Narvaes IS. Classificação e caracterização da regeneração natural em Floresta Ombrófila Mista na Floresta Nacional de São Francisco de Paula, RS [dissertation]. Santa Maria, RS; 2004. p. 143

[13] Angiosperm Phylogeny Group. An update of the Angiosperm Phylogeny Group classification for the orders and families of flowering plants: APG II. Botanical Journal of the Linnean Society. 2003;**141**:399-436

[14] Waite S. Statistical Ecology in Practise: A Guide to Analysing Environmental and Ecological Field Data. Prentice Hall: London, EN; 2000. p. 414

[15] Moreno CE. Métodos para medir la biodiversidad. 1st ed. M & T Manuales y Tesis SEA: Zaragoza; 2001. p. 84

[16] Ludwig JA, Reynolds JF. Statistical Ecology: A Primer on Methods and Computing. New York, NY: John Wiley & Sons; 1988. p. 338

[17] Cordeiro J, Rodrigues WA. Caracterização fitossociológica de um remanescente de Floresta Ombrófila Mista em Guarapuava, PR. Árvore. 2007;**31**(3):545-554

[18] Narvaes IS, Brena DA, Longhi SJ. Estrutura da regeneração natural em Floresta Ombrófila na Floresta Nacional de São Francisco de Paula, RS. Ciência Florestal. 2005;**15**(4):331-342

[19] Pielou EC. Ecological diversity. New York, NY: John Wiley; 1975. p. 165

[20] Felfili JM, Rezende RP. Conceitos e métodos em fitossociologia. Brasília, DF: UNB; 2003. p. 68

[21] Saueressig D, Moraes CM, Walk F, Inoue MT, Figueiredo Filho A. Estudo da similaridade na composição florística entre uma área de Floresta Ombrófila Mista e a regeneração natural em um plantio adulto de pinheiro-brasileiro na FLONA de Irati. In: 14 Encontro Anual de Iniciação Científica; Guarapuava, PR: UNICENTRO; 2005

[22] Kanieski MR, Longhi SJ, Araujo ACB. Quantificação da biodiversidade em floresta ombrófila mista por meio de diferentes índices alfa. Scientia Forestalis. 2010;**38**(88):567-577

[23] Schaaf LB, Figueiredo Filho A, Franklin G, Sanquetta CR, Longhi SJ. Modificações florístico-estruturais de um remanescente de floresta ombrófila mista montana no período entre 1979 e 2000. Ciência Florestal. 2006;**16**(3):271-291

[24] Sousa SGA, Carvalho J, Barddal ML. Diversidade e estrutura fitossociológica de uma floresta ciliar do rio Iraizinho, Piraquara, PR. In: UFLA, editor. Ve Simpósio Nacional sobre Recuperação de Áreas Degradadas; Lavras, MG; 2008

[25] Watzlawick LF, Sanquetta CR, Valério AF, Silvestre R, Caracterização d. composição florística e estrutura de uma Floresta Ombrófila Mista no município de General Carneiro (PR). Ambiência. 2005;**1**(2):229-237

[26] Roderjan CV, Franklin G, Saito YK, Santos EP. Caracterisation des unites phytogeographiques dans l.etat du Paraná Brasil et leur etat de conservation. Biogeographica. 2001;**77**(4):129-140

[27] Gomes JF, Longhi SJ, Araujo MM, Brena DA. Classificação e crescimento de unidades de vegetação em Floresta Ombrófila Mista, São Francisco de Paula, RS. Ciência Florestal. 2008;**18**(1):93-107

[28] Sonego RC, Backes A, Souza AF. Descrição da estrutura de uma Floresta Ombrófila Mista RS Brasil utilizando estimadores não-paramétricos de riqueza e rarefação de amostras. Acta Botânica Brasílica. 2007;**21**(4):943-955

[29] Nascimento AT, Longhi SJ, Brena DA. Estrutura e padrões de distribuição espacial de espécies arbóreas em uma amostra de Floresta Ombrófila Mista em Nova Prata RS. Ciência Florestal. 2001;**11**(1):105-119

[30] Seger CD, et al. Levantamento florístico e análise fitossociológica de um remanescente de Floresta Ombrófila Mista localizado no município de Pinhais Paraná-Brasil. Floresta. 2005;**35**(2):291-302

[31] Lingner DV, Oliveira YMM, Rosot NC, Dlugosz FL. Caracterização da estrutura e dinâmica de um remanescente de floresta com araucária no planalto catarinense. Pesquisa Florestal Brasileira. 2007;**55**:55-56

[32] Reginato M, Goldenberg R. Análise florística estrutural e fitogeográfica da vegetação em região de transição entre as Florestas Ombrófilas Mista e Densa Montana Piraquara, Paraná, Brasil. Hoehnea. 2007;**34**(3):349-364

10

Importance of Underutilized Field Crops for Increasing Functional Biodiversity

Franc Bavec, Urška Lisec and Martina Bavec

Abstract

Despite the suggestions to include two or three crops into crop rotation that is widely considered to support the richer biodiversity on fields, industrial field crop production systems are still based mainly on monoculture, where the farmers produce permanently mainly one crop. Review and analyses of different possibilities showed that more diverse functional (also important for diverse nutritional and health products of food) biodiversity of underutilized field crops needs to be established, especially if beneficial social and economic effects of introducing underutilized crops into small-scale farms are taken into account. We can conclude that functional biodiversity based on rich crop rotations associated with underutilized crops increases biodiversity in the soil and has an effect on richer and sustainable behavior of cultural plants with good balance of pests and plant diseases.

Keywords: underutilized crops, alternative crops, biodiversity, functional biodiversity, crop rotation

1. Introduction

Industrial-supported production systems with monocultures production of one crop during the years at one place are known to unbalance complete biodiversity of the fields. The main reasons for decreasing soil, plant, and landscape biodiversity are bad agricultural practices such as: (i) reduction of organic inputs in the soil, (ii) use of synthetic compounds including heavy metals as a part of fertilizers, (iii) use of high inputs of herbicides, pesticides and fungicides, and (iv) "scientific" supported GMO organisms adaptable to high inputs of herbicides. Especially, GMO plants in field crop production represent a compensation for the mentioned bad agricultural practices.

Few basic documents on agroecosystems biodiversity [1–3] gave extensive descriptions and figures about relationships of forming sustainable biodiversity. It is now clear that monoculture gives rise to decreased natural enemies of species diversity and habitat diversification, whereas rotations and cover crops can improve biodiversity [2]. An extensive overview from different aspects contains terms such as (i) prevention of culture and ethno science, (ii) knowledge about potential uses of plants, and (iii) uses of local resources for pest control, etc., but on other hand, the whole approach is done in the context of industrial-supported agriculture by synthetic pesticides and fertilizers [3]. Assessment of potential indicators for sustainability (especially biodiversity) of food production on the field level looks like built a house without strong fundament [4]. Indicators of soil quality and food quality (influenced by different crops in crop rotation) are essential for the indicator frameworks; they are strongly correlated with each other.

The main problem of dominant monoculture production (based on five main crops such as maize, wheat, soybean, rice and cassava, and potato, rye, etc., depending on specific regional circumstances) is genetic erosion of plant diversity and associated biodiversity of life cycles in the soil and behavior of the plants (pests, diseases, new components of GMO sequences in the plants, etc.). It is true and important that agricultural and biodiversity conservation sectors must work in partnership [5]. However, kind of theoretical basis is not enough for protecting biodiversity from landscape to the soil level. Especially, it needs to be in the scope of natural processes of known good agricultural practices.

In spite of discussions with different officers (Brussels, Ministry) and presented chapter [6], where we focused on agricultural measures for increasing biodiversity on the field level with supporting remarks for policy makers, "green measures" is nothing new. EU green measures (also the USA and other parts of world are not exception) in many cases smell more like "green washing" than real base for increasing rich biodiversity on the field level and produce niche products of health and nutritional foods [7].

The aim of the chapter is to analyze potential underutilized species of field crops and their benefits for introducing them into broader crop rotation instead of monoculture or two crops rotation. It is also important to show the data and idea to environmental, agronomical advisers, and policy makers for better understanding and support of rich biodiversity of crop plants in the fields.

2. Methodology

The chapter is based on extensive review of latest results and rare scientific papers (from databases Web Sci., Sci. Direct, keywords like functional biodiversity, crop rotation, and underutilized crops) particularly focusing on basic part of agricultural biodiversity with real change of crop rotation based on underutilized crops as a part of functional biodiversity. SWOT analyses of underutilized crops vs. biodiversity were also employed. As a special part of this basic change on the field instead of monoculture or two crops rotation, a special focus will be an interdisciplinary approach of underutilized crops vs. biodiversity and other environmental, social, and economic benefits.

3. Results and discussion

3.1. Review of scientific and professional bibliographies in field underutilized crops as part of functional biodiversity

According to numerous projects' (e.g., by Agrobiodiversity international, different environmental agencies like PBL Netherlands Environmental Assessment Agency [8], and others) designs and explanations of agricultural biodiversity [2, 3] from agronomic point of view, underutilized crops are the main point for real sustainable biodiversity on the fields. Considering the basic part of agricultural biodiversity, especially underutilized crops growing in organic (biological, ecological) production system without synthetic chemistry and transgenic plants are the best support for rich agricultural biodiversity. In this way, rich functional biodiversity of underutilized crop supports all parts of natural biodiversity in the soil and on the field (**Figure 1**).

Potential of underutilized crops for rich biodiversity is extensively described by different authors [9] through global and regional assessments from wild to cultivated plants. In different literatures, about 100,000 used plants, 80,000 explored plants by humans since down civilization and about 118,000 edible plants are listed. In addition, 9500 plants as a source of economics, etc., are reported at the world level. Regional types of species are diverse according to climatic conditions, but in every region, a lot of minor, underutilized, and even neglected crops also exist. It is also based on the fact that in one region, one species can be neglected, but it may be a main crop (like a chick pea).

In case of small country Slovenia, we can describe a relatively long list of potential edible and economical valuable field crops (**Table 1**). However, just a few of main crops are economically acceptable for the farmers. Because of this, agricultural biodiversity has reduced mainly to one or two crops per field. In recent years, biodiversity structure on the fields has been increasing especially on organic farms. The list in **Table 1** can be longer if herbal and vegetable crops or even their intercrops with fruit plants are included in crop rotation. In this case, in every region over the world, the increase of functional plants is nonlimited due to usable plants. Only question is whether political decision-makers support it. In case of Slovenia, we numbered more than 200 plant species in these combinations. However, many of this kind of combinations were not examples of research and specific development. In case of more traditional plants, we can represent some possibilities from more conventional intercrops to the intercrops and sole crops of underutilized plants.

Underutilized crops are better adapted to marginal soil, complex and difficult environments conditions (climate change, which will have impact on biotic and abiotic stress), and they can contribute significantly to diversification and resilience of agroecosystems [10, 11]. Underutilized crops, which are known for their drought tolerance, are the minor millets, sorghum, and amaranth [12–14]. We have to know that we need to grow population of crops, which would have the ability to maintain the satisfactory yield under changing climates. Excellent drought resistance combined with good tolerance of heat, low fertility, and a range of soils make cowpeas viable throughout the temperate climate. It makes an excellent N source and attracts many beneficial insects. Cowpeas also can be used on poor land [15].

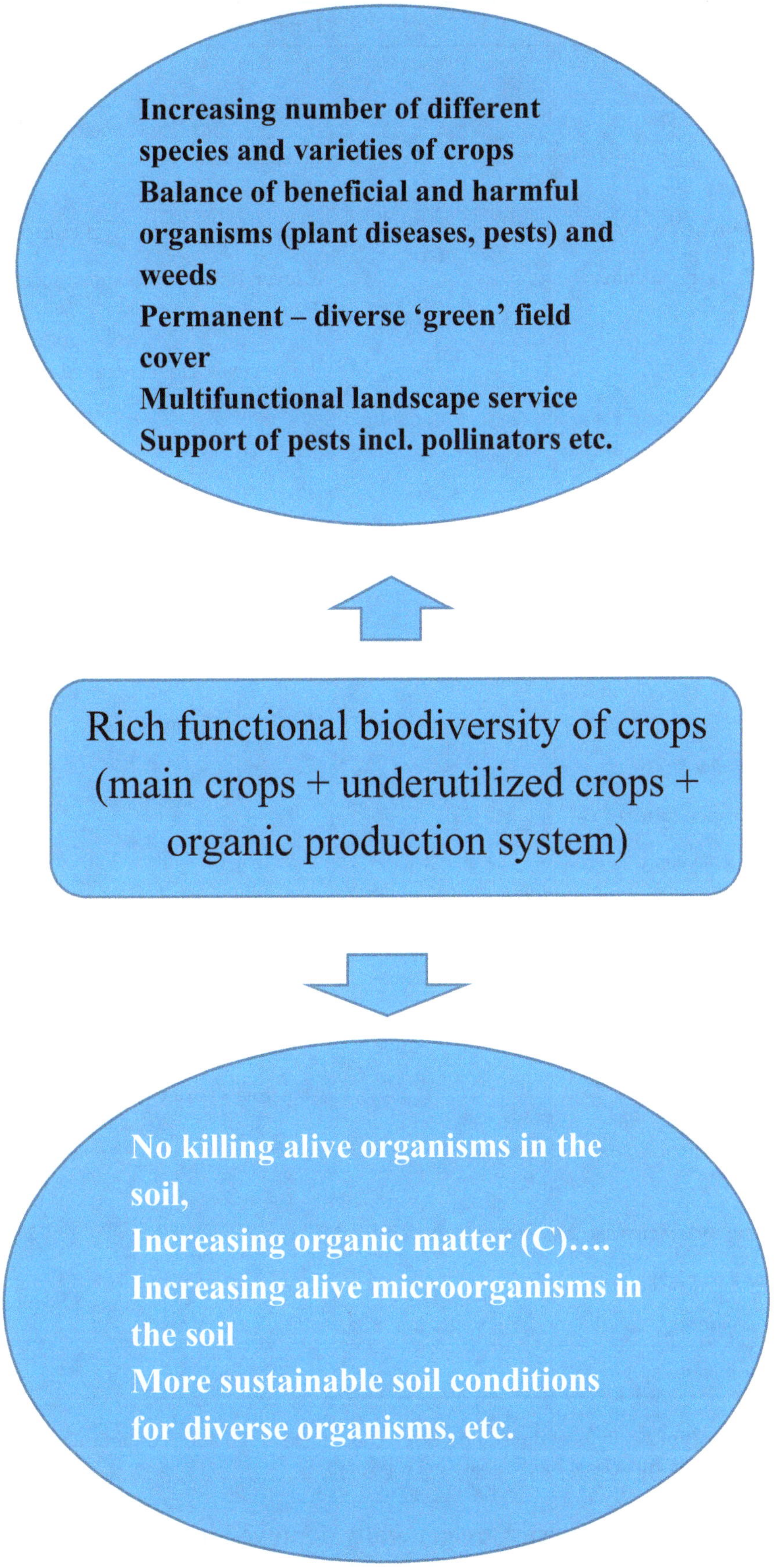

Figure 1. Some effects of underutilized crops like functional biodiversity on soil and field diversification.

Crops	Intercrops
Main crops	
Maize (*Zea mays*)	Maize × climbing bean
Cereals: wheat (*Triticum* sp.), barley (*Hordeum vulgare*), rye (*Secale cereale*)	Cereals × pea
Potatoes (*Solanum tuberosum*)	Hops × under sown cover crops
Hops (*Humulus lupulus*)—monoculture	(Clover, buckwheat, white mustard,…)
Less produced crops	
Soybean (*Glycine hispida*)	Pea × cereals, clover × maize, soybean × maize (strip sowing)
Pea (*Pisum sativum*)	
Oil (seed) pumpkins *Cucurbita pepo* group	
Oil rapeseed (*Brassica napus* var. *napus*)	
Clovers, alfalfa,…	Clover × under sown cereals
Underutilized (minor, neglected, alternative) crops	
Horse bean (*Faba bean*)	
Beans (*Phaseolus* sp.)	Maize × climbing bean
White lupine (*Lupinus albus*)	
Buckwheat (*Fagopyrum esculentum*)	
Proso millet (*Panicum miliaceum*)	
Spelt (*Triticum aestivum* ssp. *spelta*)	
Einkorn (*Triticum monococcum*)	
Emmer (*Triticum diccocum*)	
Khorosan (*Triticum turgidum* ssp. *turanicum*)	
Camelina (*Camelina sativa*)	Camelina × barley
Hemp (*Cannabis sativa*)	
Flax (*Linum usitatissium*)	
Garden poppy (*Papaver somniferum* ssp. *somniferum*)	
Safflower (*Carthamus tinctorius*)	
Quinoa (*Chenopodium quinoa*)	
Grain amaranths (*Amaranthus* sp.)	

Table 1. The list of usable (practically introduced and researched in Slovenia) and economical valuable crops including their intercrops for increasing functional biodiversity in crop rotation.

As mentioned, underutilized species are generally deployed in multicropping system [12]. If more diverse production is required, we can use different intercropping. These cropping systems are more stable and resilient than monocultures [12]. Abiotic stress can be ameliorated by intercropping, which reduces the accumulation of nitrate in the soil, permitting lower application rates of N and reducing down streams effects [16]. Altieri [2] states that in the areas where intercropping system was used, 52% less arthropods were observed compared to monoculture. There was also an increase in density of all living organisms up to 15.3%.

It is a known fact that the most effective suppression of weeds in organic and more sustainable agriculture systems is intercropping. It can provide benefits in nutrient availability and pest control, finally higher yield too [16]. The most common combination is between cereals with legumes. This enables more stable yields, greater use of land, and effectively controls the growth of weeds and soil mainly to improve the hock (fixation of nitrogen of the roots of legumes), the quantity of organic material, and the quality of the soil by the addition of atmospheric carbon.

In Serbia [17], an experiment was performed in which impact of monocultures of corn and soya compared with the intercropping of both was investigated. They examined coverage, abundance, and biomass of weeds. In the monocultures, the number of weeds was statistically different compared with other production system. The intercropping system was proved to be more effective than the estimate parameters of monoculture. Earthworm population is more affected by species-rich if more legumes are included [16]. Cropping systems increase biological diversity and prevent the onset of nematode problems as it balances a dynamic soil ecological and improves healthier soil structure with higher organic matter [15].

For soya production, temperate climates in northern part of the Alps are too low to ensure yield potential and stability. This is the reason why soya can be neglected in this region, although on the global scale, it is in the fifth place in the list of the most important main crops [14]. Amaranth is a very notorious leaf vegetable; it can be comparable to spinach. With regard to resistance to cold weather condition, quinoa is similar to peas and faba beans. However, they are not adapted to modern production techniques. Switzerland has made adjustments to test if soya can be adapted to cold climate. Hemp is known as a robust plant for producing fibers, but has some other problems such as too short fibers for the construction industries and too long fibers for textile [24]. Buckwheat has short vegetation, so it can be suitable for regions that enable stubble crop. With global warming, there will be a longer growing season. In addition, the role of underutilized crops as a buffer for climate change should be promoted [10, 14].

Different species (like millets from different genera) are important in fragile ecosystems such as those of arid and semiarid lands, mountains, steppes, and tropical forests. Crops do not require high imputes, so they can be grown in degraded and wastelands. They can increase agricultural production and enhance crop diversification. At the same time, they can contribute useful genes to breed better varieties [10].

A lot of sole or mixed populations of underutilized crops can be used as cover crops. These can cover the soil during the summer and winter, thus improving soil diversity and also prevent leaching of nutrients. They can reduce the intake of various fertilizers for the next crops and minimize the use of nonrenewable resources [18].

3.2. Underutilized crops performance of SWOT analyses

There are an extensive number of crops at the world level, continent, region, or country (see **Table 1** for Slovenia). From agronomic (**Table 2**) point, crop rotations with different underutilized crops have some opportunities and weaknesses, with more promising structure for increasing functional biodiversity (**Table 3**). In case of including underutilized crops and intercrops into crop rotation, we can increase important benefits of agronomic factors such as soil organic matter, soil structure, phosphorus mobilization, and reduction of diseases or weeds, etc. (**Figure 1**).

Strengths, opportunities, weaknesses, and threats of including underutilized crops into crop rotation have been frequently criticized over the world from supporting industrial thinking agronomists. It is partly included in the lists of presented weaknesses and threats (**Tables 2** and **3**).

Due to SWOT analyses, including comments from literature, we need to underline that underutilized crops have a good impact on preservation of the ecosystem. Grasses, sacrum, and buckwheat are good ground cover, and they effectively prevent erosion with their dense root system. They produce a lot of biomass, which ensures the subsequent increase in the organic matter with increasing soil microbiology. With their intense growth of crops, they also prevent the growth of weeds. Hardy rye, sorghum, sudangrass, and the sacrum with their excreta prevent the germination and development of weeds due to allopathy. Particularly, rye is an effective crop for weed control as it can reduce the germination of weeds by preventing access of light to the ground; consequently, the soil is colder [19].

The role of grain legumes in cropping system is nitrogen effect caused by N provision from bacterial symbiotic fixation and N sparing process that provides a long-term supply to other

Strengths	**Weaknesses**
Rich crop rotation	Loss of genetic materials
Less plant diseases, pests and weed populations	Not well organized breeding
No synthetic chemistry in process	In some cases bad adaptation to environment
No transgenic plants	Lack of whole chain knowledge
More niche product	Lack of knowledge about nutritional and health value of products
Nutritional and healthy food	
Higher prices of products than conventional	Less specialized mechanization than in conventional production
Opportunities	**Threats**
Les risks with sell prices	Low yields
Increasing employment	Unknown products as a food
Organize production whole chain (from seed to sell the product)	Opposition from intensive sector and multinational corporations
Increasing production of diverse food	
Development of new tastes by consumers	
New trading possibilities	

Table 2. SWOT analyses of crop rotations with more underutilized crops on main agronomic, social and economic parameters.

Strengths	Weaknesses
Increasing usable crops in the fields	Bad organized breeding and gene banks exploitation
Natural populations of predators and pests, natural regulation of plant diseases and weed populations	Bad genetic adaptation to growing conditions
Increasing pollinator plants	
Green covered field during longer period	
More plant rests with different effects on the soil cycles	
More microorganisms in the soil	
Opportunities	**Threats**
Preservation of honey bees	Misunderstanding of policy makers about biodiversity benefits for support
Increasing soil organic matter and available nutrients	Use underutilized crops just for genetic modification of plants
Symbiotic N fixation with legumes in crop rotation	

Table 3. SWOT analyses of functional biodiversity of underutilized crops in total agricultural biodiversity.

crops [20]. Peas, legume, has ability to fix nitrogen from the air with the help of bacteria (*Rhizobium* sp.), living on their roots. It contributes on the formation of organic substances and positive influence on the structure of the soil, as its roots exudate sugar bind soil particles to larger structures. Consequently, it retains moisture.

It is worth to mention that lupines included into crop rotation has the ability to accumulate heavy metals such as Cu, Cd, Ni, and Zn [21]. In the same time, it has a positive influence on the increase of N (it has the biggest content and uptake), the organic matter, the number of microorganisms (bacteria), and plant accessible metal "bioavailability metals" [22, 23]. According to our experience, hemp is a promising bioremediation crop as a part of crop rotation because of high production of biomass and their accumulation of heavy metals [22].

Cover crops can reduce the number of nematodes, pests, and diseases. Sudan grass is one of the good inhibitors of nematodes as it effectively reduces the formation of nematode eggs. This was evidenced in experiments on lettuce and its previous crops [25, 26].

Preissel et al. [22] compared different precrops without fertilizers application and reported that yields of cereals following grain legumes increased yield by 11–41%, and compared to oats or precrops, the benefits increase by 27–110%. Grain legumes, plants with multiple impacts on the environment can provide a stable yield and the possibility of reduce inputs (spraying, fertilization, and multiple treatments). They also indicated that grain legume intercropped with cereals is the most important in rotation system as they are the most effective natural way to prevent diseases of cereals. In the grains of cereals, the content of proteins also increased. Cropping system with legumes reduced nitrous oxide emission and caused lower nitrate-N leaching. In the same time, it had positive phytosanitary effects [23]. Alteri [2] argues the same emphasize that the polyculture (maize-beans and squash) can produce up to 4 t ha^{-1} of dry matter compared with 2 t ha^{-1} from monoculture of maize. This reflects that the presence of different crops can be efficient of using natural resources.

Diversified crop rotations may be one of the prospective methods of reducing the expansion of pests beside soil improvement. Phatak [15] proved that, by using crimson clover and vetch as cover crops, it is possible to reduce the amount of fertilizers and insecticides to 30–100%. As it is known that some crops like lupines and oilseed rape have big impact on the crop of cereals, especially on wheat. Both reduce the presence of root diseases and grass weeds. These combinations influence the effect on the final harvest index.

Pollination is an essential ecosystem service in which bumble and solitary bees play an important role. If we used crop rotations without using pesticides, we can improve production. If habitats islands were isolated from agricultural field, species richness of flower-visiting bees decreases. The same is valid for seed production. This is a proof how important is the relation between crop and pollinator diversity. It is a good example of importance of interactions among different components of biodiversity [16].

Intercropping can also have a negative effect on crops, particularly if it is not selected by the correct combinations within the intra- and interspecific varieties; consequently, there will be competition between crop plants. This is a reason why the next plant should be selected according to the characteristic of competition of the precrop.

Several vetches, clovers, and certain cruciferous crops increase high density of populations of beneficial insects such as insidious flower bugs (*Orius insidiosus*), bigeyed bugs (*Geocoris* spp.), and various lady beetles (*Coleoptera: Coccinellidae*). These predators are reproduced on nectar, pollen, thrips and aphids and are established before key pests arrived. When pests attack crops, they send chemical signals that attract beneficial insects. Lady beetles in rotation crops systems can control aphids attacking many crops [27].

Minimally tilled crimson clover or cahaba vetch before cotton planting have been successful in reducing fertilizer N down to 50% and insecticide inputs by 30–100%. A system of transplanting tomatoes, peppers, and eggplants into a killed hairy vetch or vetch/rye cover crop can influence benefits of weed, insect, and disease suppression, improved fruit quality, and overall lower production cost. Rye has the ability to reduce soil-borne diseases, nematodes, and weeds. It produces significant biomass that smothers weeds when it is left on the surface provides habitat for beneficial insects and controls weeds allelopathically through natural weed-suppressing compounds. Rye is not suitable cover where those worms are a problem (crops like corn, sweet corn, sorghum, or pearl millet) [15].

Crimson clover suppresses weeds effectively by forming a thick mulch and supports high densities of beneficial insects by providing food and habitat [15]. Buckwheat is very effective in suppressing weeds and supports high densities of beneficial insects. It is very attractive to honeybees and also sunflowers [15, 16].

After tests with oats, broccoli, white lupine, and field peas, researchers cautioned that it might take 3–5 years to effectively reduce stem lesion losses on potatoes. Cover crops with documented nematocidal properties against at least one nematode species include sorghum-sudangrass, marigold, hairy indigo, showy crotalaria, hemp, velvet bean, rapeseed, mustards, and radish. Sorghum-sudangrass and buckwheat are warm-season crops that suppress weeds by allelopathy. Cereal rye is an overwintering crop that suppresses weeds both physically and chemically. Rye residue, which is left on the soil surface, can release allelo chemicals [15].

4. Conclusions

We agreed with the statement of Frison et al. (2011) that "Diversity will be essential to improve productivity, to enhance ecosystem functions, and to provide adaptability." In addition to environmental services, underutilized crops produced in different regions are urgent for preserving food security and food quality [28, 29], especially if grown under organic farming system [30].

We conclude that only real increasing of functional biodiversity is based on underutilized crops managed in multiple crop rotation and intercrops in organic farming production system. Conservation of many underutilized crops needs to be based on practical production on small-scale farms. This kind of supporting usable functional biodiversity would be a chance for increasing positive social and economic impacts by the farmers with an extensive protection of nature. Worldwide and EU policy makers need to find better ways to develop this kind of practical and beneficial environmental development.

Author details

Franc Bavec*, Urška Lisec and Martina Bavec

*Address all correspondence to: franci.bavec@um.si

University of Maribor, Maribor, Slovenia

References

[1] Altieri MA. Biodiversity and Pest Management in Agroecosystems. New York: Haworth Press; 1994. 185 pp

[2] Altieri MA. The ecological role of biodiversity in agroecosystems. Agriculture, Ecosystems & Environment. 1999;**74**:19-31

[3] Altieri MA. Agroecology: The science of natural resources management for poor farmers in marginal environments. Agriculture, Ecosystems & Environment. 2002;**93**:1-24

[4] Štraus S. Potential Indicators for Sustainability Assessment of Food Production on the Field Level. PhD Thesis: Univ. Maribor; 2012. 195 pp

[5] Heywood V. 1999. Trends in agricultural biodiversity. Second Meeting, Montreal 2-6 September 1996

[6] Bavec F, Bavec M. Underutilized crops and intercrops in crop rotation as factors for increasing biodiversity on fields. University of Maribor, Maribor. In: Lo Y-H, Blanco JA, editors. Biodiversity in Ecosystems–Linking Structure and Function. Intech; 2015. p. 583-595

[7] Bavec F, Bavec M. Organic Production and Use of Alternative Crops. Boca Raton, New York, London. Taylor and Francis; 2006. 241 pp

[8] PBL Netherlands Environmental Agency. How sectors can contribute to sustainable use and conservation of biodiversity. CBD Technical series. 2014;**79**:17-86

[9] Padulosi S, Hodgkin T, Williams JT, Haq N. Underutilized crops: Trends, challenges and opportunities in the 21st century. In: Engels JMM et al., editors. Plant Genetic Diversity; 2002. p. 323-338

[10] Staphit B, Padulosi S, Mal B. Role of on-farm/in situ conservation and underutilized crops in the wake of climate change. Indian Journal of Plant Genetic Resources. 2010;**23**:145-156

[11] Powell B, Thilsted SH, Ickowitz A, Termote C, Sunderland T, Herforth A. Improving diets with wild and cultivated biodiversity from across the landscape. Food Security. 2015;**7**: 535-554

[12] Padulosi S, Heywood V, Hunter D, Jarvis A. Underutilized Species and Climate Change: Current Status and Outlook. Crop Adaptation to Climate Change. New York; 2011. p. 507-521

[13] Ebert AW. Potential of underutilized traditional vegetables and legumes crops to contribute to food and nutritional security income and more sustainable production system. Sustainability. 2014;**6**:319-335

[14] Stamp P, Messmer R, Wlater A. Competitive underutilized crops will depend on the state funding of breeding programmes: An opinion on the example of Europe. Plant Breeding. 2012;**131**:461-464

[15] Phatak S.C., 2016. Managing Pests. http://www.sare.org/Learning-Center/Books/Managing-Cover-Crops-Profitably-3rd-Edition/Text-Version/Managing-Pests (12 November 2016)

[16] Frison EA, Cherfas J, Hodgkin T. Agricultural biodiversity is essential for a sustainable improvement in food and nutrition security. Sustainability. 2001;**3**:238-253

[17] Dolijanović Ž, Oljača S, Kovačević D, Simić M, Dragičević V, Popović V. Weediness of a maize and soybean intercropping system. Herbologia. 2015;**15**:2-10

[18] Robačer M, Canali S, Lakenborg Kristensen H, Bavec F, Grobelnik Mlakar S, Jakop M, Bavec M. Cover crops in organic field vegetable production. Scientia Horticulturae. 2015;**208**:104-110

[19] Laub CA, Luna JM. Winter cover crop suppression practices and natural enemies of armyworm (*Lepidoptera Nuctoidae*) in no till corn. Environmental Entomology. 1992;**21**:21-49

[20] Peoples MB, Brockwell J, Herridge DF, et al. The contributions of nitrogen-fixing crop legumes to the productivity of agricultural systems. Symbiosis. 2009;**48**:1-17

[21] Fumagalli P, Comolli R, Ferre C, Ghiani A, Gentili R, Citterio S. The rotation of white lupine (*Lupinus albus* L.) with metal–accumulation plant crops: A strategy to increase the benefits of soil phytoremediation. Journal of Environmental Management. 2014;**145**:35-42

[22] Preissel S, Reckling M, Schlafke N, Zander P. Magnitude and farm economic value of grain legume pre-crop benefits in Europe: A review. Field Crops Research. 2015;**175**:64-79

[23] Reckling M, Hecker J-M, Bergkvist G, Watson CA, Zender P, Sclhlafke N, Stoddard FL, Eory V, Topp CFE, Maire J, Bachinger J. A cropping system assessment framework-evaluating effect of introducing legumes into crop rotations. European Journal of Agronomy. 2016;**76**:186-197

[24] Klavž D. Growing and phytoremediation potential of hemp (*Cannabis sativa* L.). Diploma work. University of Maribor, Maribor. 2016. 68 p

[25] Wang KH, Mc Sorley ČR, Kokalis-Burrele ČN. Effects of cover cropping, solarisation, and soil fumigation on nematode communities. Plant and Soil. 2006;**286**:229-243

[26] Bonanomi G, Filippis F, Cesarano G, Storia A, Ercolini D, Scala F. Organic farming induces changes in soil microbiota that affect agroecosystem functions. Soil Biology and Biochemistry. 2016;**103**:327-336

[27] Sharad C. Phatak and Juan Carlos Diaz-Perez. Managing Pests With Cover Crops. http://www.sare.org/Learning-Center/Books/Managing-CoverCrops-Profitably-3rd-Edition/Text-Version/Managing-Pests (Accessed 10.2.2017)

[28] Dansi A, Vodouhe R, Azokpota P, Yedomonhan H, Assogba P, Adjatin A, Loko YL, Dossou-Aminon I, Akpagana K. Diversity of the neglected and underutilized crops species of importance in Benin. The Scientific World Journal. 2012; 19 p

[29] Brussaard L, Caron P, Campbell B, Libber L, Mainka S, Rabbinge R, Babin D, Pulleman M. Reconciling biodiversity conservation and food security: Scientific challenges for a new agriculture. Environmental Sustainability. 2010;**2**:34-42

[30] Das A, Patel DP, Kumar M, Ramkrushna GI, Mukherjee A, Layek J, Nagachan SV, Buragohhhain J. Impact of seven years of organic farming on soil and produce quality and crops yields in eastern Himalayas, India. Agriculture, Ecosystems & Environment. 2017;**236**:142-153

11

Biodiversity of Amphipoda Talitridae in Tunisian Wetlands

Jelassi Raja, Khemaissia Hajer and
Nasri-Ammar Karima

Abstract

Although wetlands were remarkable habitats with their fauna and flora diversity, few studies have been devoted to the study of amphipod biodiversity in this ecosystem type. The amphipod communities of six wetland types belonging to 117 stations were studied with respect to species composition, abundance and their relationship with environmental parameters. Amphipods were collected during spring. At each station, eight quadrats of 50 × 50 cm^2 were randomly placed. Animals were preserved in alcohol at 70°C. In the laboratory, the specimens collected were identified and counted. Physicochemical parameters (organic matter, particle size, heavy metals) of sampled soils were determined. The results showed that the highest species richness was observed in lagoons with the presence of eight species namely *Orchestia montagui, Orchestia gammarellus, Orchestia mediterranea, Orchestia stephenseni, Orchestia cavimana, Platorchestia platensis, Deshayesorchestia deshayesii* and *Talitrus saltator*, whereas in the hill lakes and dams banks, no specimens were collected. The biodiversity of amphipod species depends on climatic (temperature, humidity) and edaphic (organic matter, particle size, heavy metals) factors.

Keywords: Tunisia, wetlands, neuro-inflammation, Amphipoda, diversity, environmental factors

1. Introduction

In the Mediterranean, there was a high diversity of wetlands (lagoon, lake, sebkha, wadi, hill reservoir and dam) that were of great importance in conservation of biology. They were considered among the most biologically diverse and productive ecosystems [1]. They offer a wide variety of natural habitats for plants and aquatic animals as well as semi-terrestrial and terrestrial species. The interactions of biological (plants, animals, microorganisms, etc.) and physicochemical components (granulometry, temperature, humidity, etc.) of wetlands enable them to perform many ecological functions such as shoreline stabilization and water purification. Lacaze [2] mentioned that lagoon wetlands harbour a diverse fauna, but were threatened by intense anthropogenic exploitation and pollution. As they receive continental freshwater from their catchment area, many lagoons have been subjected to severe degradation of water quality caused by pollution and/or eutrophication [3]. In Tunisia, semi-closed shallow lagoons were among the most sensitive areas to environmental stresses [4, 5].

Among wetlands, sandy beaches were more studied and characterized by the presence of a large number of invertebrates. Talitridae amphipods were among the most dominant invertebrates living on wetlands [6]. These talitrids play an important role as decomposers of organic matter and were considered as potential bio-indicators of sandy beaches quality [7, 8]. This role was estimated using genetic approach, behavioural approach as well as reproduction and spatio-temporal distribution studies [9–32].

In Tunisia, amphipod communities inhabiting wetlands bank, other than sandy beaches [21–23, 31–35] have not received much attention. Through this study, we propose a description as exhaustive as possible of the biodiversity of these communities taking into account geographical, climatic and edaphic specificities. More specifically, we addressed the following questions: (1) Does the diversity of Talitridae amphipods follow a north-south cline? (2) Is the correlation between specific diversity and wetlands type is significant?

2. Materials and methods

2.1. Study site

This study focuses on wetlands that consist of permanent or temporary areas of fresh or brackish water and adjacent lands. They include all wadis, chotts, lagoons, hill lakes, sebkhas and dams. The majority of these areas, several of which were of international importance, were found in the north, particularly near the coast. In this study, 117 stations namely *lagoons* (a stretch of salt water partially or completely separated from the open ocean by barriers of sand or coral distributed along the Tunisian coasts), *lakes* (a body of relatively still freshwater of considerable size, localized in a basin that was surrounded by land and most of them were fed and drained by rivers and streams), *sebkhas* (North African vernacular name for a shallow, salty depression. It was a common wetland type especially in semi-arid and arid climate), *wadis* (a natural stream of water of fairly large size flowing in a definite course or channel or series of diverging and converging channels), *hill lakes* (distinguished by a height >10 m and a volume >1 million m^3) and *dams* (characterized by a reservoir volume more than 3 million m^3 and a height of 15 m) were prospected (**Table 1**).

2.2. Sampling methods and laboratory procedures

Quantitative samples of amphipods were taken in spring of 2008, 2009 and 2010 in the early morning hours using quadrates method [36, 37]. In the bank of each site, eight quadrates of $50 \times 50\ cm^2$ were randomly placed. The content of each quadrat (7 cm depth) was placed in an

Stations	Wetland type	Governorate	GPS	Sediment type
1. Bizerte	Lagoon	Bizerte	37°13′8″N/009°55′1″E	Loamy sand
2. El Bcherliya	Lagoon	Bizerte	37°10′03″N/010°09′57″E	Loamy sand
3. Ghar El Melh Old harbour	Lagoon	Bizerte	37°10′04″N/010°11′40″E	Loamy sand
4. Boughaz	Lagoon	Bizerte	37°10′09″N/010°13′12″E	Sandy loam
5. Sidi Ali Mekki	Lagoon	Bizerte	37°09′50″N/010°14′45″E	Fine silt
6. Tunis North lagoon	Lagoon	Tunis	36°48′01″N/010°12′27″E	Loamy sand
7. Tunis South lagoon	Lagoon	Tunis	36°47′59″N/010°12′26″E	Sandy
8. Korba lagoon	Lagoon	Nabeul	36°38′12″N/010°54′11″E	Loamy sand
9. Tazarka lagoon	Lagoon	Nabeul	36°32′20″N/010°50′38″E	Sandy-clay-silt
10. Bhiret El Biben	Lagoon	Medenine	33°15′57″N/011°08′28″E	Sandy
11. Ichkeul	Lake	Bizerte	37°06′37″N/009°41′21″E	Loamy sand
12. Bouhnach	Lake	Ariana	36°58′57″N/010°08′56″E	Sandy loam
13. Majin Chitane	Lake	Bizerte	37°09′07″N/009°05′54″E	Sandy-clay-loam
14. El Ouafi	Sebkha	Bizerte	37°09′22″N/010°13′38″E	Sandy Silt
15. Raoud	Sebkha	Ariana	36°55′57″N/010°10′48″E	Clay
16. Ariana	Sebkha	Ariana	36°56′53″N/010°11′4″E	Sandy clay
17. Kalaat Andalous	Sebkha	Ariana	37°05′06″N/010°10′16″E	Clay
18. Sliman	Sebkha	Nabeul	36°42′02″N/010°27′37″E	Sandy
19. Maâmoura	Sebkha	Nabeul	36°28′2″N/010°48′21″E	Sandy
20. Sidi Khlifa	Sebkha	Sousse	36°14′20″N/10°26′15″E	Clay
21. Assa Jriba	Sebkha	Sousse	36°0′46″N/10°25′36″E	Clay
22. Halk El Menzel	Sebkha	Sousse	36°0′23″N/10°27′15″E	Loamy sand
23. Sousse	Sebkha	Sousse	35°47′45″N/10°38′48″E	Sandy silt loam
24. Monastir	Sebkha	Monastir	35°46′21″N/10°46′47″E	Loamy sand
25. Argoub	Sebkha	Gabès	33°38′20″N/10°16′56″E	Clay
26. Khalfallah	Sebkha	Medenine	33°26′59″N/010°56′32″E	Clay
27. Gorgabiya	Sebkha	Medenine	33°23′45″N/10°54′54″E	Sandy
28. Moknine	Sebkha	Monastir	35°37′13″N/10°55′17″E	Clay loam
29. Gargour	Sebkha	Sfax	34°37′32″N/10°38′22″E	Sandy clay
30. Sidi El Hani	Sebkha	Sousse	35°32′14″N/010°18′35″E	Sandy Silt
31. Kalbiya	Sebkha	Kairouan	35°54′25″N/010°17′08″E	Silty
32. Metbasta	Sebkha	Kairouan	35°45′14″N/010°06′57″E	Silty
33. M′Habbil	Sebkha	Medenine	33°24′47″N/010°51′59″E	Clay
34. Kairouan	Sebkha	Kairouan	35°44′6″N/010°6′52″E	Silty
35. Mchiguig	Sebkha	Sfax	34°58′58″N/010°03′06″E	Sandy Silt

Stations	Wetland type	Governorate	GPS	Sediment type
36. Thrayaa	Sebkha	Gabès	34°10′10″N/010°00′47″E	Sandy silt loam
37. Gataaya	Sebkha	Kébili	33°41′44″N/008°53′44″E	Sandy clay
38. Jemna	Sebkha	Kébili	33°34′48″N/009°00′15″E	Clay
39. Blidette Sguira	Sebkha	Kébili	33°35′18″N/008°51′06″E	Sandy silt loam
40. Blidette Kbira	Sebkha	Kébili	33°34′27″N/008°51′37″E	Sandy silt loam
41. Guidma	Sebkha	Kébili	33°25′44″N/008°47′45″E	Sandy clay
42. Golaa	Sebkha	Kébili	33°31′18″N/008°57′26″E	Sandy clay
43. Zarzara	Sebkha	Kébili	33°31′07″N/008°56′30″E	Clay
44. El Korsi	Wadi	Bizerte	37°11′12″N/009°46′52″E	Sandy loam
45. Tinja	Wadi	Bizerte	37°10′10″N/009°45′26″E	Loamy sand
46. Lebna wadi Estuary	Wadi	Nabeul	36°38′58″N/010°54′57″E	Sandy loam
47. Khniss	Wadi	Monastir	35°43′13″N/010°48′57″E	Sandy
48. Lakaarit	Wadi	Gabès	34°06′29″N/009°58′55″E	Sandy
49. El Fared	Wadi	Gabès	33°44′59″N/010°12′31″E	Sandy-clay-silt
50. Majerda	Wadi	Bizerte	37°05′03″N/010°08′17″E	Loamy sand
51. Joumin	Wadi	Bizerte	37°0′37″N/009°41′59″E	Sandy
52. Sidi Bou Ali	Wadi	Sousse	35°58′8″N/010°27′20″E	Sandy
53. Hamdoun	Wadi	Monastir	35°46′51″N/010°40′48″E	Sandy loam
54. Zerkine	Wadi	Gabès	33°41′22″N/010°15′12″E	Sandy
55. Zigzaw	Wadi	Gabès	33°35′40″N/010°18′42″E	Clay
56. Zas	Wadi	Medenine	33°30′53″N/010°20′28″E	Clay
57. Koutine	Wadi	Medenine	33°26′34″N/010°23′9″E	Silty
58. Hessi Amor	Wadi	Medenine	33°21′47″N/010°37′14″E	Clay
59. Bouhamed	Wadi	Sidi Bouzid	33°18′6″N/010°44′5″E	Silty
60. Demna	Wadi	Gabès	33°56′27″N/010°1′35″E	Loamy sand
61. Maleh	Wadi	Gabès	34°0′2″N/009°59′57″E	Loamy sand
62. Widran	Wadi	Sfax	34°31′7″N/010°4′17″E	Clay
63. Zit	Wadi	Zaghouan	36°27′01″N/010°16′43″E	Sandy
64. El Harat	Wadi	Zaghouan	36°21′50″N/010°18′34″E	Sandy
65. Lassoued	Wadi	Siliana	36°24′20″N/010°12′37″E	Sandy loam
66. Sidi Hmid	Wadi	Zaghouan	36°24′21″N/009°58′56″E	Silty
67. Bouthiben	Wadi	Zaghouan	36°22′16″N/009°54′0″E	Sandy
68. El Kbir wadi	Wadi	Siliana	36°13′26″N/009°44′49″E	Loamy sand
69. Siliana	Wadi	Siliana	36°12′03″N/009°42′57″E	Sandy-clay-silt

Stations	Wetland type	Governorate	GPS	Sediment type
70. El Kbir	Wadi	Siliana	36°07′11″/009°35′28″E	Sandy-clay-silt
71. Bargou	Wadi	Siliana	36°05′25″N/009°33′48″E	Loamy sand
72. Massouj	Wadi	Siliana	36°04′57″N/009°22′30″E	Fine silt
73. Saboun	Wadi	Siliana	35°52′11″N/009°11′37″E	Silty
74. Zguifa	Wadi	Siliana	35°45′55″N/009°01′22″E	Loamy sand
75. Raguey	Wadi	Jendouba	36°27′51″N/008°23′27″E	Sandy loam
76. Mazbla	Wadi	Jendouba	36°29′11″N/008°18′28″E	Sandy loam
77. El Maleh	Wadi	Ariana	36°58′41″N/010°09′55″E	Sandy loam
78. Lanj	Wadi	Jendouba	36°34′46″N/008°30′25″E	Sandy
79. Lahmam	Wadi	Jendouba	36°32′55″N/008°26′53″E	Loamy sand
80. Soufi	Wadi	Jendouba	36°29′20″N/008°23′49″E	Sandy loam
81. Menzel Tmim	Wadi	Nabeul	36°42′26″N/010°43′27″E	Sandy loam
82. El Widyen	Wadi	Nabeul	36°47′03″N/010°53′39″E	Sandy
83. Sliman	Wadi	Nabeul	36°41′36″N/010°28′53″E	Loamy sand
84. Lebna	Wadi	Nabeul	36°39′13″N/010°54′31″E	Sandy loam
85. Houith	Hill lake	Bizerte	37°4′59″N/009°58′5″E	Loamy sand
86. Morra	Hill lake	Bizerte	37°05′53″N/009°59′08″E	Sandy
87. Bnt Liba	Hill lake	Bizerte	37°05′52″N/009°59′08″E	Sandy-clay-silt
88. Ghar Ettine	Hill lake	Bizerte	37°04′02″N/009°15′53″E	Sandy
89. Sidi Daoued	Hill lake	Bizerte	37°03′14″N/009°23′47″E	Sandy
90. Khelifa wadi	Hill lake	Zaghouan	36°13′40″N/009°47′13″E	Sandy
91. Jetta	Hill lake	Siliana	35°59′44″N/009°26′48″E	Sandy
92. Ain Ben Ali	Hill lake	Siliana	36°03′47″N/009°17′35″E	Sandy
93. Zrab wadi	Hill lake	Siliana	36°02′8″N/009°16′54″E	Sandy
94. Khalsi	Hill lake	Siliana	35°57′10″N/009°10′32″E	Sandy
95. Jdaïda wadi	Hill lake	Siliana	35°53′53″N/009°11′12″E	Loamy sand
96. Ettal wadi	Hill lake	Siliana	35°53′20″N/009°10′54″E	Sandy loam
97. Ksayir Hamdoun	Hill lake	Siliana	35°48′10″N/009°03′57″E	Sandy
98. Ouled Ali	Hill lake	Siliana	35°50′58″N/009°09′31″E	Sandy
99. Zraybiya	Hill lake	Jendouba	36°28′25″N/008°21′29″E	Loamy sand
100. At 5km d'El Kssour	Hill lake	Kef	35°52′05″N/008°55′52″E	Clay
101. Bni Mtir	Dam	Jendouba	36°44′47″N/008°44′19″E	Sandy loam
102. Sidi Barrak	Dam	Béja	37°00′52″N/009°06′12″E	Sandy
103. El Hma	Dam	Ben Arous	36°35′16″N/010°18′24″E	Sandy clay
104. Bakbaka	Dam	Ben Arous	36°34′35″N/010°20′17″E	Loamy sand

Stations	Wetland type	Governorate	GPS	Sediment type
105. Bnt Jedidi	Dam	Nabeul	36°25′09″N/010°27′26″E	Sandy-clay-silt
106. Ermal wadi	Dam	Sousse	36°19′50″N/010°21′29″E	Sandy loam
107. Jneyhiya	Dam	Siliana	36°12′25″N/009°44′20″E	Sandy loam
108. Siliana	Dam	Siliana	36°07′57″N/009°21′14″E	Loamy sand
109. Lakhmas	Dam	Siliana	35°59′55″N/009°28′15″E	Sandy-clay-silt
110. El Gattar	Dam	Siliana	36°01′47″N/009°15′56″E	Sandy
111. Cheikh El Maïz	Dam	Siliana	36°01′15″N/009°15′8″E	Sandy
112. El Kharroub wadi	Dam	Siliana	36°01′43″N/009°15′8″E	Sandy
113. Mchaker wadi	Dam	Siliana	35°58′57″N/009°10′20″E	Sandy loam
114. Ermal	Dam	Siliana	35°49′21″N/009°07′33″E	Loamy sand
115. Mallègue	Dam	Kef	36°18′48″N/008°42′21″E	Sandy loam
116. Kasseb	Dam	Béja	36°45′36″N/009°0′5″E	Sandy
117. Ermal	Dam	Siliana	36°23′54″N/010°04′52″E	Loamy sand

Table 1. Localization of the studied stations.

individual bag, and then the animals were sorted by hand. Twenty minutes were devoted to each quadrat. Humidity and temperature of air and soil were measured *in situ* at each site. At the laboratory, amphipod specimens were preserved in 70% ethanol. Then, they were identified, counted and sexed. The identification of these species was carried out under Leica MS 5 binocular microscope, using the key of Ruffo [38].

2.3. Soil analysis

The particle size, organic matter and heavy metals of soil samples taken from 117 stations were analysed. Grain size distribution of these composite samples was analysed using different sieves in descending order (from 2 to 25 μm).

A subsample was brought to the inductively coupled plasma-mass spectrometry (ICP-MS) laboratory at University of Kiel and sieved to obtain the <250-μm grain size fraction which was then dried and milled [39]. Heavy metals were extracted from a 250-mg sample of powder with 10 mL 7 N nitric acid on a hot plate at 80°C (2.5 h). The solution was made up to 20 mL, centrifuged at 3500 rpm for 15 min, and the supernatant transferred to a 20-mL sample vial. The metals vanadium (V), chromium (Cr), manganese (Mn), cobalt (Co), nickel (Ni), copper (Cu), zinc (Zn), arsenic (As), cadmium (Cd), tin (Sn), thallium (Tl), lead (Pb), lithium (Li), rubidium (Rb) and strontium (Sr) were analysed by inductively coupled plasma-mass spectrometry (ICP-MS). Average analytical reproducibility was estimated from replicate analyses of some samples and was found to be better than 2% Relative Standard Deviation (RSD) (1 sigma relative standard deviation) for all elements. The accuracy of analytical results was monitored by analysing certified reference materials (CRM): GSMS-2 (marine sediment; Chinese Academy of Geological Sciences, PR China) and Reference material, coastal sediment (PACS-1) (coastal sediment;

National Research Council Canada (NRCC) Canada) as unknowns along with the samples. Organic matter content was determined by weighing before and after ashing at 450°C for 3 h at the University of Salzburg.

2.4. Data analysis

To compare the amphipod community structure among stations, different faunistic parameters were calculated using quantitative data such as species richness, relative species abundance, etc. Mean density of the amphipod community at each station and the mean density of each species at each station were expressed as number of individuals per m^2. Species diversity and evenness were calculated by the Shannon-Weaver index and Pielou's evenness index [40], respectively. The degree of similarity between sampling stations was evaluated using similarity cluster dendrograms. The analysis above was performed with the PRIMER software package [41]. Principal component analysis of amphipod distribution and site characteristics was performed using Xlstat software.

3. Results

3.1. Temperature, humidity, organic matter and grain size

Temperature (°C) and humidity (%) were measured *in situ* in different wetland types. The mean values for these two parameters varied between 22.453 ± 2.797°C in dams, 27.387 ± 5.289°C in sebkhas, 51.243 ± 18.627% in sebkhas and 65.50 ± 12.388% in lagoons (**Figure 1A** and **B**).

The percentage of organic matter differs between and within wetland types (**Figure 1C**). The highest values were observed in the banks of Bizerte lagoon (9.46%), Majin Chitane (12.23%), Halk Menzel (16.13%), Bargou wadi (20.66%), Ouled Ali (17.62%) and Kasseb (12.64%) (**Figure 1C**).

An heterogeneity in grain size nature was observed between stations ranging from sandy substrates, loamy sand, sandy loam, sandy silt, sandy-clay, silty clay, clay-loam, sandy-clay-silt, sandy-clay-loam, sandy-silt-loam, fine silt to clay substrates (**Table 1**).

3.2. Heavy metals

In the lagoon, the highest concentrations for the majority of heavy metals, vanadium, nickel, zinc, arsenic, cadmium, thallium and lead were recorded in the northern lagoon of Tunis. The bank of Bizerte lagoon was characterized by the highest concentrations of chromium (26.393 ppm) and manganese (281.748 ppm). While the highest copper content (39.098 ppm) was observed in El Bcherliya. The Korba lagoon revealed the highest concentration in cobalt and rubidium with 8.311 and 15.814 ppm, respectively. Bhiret El Biben was characterized by the highest concentration of lithium (29.087 ppm), strontium (2101.549 ppm) and tin (7.340 ppm). In addition, the lowest concentration for all the heavy metals studied was recorded in the bank of Sidi Ali Mekki lagoon. The different metals analysed in these lagoons do not exceed the maximum tolerate values [42] except lead that exceeds 100 ppm in the northern lagoon of Tunis (133.556 ppm).

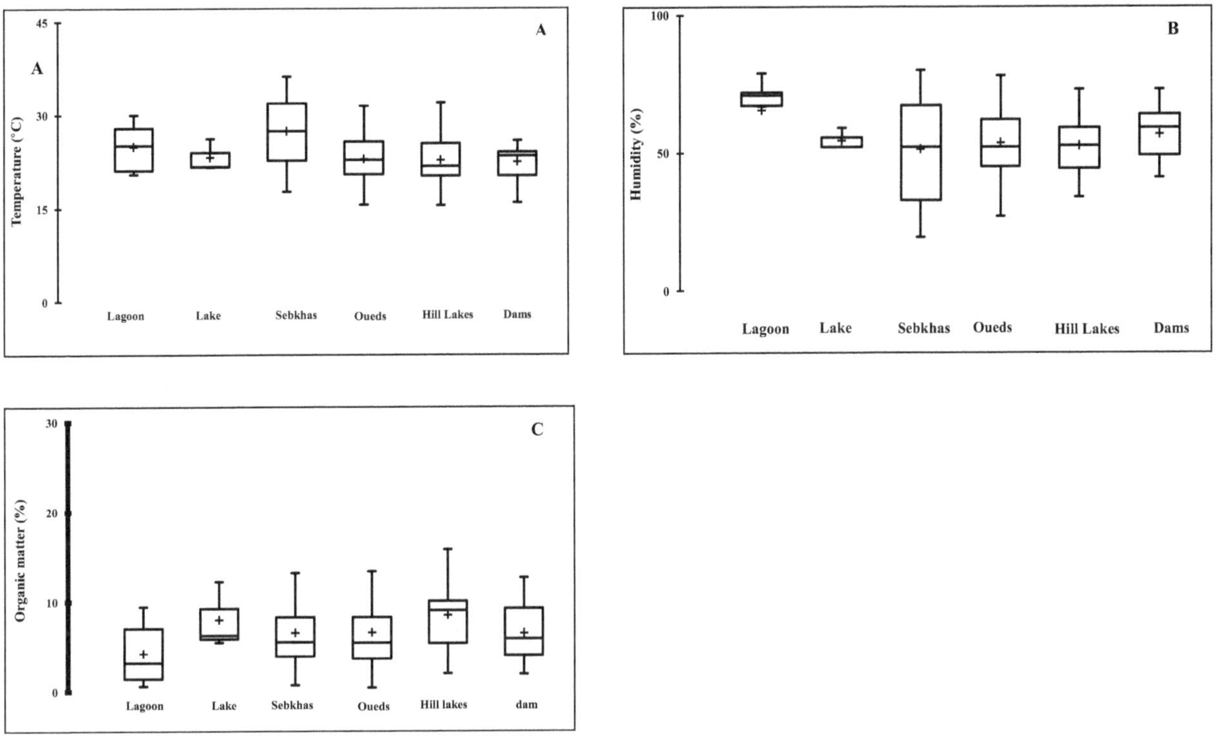

Figure 1. Environmental factors (A: Temperature (°C), B: Humidity (%), C: Organic matter (%)) measured at each wetland types.

In the second type of wetlands, lakes, the highest contents of manganese (1806 ppm), zinc (131.955 ppm), arsenic (4.211 ppm), cadmium (0.678 ppm), thallium (0.170 ppm) and lead (47.060) were observed in Ichkeul lake. Furthermore, Bouhnach lake was characterized by the important contents of lithium (7.735 ppm), vanadium (23.893 ppm), chromium (15.711 ppm), cobalt (7.497 ppm), nickel (16.567 ppm), strontium (643.783 ppm) and tin (0.140 ppm) and it was rather Majin Chitan lake that presented the highest concentration of copper (15.577 ppm) and rubidium (11.632 ppm). According to Henin [42], these stations were not contaminated since the content of heavy metals does not exceed the maximum tolerated values.

Concerning sebkhas, the highest concentrations in vanadium, rubidium and thallium were recorded in Sebkha Halk Menzel (36.357, 19.239 and 0.140 ppm, respectively). For chromium, manganese and tin, the highest concentrations were recorded in sebkhas of Sidi Khlifa (554.628 ppm), Moknine (387.880 ppm) and Sousse (2.793 ppm), respectively. The highest concentration of cobalt, nickel, strontium and cadmium was, respectively, observed in the Halk Menzel (10.460 ppm), Raoued (23.106 ppm), Sidi El Hani (3305.249 ppm) and Golla (0.422 ppm). Concerning copper, arsenic and lithium, the highest concentration was, respectively, recorded in sebkha Ariana (23.238 ppm), sebkha Golla (14.507 ppm) and sebkha Kairouan (41.861 ppm), and it was rather the sebkha Monastir that showed the highest concentration of zinc (86.453 ppm) and lead (48.741 ppm). According to these results and taking into account the tolerance thresholds, no sebkha was considered polluted with the exception of sebkha Sidi Khlifa that was considered as polluted by chromium which exceeds the maximum tolerated value (150 ppm) [42].

In wadis, the highest concentrations of vanadium (44.619 ppm), chromium (40.413 ppm), zinc (147.822 ppm) and lead (303.910 ppm) were recorded in Lahmam wadi. The highest concentration of cobalt (19,723 ppm), nickel (29,283 ppm), rubidium (27,016 ppm), and thallium (0.183 ppm) were found in Zit wadi. Whereas for lithium, manganese, copper, arsenic, strontium, cadmium and tin, the highest concentrations were, respectively, recorded in Soufi wadi (26.527 ppm), Mazbla wadi (644.069 ppm), Joumin wadi (19.448 ppm), Lanj wadi (9.380 ppm), Khniss wadi (1410.100 ppm), Bargou wadi (1.412 ppm) and El Korsi (1.019 ppm). According to Henin [42], the different prospected wadis were not polluted except for Joumin, Lassoued and Bargou wadis, which were considered as polluted with cadmium whose percentage exceeds the maximum tolerated value (0.7 ppm) as well as Lahmam wadi in which a lead concentration exceeded 100 ppm.

Sixteen hill lakes and 17 dams belonging to different bioclimatic stages were prospected. In the banks of these closed and artificial ecosystems, no amphipod was found.

Concerning heavy metals, in hill lakes, the highest concentrations of vanadium (46.795 ppm), cobalt (14.661 ppm), nickel (30.362 ppm), copper (16.611 ppm) and lead (23.047 ppm) were observed in Sidi Daoued hill lake. The highest concentration of zinc (88.804 ppm), arsenic (4.590 ppm), cadmium (3.031 ppm), thallium (0.176 ppm) was found in Zrab wadi hill lake. The khlifa wadi hill lake was characterized by the important content of lithium (24.891 ppm) and strontium (930.812 ppm) and it was rather Ouled Ali hill lake that presented the highest concentration of rubidium (16.526 ppm) and tin (0.629 ppm). Hill lakes of Khalsi and Ksayir Hamdoun were characterized by the highest concentration of chromium (43.394 ppm) and manganese (530.039 ppm). This analysis of heavy metals revealed that only Ain Ben Ali, Zad and Khalsi hills lakes were contaminated by the cadmium. Concerning dams, our results showed that Sidi Barrak dam was characterized by the highest concentration of majority of heavy metals, namely manganese (1060.291 ppm), cobalt (14.085 ppm), copper (22.840 ppm), zinc (151.90 ppm), arsenic (6.246 ppm), thallium (0.544 ppm) and lead (166.067 ppm); while Kasseb dam showed the highest concentration of vanadium (37.377 ppm), chromium (41.476 ppm), nickel (32.579 ppm) and rubidium (18.714 ppm). The highest concentration in lithium (15.013 ppm), cadmium, (5.426 ppm) and tin (0.137 ppm) was found in Gattar dam; while that of strontium (731.645 ppm) was observed in Jneyhiya dam.

3.3. Species richness

Eight species of amphipoda Talitridae, namely *Orchestia montagui* Audouin, 1826, *Orchestia mediterranea* Costa, 1853, *Orchestia gammarellus* (Pallas, 1766), *Orchestia stephenseni* Cecchini, 1928, *Orchestia cavimana* Heller, 1865, *Platorchestia platensis* (Kroyer, 1845), *Deshayesorchestia deshayesii* (Audouin, 1826) and *Talitrus saltator* (Montagu, 1808) were collected in different wetlands.

Species richness (S) varied between stations of the same as well as the different types of wetlands.

In lagoons, species richness varied between one species in El Bcherliya and eight species in the bank of Bizerte lagoon near Menzel Jmil. The differences observed between lagoons were highly significant ($F = 5.317$; $df = 9$; $p < 0.0001$). In the bank of lakes, amphipods were collected only in Ichkeul lake ($S = 5$).

Concerning sebkhas, among 30 sebkhas studied, talitrids were found in only four sebkhas namely: El Ouafi, Maâmoura, Moknine and Gargour. Species richness was equal to one species in the bank of sebkhas El Ouafi and Maâmoura namely *Orchestia gammarellus* and *Talitrus saltator*, respectively. In two other sebkhas, *Orchestia gammarellus* and *Orchestia mediterranea* were collected.

For wadis, individuals were collected only in six wadis among the 41 stations prospected. Species richness varies from one (El Fared wadi, Laakarit wadi, Khniss wadi and Lebna wadi) to six species (El Korsi). In hill lakes and dams, no species were collected.

3.4. Relative abundance and density

A total of 340 specimens of amphipoda Talitridae were collected in lagoons. The bank of Bizerte lagoon revealed statistically the most important relative abundance of amphipod community (36.04%) (Anova test: $F = 5.330$, $df = 9$, $p < 0.0001$). Moreover, in this station, *Orchestia mediterranea* was the most abundant species (25.7%). However, in the banks of El Bcherliya, the Ghar El Melh old harbour, Tunis north and south lagoons, it was rather *Orchestia gammarellus* that dominated. These two species have the same relative abundance (46.7%) in bank of Sidi Ali Mekki lagoon. In Bhiret El Biben lagoon, *Orchestia montagui* was the most abundant species (28.3%). The Anova test revealed that differences between the different lagoons were highly significant ($F = 7.922$; $df = 7$; $p < 0.0001$). The mean community density varied between 0.5 ind.m^{-2} in the bank of El Bcherliya and 241.5 ind.m^{-2} in that of Bizerte lagoon. Furthermore, our results showed that *Orchestia mediterranea* presented the most important density in the bank of Bizerte lagoon (62 ind.m^{-2}). Whereas, in the bank of El Bcherliya, Ghar El Melh old harbour, Tunis north and south lagoons, it was *Orchestia gammarellus* that exhibited the largest density with, respectively, 0.5, 19, 34 and 34.5 ind.m^{-2}. These two species were recorded with the same mean density in the bank of Sidi Ali Mekki lagoon (3.5 ind.m^{-2}). In lakes, 170 individuals were collected where *Orchestia mediterranea* presented the highest abundance (26.5%) and density (22.5 ind.m^{-2}).

In sebkhas, 352 specimens of amphipods were collected. Sebkha Gargour revealed the highest relative abundance (50%) followed by sebkha Moknine which abundance was equal to 34.66%. However, in the bank of the two other sebkhas, the abundance was relatively low in Mâamoura with 15.06% and very low in sebkha El Ouafi with 0.28%. The Anova test revealed a highly significant difference in relative abundance between these sebkhas ($F = 8.288$, $df = 29$, $p < 0.0001$). The relative abundance of *Orchestia gammarellus* and *Talitrus saltator* were maximal (100%), respectively, in sebkha El Ouafi and Mâamoura since each sebkha harbours only one species. In Moknine, abundance was greater for *Orchestia gammarellus* (53.3%) than in *Orchestia mediterranea* (46.7%) and inversely in the sebkha Gargour where the highest abundance was recorded for *Orchestia mediterranea* with 59.1%. In addition, no significant difference in relative abundance between species was found (Anova test: $F = 1.461$, $df = 2$, $p = 0.233$). The global mean density oscillated between 0.5 ind.m^{-2} in sebkha El Ouafi and 88 ind.m^{-2} in sebkha Gargour. The study of the mean density per species showed a very low density of

Orchestia gammarellus in sebkha Ouafi (0.5 ind.m^{-2}); this density became more pronounced in sebkha Moknine and Gargour with, respectively, 32.5 and 36 ind.m^{-2}. In these two last stations, *Orchestia mediterranea* had a density of 28.5 and 52 ind.m^{-2}, respectively.

Concerning wadis, 558 individuals were found. The most important global mean density was observed in the bank of El Korsi. *Orchestia mediterranea,* species living in allopatry in Khniss, Laakarit and El Fared wadis showed a density, respectively, equal to 49.5, 47.5 and 0.5 ind.m^{-2}; whereas, where it was in sympatry, its density was equal to 16 (El Korsi) and 7.5 ind.m^{-2} (Tinja). Furthermore, *Talitrus saltator,* which was the only amphipod collected in Lebna wadi estuary (67 ind.m^{-2}), showed a relatively lower density in El Korsi (13 ind.m^{-2}) and Tinja (6 ind.m^{-2}).

3.5. Diversity

According to the Simpson index (Is), the most important diversity was observed in the Bizerte lagoon where we noted the highest value which tends towards the specific richness (6.059) and the community was more balanced in Boughaz.

In Ghar El Melh old harbour, we obtained the lowest diversity compared to that observed in Boughaz. This result could be explained by the fact that this index does not consider rare species into account.

The Shannon-Weaver (H') index ranged from 1.287 in the bank of Sidi Ali Mekki lagoon to 2.771 in the bank of Bizerte lagoon where the diversity was relatively significant. This index, which takes into account the rare species, was often accompanied by the equitability index, which was more or less insensitive to specific richness. It ranged from 0.812 (Sidi Ali Mekki) to 0.996 (Boughaz) where the community was more balanced.

In the banks of different sebkhas, we did not observed a great diversity, so the analysis of diversity was not carried out.

In wadis, results showed that the Simpson index varies between 1 in the Lebna wadi estuary, Khniss, Laakarit and El Fared wadis and 5.78 in El Korsi station where we found the highest species richness (S = 6). The Shannon-Weaver index confirmed the previous index showing that the most important diversity was observed in El Korsi station (H' = 2.56). Moreover, the study of the equitability index showed that the community was more balanced in this station (J' = 0.99) where species were equitably distributed.

3.6. Amphipod distribution according to environmental factors and wetland types

To better understand the species distribution in the different wetland types and to elucidate the parameters involved in their distribution, a canonical correspondence analysis was carried out (**Figure 2**). The first three axes, F1, F2 and F3 extract, respectively, 71.43, 20.23 and 5.55% of the variance. The two species, *Orchestia mediterranea* and *Orchestia gammarellus* that dominated the majority of lagoons and sebkhas were positively correlated with the strontium content and negatively with concentrations of vanadium, chromium, manganese, cobalt, nickel, copper, zinc, arsenic, rubidium, cadmium, thallium and lead. However,

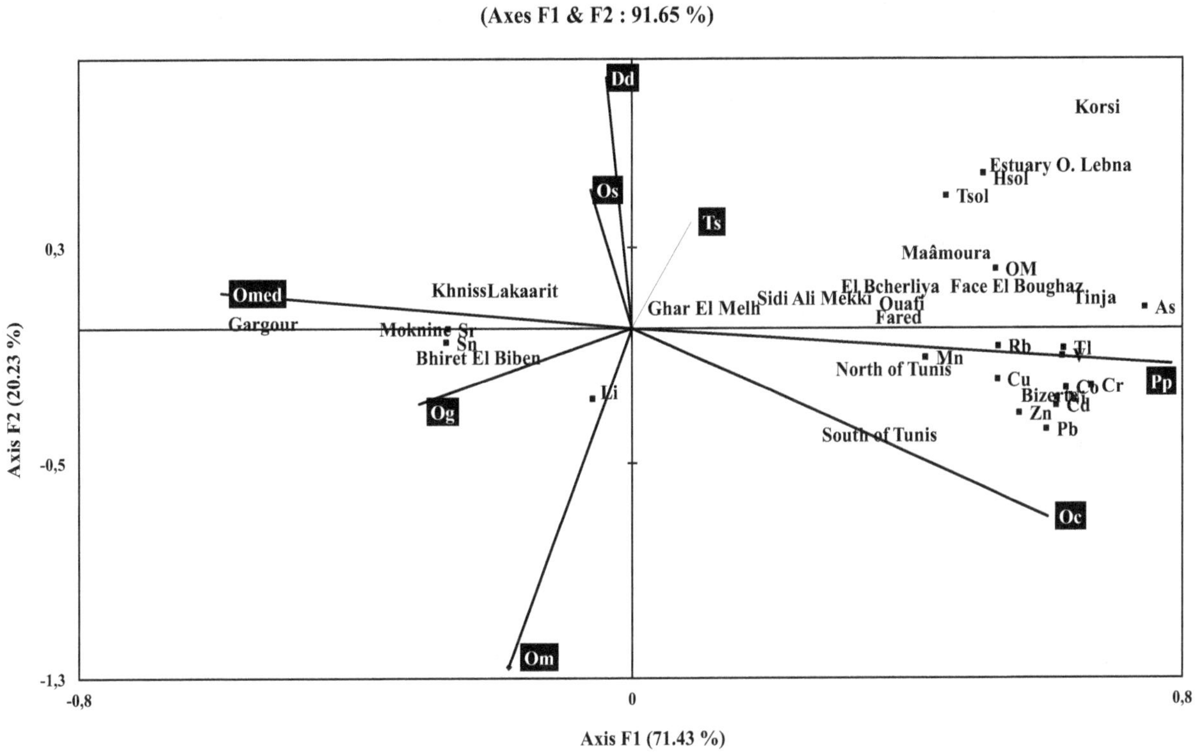

Figure 2. Canonical correspondence analysis (CCA) performed on the abiotic parameters. Om: *Orchestia montagui*, Og: *Orchestia gammarellus*, Omed: *Orchestia mediterranea*, Os: *Orchestia stephenseni*, Oc: *Orchestia cavimana*, Pp: *Platorchestia platensis*, Ts: *Talitrus saltator*, Dd: *Deshayesorchestia deshayesii*, Hsol: soil humidiy, Tsol: soil temperature, Gra: granulometry, V: vanadium, Cr: chromium, Mn: manganese, Co: cobalt, Ni: nickel, Cu: copper, Zn: zinc, As: arsenic, Cd: cadmium, Sn: tin, Tl: thallium, Pb: lead, Li: lithium, Rb: rubidium and Sr: strontium.

Talitrus saltator, abundant species in El Korsi and the estuary of Lebna wadi, as well as *Orchestia stephenseni* and *Deshayesorchestia deshayesii* were positively correlated with temperature and humidity. In the third axis, *Orchestia cavimana* and *Platorchestia platensis* were found positively correlated with organic matter and negatively with lithium and tin content.

4. Discussion

The study of the biodiversity talitrid populations in six types of wetlands revealed differences between these types.

The different prospections carried out in wetlands showed globally that the most important species richness was observed in lagoons. Moreover, no individual was collected in hill lakes and dams. Among lagoons studied, the bank of Bizerte lagoon was the most diverse one with eight species. This result was confirmed by several diversity indices performed in the present study.

Our hypothesis concerning the existence of a north/south diversity gradient was confirmed only for wadis. Furthermore, we did not reveal any significant difference concerning the vulnerability of lentic wetlands compared to the lotic type. Concerning Oniscidean group collected from the banks of Tunisian wetlands where many species were found in sympatry with amphipods, a positive correlation between species richness and altitudinal gradient has been highlighted [43]. The same authors showed that species richness differs significantly depending on wetland types or bioclimatic zones.

A total of 2420 amphipods belonging to different species were determined in all prospected wetlands; more than half of the specimens were collected in lagoons (N = 1340) with a mean density of 241.5 ind.m^{-2} observed in the bank of the Bizerte lagoon; *Orchestia mediterranea* showed the most important density in this lagoon. In sebkhas and wadis, the highest densities were recorded, respectively, in sebkha Gargour and El Korsi where *Orchestia mediterranea* and *Platorchestia platensis* were characterized by the highest density, respectively. Studying *Talitrus saltator* and *Britorchestia brito* populations in Zouara beach, Charfi-Cheikhrouha et al. [26] determined a mean density equal to 262.94 ± 85 ind.m^{-2}. These authors showed that the density increased in autumn and winter and reached a maximum in March for *Talitrus saltator*; while for *Britorchestia brito,* this density increased from the middle of March and peaked in October. In Algeria, in the bay of Bou Ismail, *Orchestia montagui* and *Deshayesorchestia deshayesii* reached more than 45,000 ind.m^{-2} [44]. In the Bou Regreg estuary, *Orchestia mediterranea* showed densities ranging from 3380 (February) to 7000 ind.m^{-2} (August) [45]. Studying the spatio-temporal distribution of amphipods in different wetlands in Tunisia, Jelassi [46] highlighted that the most important densities were observed during spring.

The diversity of the different talitrid species was related to the presence of different parameters. This relation depends on the wetland type; for example, in lagoons, the sandhopper *Talitrus saltator* was correlated with climatic (temperature, humidity) as well as edaphic factors (organic matter, granulometry, heavy metals of soil) whereas in sebkhas and wadis, this species was correlated only with edaphic factors. These results were also observed for other species. In this context, several studies have investigated the role of environmental factors and have revealed the influence of some factors rather than others. Jelassi et al. [31] have shown that talitrid abundance in the bank of Bizerte lagoon was closely related to air temperature. Bouslama et al. [47] showed that temperature was the important factor influencing the zonation whose augmentation induces the displacement or the migration of *Talitrus saltator* population from the top to the bottom of the beach. This result was similar to that found by Fallaci et al. [48], who indicated that mean zonation of this species was influenced by temperature during its activity period. Other authors such as Colombini et al. [49] confirmed the importance of sediment parameters in the selection of specific distribution area especially for young individuals.

Our results showed that the two species *Orchestia cavimana* and *Platorchestia platensis* were correlated with organic matter. Jelassi et al. [32] highlighted that air and soil temperature were the best predictors for *O. stephenseni* abundance that negatively corresponded with the proportion of fine sand fraction and organic matter content of the soil. *O. montagui* and *O. cavimana* abundances corresponded positively with air humidity and the soil lithium and rubidium contents, but negatively with the soil tin content and the proportion of the silt and clay fraction. *D. deshayesii* and *P. platensis* did not exhibit any clear correspondence with station characteristics.

According to Williams [50], the relationship between population movements and trophic preferences does not seem to be a major parameter in the structuring of zonation despite the important mobility of the sandhopper *Talitrus saltator*, which would induce a greater choice of nutrient sources. Studying the biodiversity of amphipods in some coastal lagoons in Tunisia, Jelassi et al. [33] showed that the most important species richness observed in the bank of Bizerte lagoon would be related to the presence of important vegetation in spring as well as the *Cymodocea nodosa* leaf litter and a high percentage of organic matter.

Attention was also given to biodiversity and biogeography for Oniscidean communities living in sympatry with amphipods in different wetland types prospected in the present study. Khemaissia et al. [43] showed that *Porcellio lamellatus, Tylos europaeus, Armadilloniscus ellipticus, Armadillo officinalis, Porcellio sexfasciatus* and *Chaetophiloscia elongata*, abundant species in the banks of lagoons, were associated with sodium content, pH and temperature of soil. However, other species such as *Armadillidium pelagicum, Armadillidium sulcatum, Armadillidium vulgare, Armadillidium boukornini, Armadillidium tunisiense* and *Porcellio dominici* were abundant around dams and hill reservoirs and were positively associated with elevation. The distribution of *Leptotrichus panzerii* and *Armadillidium granulatum*, in the sebkhas, was correlated with calcium content and humidity of soil.

Through these results, we did not reveal any significant difference regarding the vulnerability of lentic type wetlands compared to the lotic type wetlands. In order to test this hypothesis, the number of this last wetland type (lotic type) should be multiplied.

Acknowledgements

The present study was funded by the Research Unit of Bio-ecology and Evolutionary Systematics (UR11ES11), Faculty of Science of Tunis, University of Tunis El Manar. We would like to thank Dr. Dieter Garbe-Schönberg (Institut für Geowissenschaften, ICP-MS Labor, Universität zu Kiel, Kiel, Germany) and Dr. Martin Zimmer (Leibniz Center for Tropical Marine Ecology Fahrenheitstr. Bremen, Bermany) for assistance with soil analysis.

Disclosure statement

No potential conflict of interest was reported by the authors.

Author details

Jelassi Raja[1,2]*, Khemaissia Hajer[2] and Nasri-Ammar Karima[2]

*Address all correspondence to: djelassi.raja@gmail.com

1 National Institute of Sciences and Technology of the Sea, Salammbô, Tunis, Tunisia

2 University of Tunis El Manar, Faculty of Sciences of Tunis, Research Unit of Bio-ecology and Evolutionary Systematics El Manar II, Tunis, Tunisia

References

[1] Medail F, Quezel P. Biodiversity hotspots in the Mediterranean basin: Setting global conservation priorities. Conservation Biology. 1999;**13**:1510-1513

[2] Lacaze JC. L'eutrophisation des eaux marines et continentales: Causes, manifestations, conséquences et moyens de lutte. Paris: Ellipse; 1996

[3] Benrejeb A, Romdhane MS. Impact des perturbations anthropiques sur l'évolution du phytoplancton de la lagune de Boughrara, Tunisie. Bulletin de l'Institut National des Sciences et Technologie de la Mer de Salammbô. 2002;**29**:65-75

[4] Romdhane MS. Les îles et les lagunes de Tunisie entre l'aménagement et la protection. Effets des changements globaux sur les écosystèmes marins et les habitats côtiers. Colloque PRICAT 6 RIGEDMER, Sousse. 2001

[5] Turki S, Hamza A. Le phytoplankton toxique. Effets des changements globaux sur les écosystèmes marins et les habitats côtiers. Colloque PRICAT 6 RIGEDMER, Sousse. 2001

[6] McLachlan A, Wooldridge T, Dye AH. The Ecology of sand beaches in South Africa. South African Journal of Zoology. 1981;**16**:219-231

[7] Griffiths CL, Stenton-Dozey JME, Koop K. Kelp wrack and the flow of energy through a beach ecosystem. In: McLachlan A, Erasmus T, editors. Sandy Beaches as Ecosystem. The Hague: W. Junk Publishers; 1983. pp. 547-556

[8] Ketmaier V, Scapini F, De Matthaeis E. Exploratory analysis of talitrid population genetics as an indicator of the quality of sandy beaches. Estuarine Coastal and Shelf Science. 2003;**58S**:159-167

[9] Conceiçao MB, Bishop JDD, Thorpe JP. Genetic relationships between ecologically divergent species of talitrid amphipod (Crustacea). Marine Ecology Progress Series. 1998;**165**:225-233

[10] Bulnheim HP, Schwenzer DE. Allozyme variation and genetic divergence in populations of *Talitrus saltator* (Crustacea: Amphipoda) around the Atlantic coast, the Azores and the Canary Islands. Cahiers de Biologie Marine. 1999;**40**:185-194

[11] De Matthaeis E, Davolos D, Cobolli M, Ketmaier V. Isolation by distance in equilibrium and non-equilibrium populations of four tatlitrid species in the Mediterranean Sea. Evolution. 2000;**54**:1606-1613

[12] De Matthaeis E, Ketmaier V, Davolos D, Schembri PJ. Patterns of genetic diversity in Mediterranean supralittoral amphipods (Crustacea, Amphipoda). Polskie Archiwum Hydrobiologii. 2000;**47**:351-361

[13] Bouslama MF, Charfi-Cheikhrouha F, De Matthaeis E. Flux génique et structure géographique de quelques populations naturelles de *Talitrus saltator* (Montagu, 1808). Bulletin de l'Association Tunisienne des Sciences de la Mer. 2001;**5**:90-93

[14] Gérard JF, Vancassel M, Laffort B. Spread of phenotypic plasticity or genetic assimilation: The possible role of genetic constraints. Journal of Theoretical Biology. 1993;**164**:341-349

[15] Scapini F. Variation in scototaxis and orientation adaptation of *Talitrus saltator* populations subjected to different ecological constraints. Estuarine Coastal and Shelf Science. 1997;**44**:139-146

[16] Nasri-Ammar K, Morgan E. Variation saisonnière du rythme de l'activité locomotrice de *Talitrus saltator* issu de la plage de Korba (Cap Bon, Tunisie). Bulletin de la Sociéte Zoologique de France. 2005;**130**(1):19-29

[17] Nasri-Ammar K, Morgan E. Seasonality of the endogenous activity rhythm in *Talitrus saltator* (Montagu) from a sandy beach in north-eastern Tunisia. Biological Rhythm Research. 2006;**37**:479-488

[18] Rossano C, Morgan E, Scapini F. Variation of the locomotor activity rhythms in three species of talitrid amphipods, *Talitrus saltator, Orchestia montagui,* and *Orchestia gammarellus,* from various habitats. Chronobiology International. 2008;**25**(4):511-532

[19] Ayari A, Nasri-Ammar K. Seasonal variation of the endogenous rhythm in two sympatrics amphipod: *Talitrus saltator* and *Deshayesorchestia deshayesii* from Bizerte beach (North of Tunisia). Biological Rhythm Research. 2012;**43**(5):515-526

[20] Ayari A, Nasri-Ammar K. Locomotor rhythm phenology of *Talitrus saltator* from two geomorphologically different beaches of Tunisia: Bizerte (North of Tunisia) and Gabes gulf (South of Tunisia). Biological Rhythm Research. 2012;**43**(2):113-123

[21] Jelassi R, Nasri-Ammar K. Seasonal variation of locomotor activity rhythm of *Orchestia montagui* in the supralittoral zone of Bizerte lagoon (North of Tunisia). Biological Rhythm Research. 2013;**44**:718-729

[22] Jelassi R, Ayari A, Nasri-Ammar K. Seasonal variation of locomotor activity rhythm of *Orchestia gammarellus* in the supralittoral zone of Ghar Melh lagoon (North-East of Tunisia). Biological Rhythm Research. 2013;**44**:956-967

[23] Jelassi R, Akkari-Ayari A, Bohli-Abderrazak D, Nasri-Ammar K. Endogenous locomotor activity rhythm of two sympatric species of Talitrids (Crustacea, Amphipoda) from the supralittoral zone of Bizerte lagoon (Northern Tunisia). Biological Rhythm Research. 2013;**44**:265-275

[24] Borgioli C, Martelli L, Porri F, D'Elia A, Marchetti GM, Scapini F. Orientation in *Talitrus saltator* (Montagu): Trends in intrapopulation variability related to environmental and intrinsic factors. Journal of Experimental Marine Biology and Ecology. 1999;**238**:29-47

[25] Scapini F, Porri F, Borgioli C, Martelli L. Solar orientation of adult and laboratory-born juvenile sandhoppers: Inter- and intra-population variation. Journal of Experimental Marine Biology and Ecology. 1999;**238**:107-126

[26] Charfi-Cheikhrouha F, El Gtari M, Bouslama MF. Distribution and reproduction of two sandhoppers, *Talitrus saltator* and *Talorchestia brito* from Zouaraa beach-dune system (Tunisia). Polskie Archiwum Hydrobiologii. 2000;**43**:621-629

[27] Gonçalves SC, Marques JC, Pardal MA, Bouslama MF, El Gtari M, Charfi-Cheikhrouha F. Comparison of the biology, dynamics, and secondary production of *Talorchestia brito* (Amphipoda, Talitridae) in Atlantic (Portugal) and Mediterranean (Tunisia) populations. Estuarine Coastal and Shelf Science. 2003;**58**:901-916

[28] Marques JC, Goncalaves SC, Pardal MA, Chelazzi L, Colombini I, Fallaci M, Bouslama MF, El Gtari M, Charfi-Cheikhrouha F, Scapini F. Comparison of *T. Saltator* (Amphipoda, Talitridae) biology, dynamics and secondary production in Atlantic (Portugal) and Mediterranean (Italy and Tunisia) populations. Estuarine Coastal and Shelf Science. 2003;**58**:127-148

[29] Bouslama MF, Neto JM, Charfi-Cheikhrouha F, Marques JC. Biology, population dynamics and secondary production of *Talitrus saltator* (Amphipoda, Talitridae) at Korba beach (east coast of Tunisia). Crustaceana. 2007;**80**:1103-1119

[30] Gambineri S, Scapini F. Importance of orientation to the sun and local landscape features in young inexpert *Talitrus saltator* (Amphipoda: Talitridae) from two Italian beaches differing in morphodynamics, erosion or stability. Estuarine Coastal and Shelf Science. 2008;**77**:357-368

[31] Jelassi R, Khemaissia H, Nasri-Ammar K. Intra-annual variation of the spatiotemporal distribution and abundance of Talitridae and Oniscidea (Crustacea, Peracarida) at Bizerte Lagoon (northern Tunisia). African Journal of Ecology. 2012;**50**:381-392

[32] Jelassi R, Zimmer M, Khemaissia H, Garbe-Schönberg D, Nasri-Ammar K. Amphipod diversity at three Tunisian lagoon complexes in relation to environmental conditions. Journal of Natural History. 2013;**47**(45-46):2849-2868

[33] Jelassi R, Khemaissia H, Zimmer M, Garbe-Schönberg D, Nasri-Ammar K. Biodiversity of Talitridae family (Crustacea, Amphipoda) in some Tunisian coastal lagoons. Zoological Studies. 2015;**54**(17):1-10. DOI: 10.1186/s40555-014-0096-1

[34] Jelassi R, Bohli-Abderrezek D, Ayari A, Nasri-Ammar K. Effects of light pulses on the locomotor activity rhythm of *Orchestia montagui* (Amphipoda, Talitridae). Biological Rhythm Research. 2016;**48**:43-55

[35] Jelassi R, Khemaissia H, Zimmer M, Garbe-Schönberg D, Nasri-Ammar K. Influence of environmental conditions on the distribution of Amphipoda, Talitridae, in the lagoon complex of Ghar El Melh (north-east of Tunisia). African Journal of Ecology. 2017. DOI: 10.1111/aje.12375

[36] Achouri MS, Hamaied S, Charfi-Cheikhrouha F. The diversity of terrestrial Isopoda in the Berkoukech area, Kroumirie, Tunisia. Crustaceana. 2008;**81**(8):917-929

[37] Hamaïed-Melki S, Achouri MS, Aroui O, Bohli D, Charfi-Cheikhrouha F. Terrestrial isopod diversity in the wadi Moula-Bouterfess catchment area (Kroumirie, north-west of Tunisia). African Journal of Ecology. 2010;**49**:31-39

[38] Ruffo S. The Amphipoda of Mediterranean. Part IV: Localities and map-Agenda to parts 1-3-Key to families-Ecology-Faunistics and zoogeography. Monaco: Mémoire de l'Institut Océanographique; 1993. p. 13

[39] Bat L, Raffaelli D. Effects of gut sediment contents on heavy metal levels in the amphipod *Corophium volutator* (Pallas). Turkish Journal of Zoology. 1999;**23**:67-71

[40] Pielou EC. The measurement of diversity in different types of biological collections. Journal of Theoretical Biology. 1966;**13**:131-144

[41] Clarke KR, Warwick RM. Change in Marine Communities: An Approach to Statistical Analysis and Interpretation. 2nd ed. Plymouth: Marine Laboratory; 2001. p. 176

[42] Henin S. Les éléments traces dans les sols. Science du sol. 1983;**2**:67-71

[43] Khemaissia H, Raja Jelassi R, Touihri M, Souty-Grosset C, Nasri-Ammar K. Diversity of terrestrial isopods in the northern Tunisian wetlands. African Journal of Ecology. 2016. DOI: 10.1111/aje.12337

[44] Louis M. Etude d'un peuplement mixte d'*Orchestia montagui* Audouin et d'*Orchestia deshayesii* Audouin dans la baie de Bou Ismail. Bulletin d'Ecologie. 1980;**11**:97-111

[45] ElKaïm B, Irlinger JP, Pichard S. Dynamique de la population d'*Orchestia mediterranea* L. (Crustacé, Amphipode) dans l'estuaire de Bou Regreg (Maroc). Canadian Journal of Zoology. 1985;**63**:2800-2809

[46] Jelassi R. Eco-éthologie des peuplements d'Amphipodes au niveau des zones humides de la Tunisie [Thèse de doctorat en Biologie]. Faculté des Sciences de Tunis, Université de Tunis El Manar; 2014. p. 328

[47] Bouslama MF, El Gtari M, Charfi-Cheikhrouha F. Impact of environmental factors on zonation, abundance, and other biological parameters of two Tunisian populations of *Talitrus saltator* (Amphipoda, Talitridae). Crustaceana. 2009;**82**(2):141-157

[48] Fallaci M, Colombini I, Lagar M, Scapini F, Chelazzi L. Distribution patterns of different age classes and sexes in a Tyrrhenian population of *Talitrus saltator* (Montagu). Marine Biology. 2003;**142**:101-110

[49] Colombini I, Aloia A, Bouslama MF, El Gtari M, Fallaci M, Ronconi L, Scapini F, Chelazzi L. Small-scale spatial and seasonal differences in the distribution of beach arthropods on the northern Tunisian coasts. Are species evenly distributed along the shore? Marine Biology, Berlin. 2002;**140**:1001-1012

[50] Williams JA. Burrow-zone distribution of the supralittoral Amphipod *Talitrus saltator* on Derbyhaven beach, Isle of Man-a possible mechanism for regulating desiccation stress? Journal of Crustacean Biology. 1995;**15**:466-475

Permissions

All chapters in this book were first published in SSB, by InTech Open; hereby published with permission under the Creative Commons Attribution License or equivalent. Every chapter published in this book has been scrutinized by our experts. Their significance has been extensively debated. The topics covered herein carry significant findings which will fuel the growth of the discipline. They may even be implemented as practical applications or may be referred to as a beginning point for another development.

The contributors of this book come from diverse backgrounds, making this book a truly international effort. This book will bring forth new frontiers with its revolutionizing research information and detailed analysis of the nascent developments around the world.

We would like to thank all the contributing authors for lending their expertise to make the book truly unique. They have played a crucial role in the development of this book. Without their invaluable contributions this book wouldn't have been possible. They have made vital efforts to compile up to date information on the varied aspects of this subject to make this book a valuable addition to the collection of many professionals and students.

This book was conceptualized with the vision of imparting up-to-date information and advanced data in this field. To ensure the same, a matchless editorial board was set up. Every individual on the board went through rigorous rounds of assessment to prove their worth. After which they invested a large part of their time researching and compiling the most relevant data for our readers.

The editorial board has been involved in producing this book since its inception. They have spent rigorous hours researching and exploring the diverse topics which have resulted in the successful publishing of this book. They have passed on their knowledge of decades through this book. To expedite this challenging task, the publisher supported the team at every step. A small team of assistant editors was also appointed to further simplify the editing procedure and attain best results for the readers.

Apart from the editorial board, the designing team has also invested a significant amount of their time in understanding the subject and creating the most relevant covers. They scrutinized every image to scout for the most suitable representation of the subject and create an appropriate cover for the book.

The publishing team has been an ardent support to the editorial, designing and production team. Their endless efforts to recruit the best for this project, has resulted in the accomplishment of this book. They are a veteran in the field of academics and their pool of knowledge is as vast as their experience in printing. Their expertise and guidance has proved useful at every step. Their uncompromising quality standards have made this book an exceptional effort. Their encouragement from time to time has been an inspiration for everyone.

The publisher and the editorial board hope that this book will prove to be a valuable piece of knowledge for researchers, students, practitioners and scholars across the globe.

List of Contributors

Wondmagegn Tafesse Tirkaso and Ing-Marie Gren
Department of Economics, Swedish University of Agricultural Sciences, Uppsala, Sweden

Leonard Sandin, Joel Segersten, David Spjut and Erik Degerman
Department of Soil and Water, Swedish University of Agricultural Sciences, Uppsala, Sweden

Radji Raoufou and Kokou Kouami
Laboratory of Forest Research, Faculty of Science, University of Lomé, Lomé, Togo

Victor L. Finot
Department of Animal Production, Faculty of Agronomy, University of Concepción, Chile

Carlos M. Baeza, Eduardo Ruiz, Oscar Toro and Pedro Carrasco
Department of Botany, Faculty of Natural Sciences and Oceanography, University of Concepción, Chile

Olga Maria Correia Chitas Ameixa
Biology Department and CESAM, University of Aveiro, Aveiro, Portugal
CE3C, Centre for Ecology, Evolution and Environmental Changes, Azorean Biodiversity Group, Department of Biology, University of the Azores, Portugal

António Onofre Soares
CE3C, Centre for Ecology, Evolution and Environmental Changes, Azorean Biodiversity Group, Department of Biology, University of the Azores, Portugal

Amadeu M.V.M. Soares and Ana I. Lillebø
Biology Department and CESAM, University of Aveiro, Aveiro, Portugal

Alfred Francis Murye
Centre for Environmental Management, University of the Free State, Bloemfontein, South Africa

André J. Pelser
Department of Sociology, University of the Free State, Bloemfontein, South Africa

Olga L. Kupika and Edson Gandiwa
Department of Wildlife Ecology and Conservation, Chinhoyi University of Technology, Chinhoyi, Zimbabwe

Shakkie Kativu
Department of Biological Sciences, Faculty of Science, University of Zimbabwe, Harare, Zimbabwe

Godwell Nhamo
Institute for Corporate Citizenship, University of South Africa, Pretoria, South Africa

Basáñez Muñoz Agustín de Jesús, Serrano Solis Arturo, Cuervo López Liliana, Capistrán Barradas Ascención and Naval Ávila Celina
Cuerpo Académico Manejo de Ambientes Marinos y Costeros, Facultad de Ciencias Biológicas y Agropecuarias-Tuxpan, Universidad Veracruzana, Veracruz, México

Martínez Cortés Esmeralda
Maestría en Manejo de Ecosistemas Marinos y Costeros, Facultad de Ciencias Biológicas y Agropecuarias-Tuxpan, Universidad Veracruzana, Veracruz, México

Javier Ferreiro da Costa and Pablo Ramil-Rego
IBADER (Instituto de Biodiversidade Agraria e Desenvolvemento Rural)-GI-1934 TB, Universidade de Santiago de Compostela (USC), Lugo, Spain

Maria Raquel Kanieski and Philipe Ricardo Casemiro Soares
Santa Catarina University State, Lages, Santa Catarina, Brazil

Solon Jonas Longhi
Santa Maria Federal University, Santa Maria, Rio Grande do Sul, Brazil

Franc Bavec, Urška Lisec and Martina Bavec
University of Maribor, Maribor, Slovenia

Jelassi Raja
National Institute of Sciences and Technology of the Sea, Salammbô, Tunis, Tunisia
University of Tunis El Manar, Faculty of Sciences of Tunis, Research Unit of Bio-ecology and Evolutionary Systematics El Manar II, Tunis, Tunisia

Khemaissia Hajer and Nasri-Ammar Karima
University of Tunis El Manar, Faculty of Sciences of Tunis, Research Unit of Bio-ecology and Evolutionary Systematics El Manar II, Tunis, Tunisia

Index

P

R

S

T

U

W

www.ingramcontent.com/pod-product-compliance
Lightning Source LLC
LaVergne TN
LVHW080913210926
848485LV00004B/648